Lecture Notes in Computer Science   13712

More information about this series at https://link.springer.com/bookseries/558

Silvia Lizeth Tapia Tarifa · José Proença (Eds.)

Formal Aspects of Component Software

18th International Conference, FACS 2022
Virtual Event, November 10–11, 2022
Proceedings

 Springer

Editors
Silvia Lizeth Tapia Tarifa ⓘ
University of Oslo
Oslo, Norway

José Proença ⓘ
CISTER Research Centre, ISEP
Porto, Portugal

ISSN 0302-9743   ISSN 1611-3349 (electronic)
Lecture Notes in Computer Science
ISBN 978-3-031-20871-3   ISBN 978-3-031-20872-0 (eBook)
https://doi.org/10.1007/978-3-031-20872-0

This Springer imprint is published by the registered company Springer Nature Switzerland AG
The registered company address is: Gewerbestrasse 11, 6330 Cham, Switzerland

Preface

This volume contains the papers presented at the 18th International Conference on Formal Aspects of Component Software (FACS 2022), held online during November 10–11, 2022.

FACS aims to bring together practitioners and researchers in the areas of component software and formal methods in order to promote a deeper understanding of how formal methods can or should be used to make component-based software development succeed. The component-based software development approach has emerged as a promising paradigm to transport sound production and engineering principles into software engineering and to cope with the ever increasing complexity of present-day software solutions. However, many conceptual and technological issues remain in component-based software development theory and practice that pose challenging research questions. Moreover, the advent of digitalization and industry 4.0 that requires better support from component-based solutions, e.g., cloud computing, cyber-physical and critical systems, and the Internet of Things, has brought to the fore new dimensions. These include quality of service, safety, and robustness to withstand inevitable faults, which require established concepts to be revisited and new ones to be developed in order to meet the opportunities offered by these supporting technologies.

We received 25 submissions from authors in 13 countries, out of which the Program Committee selected 13 papers. All submitted papers were reviewed, on average, by three referees in a single blind manner. The conference proceedings were made available at the conference. The proceedings includes all the final versions of the papers that took into account the comments received by the reviewers. Authors of selected accepted papers will be invited to submit extended versions of their contributions to appear in a special issue of ACM's Formal Aspects of Computing: Applicable Formal Methods (FAC) journal.

We would like to thank all researchers who submitted their work to the conference, to the Steering Committee members who provided precious guidance and support, to all colleagues who served on the Program Committee, and to the external reviewers, who helped us to prepare a high-quality conference program. Particular thanks to the invited speakers, Christel Baier from TU Dresden in Germany, Renato Neves from the University of Minho in Portugal, Ina Schaefer from the Karlsruhe Institute of Technology in Germany, and Volker Stolz from the Western Norway University of Applied Sciences in Norway, for their efforts and dedication to present their research and to share their perspectives on formal methods for component software at FACS. We are extremely grateful for the help in managing practical arrangements from the local organizers at the University of Oslo, in particular to Rudolf Schlatte. We also thank the Portuguese Foundation for Science and Technology (FCT project NORTE-01-0145-FEDER-028550), the Research Council of Norway (project ADAPt with No. 274515), the ECSEL Joint Undertaking (under grant agreement No. 876852), and Springer for their sponsorship.

November 2022   Silvia Lizeth Tapia Tarifa
                                                                          José Proença

Organization

Program Chairs

Silvia Lizeth Tapia Tarifa	University of Oslo, Norway
José Proença	Polytechnic Institute of Porto, Portugal

Steering Committee

Farhad Arbab	CWI and Leiden University, The Netherlands
Kyungmin Bae	Pohang University of Science and Technology, South Korea
Luís Soares Barbosa	University of Minho, Portugal
Sung-Shik Jongmans	Open University and CWI, The Netherlands
Zhiming Liu	Southwest University, China
Markus Lumpe	Swinburne University of Technology, Australia
Eric Madelaine	Inria Sophia Antipolis, France
Peter Csaba Ölveczky	University of Oslo, Norway
Corina Pasareanu	CMU, USA
José Proença	Polytechnic Institute of Porto, Portugal
Gwen Salaün	Université Grenoble Alpes, France
Anton Wijs	Eindhoven University of Technology, The Netherlands

Program Committee

Farhad Arbab	CWI and Leiden University, The Netherlands
Kyungmin Bae	Pohang University of Science and Technology, South Korea
Guillermina Cledou	University of Minho, Portugal
Peter Csaba Ölveczky	University of Oslo, Norway
Brijesh Dongol	University of Surrey, UK
Clemens Dubslaff	Technische Universität Dresden, Germany
Marie Farrell	National University of Ireland Maynooth, Ireland
Samir Genaim	IMDEA, Spain
Fatemeh Ghassemi	University of Tehran, Iran
Ludovic Henrio	Inria Lyon, France
Sung-Shik Jongmans	Open University and CWI, The Netherlands
Olga Kouchnarenko	University of Franche-Comté, France

Ivan Lanese	University of Bologna, Italy
Zhiming Liu	Southwest University, China
Mieke Massink	CNR ISTI, Italy
Jacopo Mauro	University of Southern Denmark, Denmark
Hernán Melgratti	University of Buenos Aires, Argentina
Catuscia Palamidessi	Inria Saclay, France
Corina Pasareanu	CMU, USA
Violet Ka I Pun	University of Oslo, Norway
Gwen Salaün	Université Grenoble Alpes, France
Camilo Rocha	Pontificia Universidad Javeriana Cali, Colombia
Luís Soares Barbosa	University of Minho, Portugal
Emilio Tuosto	Gran Sasso Science Institute, Italy
Shoji Yuen	Nagoya University, Japan
Anton Wijs	Eindhoven University of Technology, The Netherlands

Organizing Committee

Rudolf Schlatte	University of Oslo, Norway
Silvia Lizeth Tapia Tarifa	University of Oslo, Norway
José Proença	Polytechnic Institute of Porto, Portugal

Additional Reviewers

Edixhoven, Luc
Fabbretti, Giovanni
Faqrizal, Irman
Ferres, Bruno
Guanciale, Roberto
Imai, Keigo
Liu, Jie
Senda, Ryoma
Zhang, Yuanrui

Contents

Types and Choreographies

Modelling and Verification

Compositional Simulation of Abstract State Machines for Safety Critical Systems

Silvia Bonfanti[1]([✉]) [ID], Angelo Gargantini[1] [ID], Elvinia Riccobene[2] [ID], and Patrizia Scandurra[1] [ID]

[1] University of Bergamo, Bergamo, Italy
{silvia.bonfanti,angelo.gargantini,patrizia.scandurra}@unibg.it
[2] Università degli Studi di Milano, Milano, Italy
elvinia.riccobene@unimi.it

Abstract. Model-based simulation is nowadays an accepted practice for reliable prototyping of system behavior. To keep requirements complexity under control, system components are specified by separate models, validated and verified in isolation from the rest, but models have to be subsequently integrated and validated as a whole. For this reason, engines for orchestrated simulation of separate models are extremely useful.

In this paper, we present a compositional simulation technique for managing the co-execution of Abstract State Machines (ASMs) communicating through I/O events. The proposed method allows the co-simulation of ASM models of separate subsystems of a Discrete Event System in a straight-through processing manner according to a predefined orchestration schema.

We also present our experience in applying and validating the proposed technique in the context of the MVM (Mechanical Ventilator Milano) system, a mechanical lung ventilator that has been designed, successfully certified, and deployed during the COVID-19 pandemic.

Keywords: Models composition · Models co-simulation · Abstract state machines · Mechanical ventilator milano

1 Introduction

Model-based simulation is a widely accepted technique for prototyping the behavior of Discrete Event Systems (DESs), and the only practical alternative for understanding the behavior of complex and large systems [9,27]. However, due to requirements complexity and system size, models are becoming unmanageable in the engineering process and are achieving unprecedented levels of scale and complexity in many fields [9]. Thus, it is becoming increasingly important to efficiently and effectively manage a system model as a composition of different sub-models to analyze separately and then integrate [12,25].

S. L. Tapia Tarifa and J. Proença (Eds.): FACS 2022, LNCS 13712, pp. 3–19, 2022.
https://doi.org/10.1007/978-3-031-20872-0_1

Towards this direction, this paper presents a *compositional model-based simulation* technique to combine the simulation of formal models representing independent and interacting subsystems of a discrete-event system under prototyping. To illustrate the proposed approach, we adopt Abstract State Machines (ASMs) [8,15] as executable state-based formal modeling language for DESs. The approach allows the co-simulation of ASM models of separate subsystems in a *straight-through processing* manner [5] according to a predefined orchestration schema. Coordination/orchestration operators allow the definition of composition patterns for simulating models. The proposed approach is supported by a simulation composer engine, called `AsmetaComp`, which exploits the simulation tool `AsmetaS@run.time` [23] to execute ASMs as *living models* at runtime [10,26].

The paper also presents our experience in validating the proposed method on the Mechanical Ventilator Milano (MVM), a mechanical lung ventilator developed for COVID-19 patients. The MVM is an adequate case study since it is a safety-critical system and it is made of different interacting hardware/software subsystems with human-in-the-loop, which have been developed by different teams. Modeling the entire system in a single ASM model would be impractical and produce a complex model to manage, analyze and share with all groups. Our compositional modeling process started from a formal specification of the MVM subsystems as separate ASM models, which are validated and verified independently. The compositional simulation technique was then used to compose by orchestration the subsystems models according to established communication flows and I/O events exchanged among the MVM components. Therefore, we validated by scenarios simulation the composed model of the MVM system by checking if it performs in accordance with the expected outcomes from the requirements. The results of such modeling and validation approach helped us to clarify and improve the MVM component interfaces and the communication protocol, while taming the system model complexity.

The contributions of this paper can be summarized as follows: (i) we promote a practical and rigorous compositional modeling and simulation method for DESs through formal state-based models (like ASMs) supported by V&V (validation and verification) tools (like the ASMETA toolset for ASMs [1,8]); (ii) we provide the concept of I/O ASM and define a compositional execution semantics of orchestrated I/O ASMs; (iii) we report our experience in applying the proposed method, through the supporting simulation composer engine `AsmetaComp`, to a safety-critical case study in the health-care domain (the MVM).

This paper is organized as follows. Section 2 presents the MVM case study used to show the proposed approach. Section 3 provides basic concepts about ASMs and the ASMETA toolset. Section 4 presents the proposed compositional model-based simulation technique. Section 5 presents the application of the proposed technique to the MVM case study and describes the features of the composition engine supporting the proposed technique. Section 6 reports our lesson learned from such a modeling and validation experience with the MVM system. Finally, Sect. 7 shows related work and Sect. 8 concludes the paper with future directions for its extensions.

2 The Mechanical Ventilator Milano (MVM) Case Study

During the COVID-19 pandemic, our research team was involved in the design, development, and certification of a mechanical lung ventilator called MVM (Mechanical Ventilator Milano)[1] [7]. This section introduces the MVM case study by briefly outlining its purpose, the problem of the case that we address, and an informal description of the behavior of the main MVM components.

2.1 Problem Context

Due to time constraints and lack of skills, no formal method was applied to the MVM project. However, later we had planned to assess the feasibility of developing (part of) the ventilator by using a component-based formal specification development. In this project, instead of creating one single system model, we split the system into several subsystems and loosely coupled their formal specifications and analysis. Right from the very beginning, the sub-models were developed by different groups and for different engineering domains and target platforms, in order to speed up the overall formal development process from requirements to code.

Then for the purpose of integrating all sub-models and simulating the whole system, we have adopted the compositional model simulation technique presented here, and used the supporting tool `AsmetaComp` to put it in practice. This technique provided us a high degree of flexibility, since it allowed us to effectively develop the sub-models separately, then to couple them via well-defined input and output interfaces, and examine the behavior of the overall integrated system according to a specific orchestrated co-simulation schema.

Figure 1 shows a high-level overview of the main architecture components of the MVM system using a UML-like notation.

The core components of the MVM system are the software *MVMController* that manages the lung ventilation based on user inputs (acquired through the *GUI* component) and patient parameters, and the *Supervisor* software that monitors the overall system behavior and ensures that the machine *HW* operates safely. The wires between components (the solid lines with arrows) in Fig. 1 represent the information flow between components: an arrow incoming is the input for the component, an arrow outgoing is the output for the component. The wires are labeled with the name of the UML interface representing the signals exchanged between components, e.g., *ISC* shows the signals sent from the supervisor to the controller: the watchdog status and the status of self-test mode. In [11], we already reported our practical experience in using ASMs/ASMETA for modeling, analyzing, and encoding only the MVM software controller. For the purpose of this paper, the model of the supervisor and the hardware subsystems have also been developed. In addition, this paper describes the integration of all sub-system models via I/O interfaces and their orchestrated co-simulation.

---

[1] https://mvm.care/.

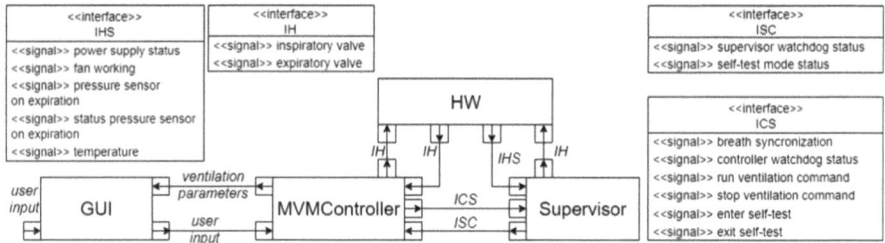

Fig. 1. Main components of the MVM system and signals exchanged

2.2 System Behavioral Description

MVM provides ventilation support in pressure-mode (i.e. the ventilator uses the pressure as a variable to control the respiratory cycle), for patients that are in intensive therapy and that require mechanical ventilation.

MVM supports two ventilation modes: *Pressure Controlled Ventilation* (PCV) and *Pressure Support Ventilation* (PSV). PCV mode is used for patients that are not able to start breathing on their own. The duration of the respiratory cycle is kept constant and set by the doctor; and the pressure changes between the target inspiratory pressure and the positive end-expiratory pressure. In PSV mode, the respiratory cycle is controlled by the patient. A new inspiration is initiated when a sudden pressure drop occurs, while expiration starts when the patient's inspiratory flow drops below a set fraction of the peak flow. If a new inspiratory phase is not detected within a certain amount of time (apnea lag), MVM will automatically switch to the PCV mode because it is assumed that the patient is not able to breathe alone. The air enters/exits through two valves: an input valve (opened during inspiration) and an output valve (opened during expiration). If the ventilator is not running, the input valve is closed and the output valve is opened to allow the patient to breath thanks to two relief valves; this configuration is called *safe-mode*.

Before starting the ventilation, the MVM controller passes through three phases. The *start-up* in which the controller is initialized with default parameters, *self-test* which ensures that the hardware is fully functional, and *ventilation off* in which the controller is ready for ventilation when requested.

The supervisor monitors the whole system behavior. After *initialization* and *startup* phases, the *self-test* procedure that checks all the hardware components is performed. When the ventilator is running, it is in *ventilation on* phase, in which it monitors all the ventilation parameters to avoid harm to the patient. If the supervisor detects a dangerous situation, the ventilator and the valves are moved to *safe-mode*. When the ventilator is not running, it is in *ventilation off* phase and the valves are in *safe-mode*.

3 Preliminary Concepts on ASMs and ASMETA

ASMs [14] are an extension of Finite State Machines. Unstructured control states are replaced by *states* comprising arbitrarily complex data (i.e., domains of

objects with functions defined on them); state *transitions* are expressed by transition rules describing how the data (state function values saved into *locations*)
change from one state to the next.
An ASM model is structured in terms of:

- The **signature** section where domains and functions are declared. The model
  interface with its environment is specified by *monitored* functions that are
  written by the environment and read by the machine, and by *out* functions
  that are written by the machine and read by the environment; *controlled*
  functions are the internal functions used by the machine (read in the current
  state and updated in the next state).
- The **definitions** section where all transition rules and possible invariants are specified. *Transition rules* have different constructors depending
  on the update structure they express, e.g., guarded updates (**if-then**,
  switch-case), simultaneous parallel updates (**par**), etc. The *update* rule
  $f(t_1, \ldots, t_n) := v$, being f an n-ary function, t_i terms, and v the new value
  of $f(t_1, \ldots, t_n)$ in the next state, is the basic unit of rules construction. State
  invariants are first order formulas that must be true in each computational
  state.
- The **main rule** that is, at each state, the starting point of the computation;
  it, in turns, calls all the other transitions rules.
- The **default init** section where initial values for the *controlled* functions
  are defined.

An ASM *run* is defined as a finite or infinite sequence $S_0, S_1, \ldots, S_n, \ldots$ of
states: starting from an initial state S_0, a *run step* from S_n to S_{n+1} consists in
firing, in parallel, all transition rules and leading to simultaneous updates of a
number of locations. In case of an inconsistent update (i.e., the same location is
updated to two different values by firing transition rules) or invariant violations,
the model execution fails, but the model is kept alive by restoring the state in
which it was before the failing step (*model roll-back*). In the sequel, we shortly
write *model succeeds* or *fails* to mean a model performing either a successful run
step or a failing one.

The development process from formal requirement specification to code generation is supported by ASMETA [1,8], a set of tools around the ASM formal
method. ASMs can be understood as executable pseudo-code or virtual machines
working over abstract data structures at any desired level of abstraction.

3.1   Modeling Example

As an example of ASM model, Code 1 reports excerpts of the ASM **MVMcontroller**
modeling the controller using the **AsmetaL** textual modeling language.

In the section **signature** (see Lines 2–11), respirationMode is a monitored
function specifying the input command received by the controller from the GUI,
as part of the *user inputs*, to change its mode of operation (PCV or PSV), while
the controlled function **state** represents the controller status. The out functions

```
1   asm MVMcontroller                              rule r_runPCV =
2   signature:                                       par
3   enum domain States = {STARTUP | SELFTEST          if phase = INSPIRATION then r_runPCVInsp[] endif
4   | VENTILATIONOFF | PCV_STATE | PSV_STATE}          if phase = EXPIRATION then r_runPCVExp[] endif
5      enum domain Modes = {PCV | PSV}              endpar
6      ...                                          rule r_runPCVInsp =
7   dynamic controlled state: States                 par
8   dynamic monitored respirationMode: Modes           if not stopVentilation then
9   dynamic monitored stopRequested: Boolean             if stopRequested then stopVentilation := true endif
10  dynamic out iValve: ValveStatus                    endif
11  dynamic out oValve: ValveStatus ...                if expired(timerInspirationDurPCV) then
12  definitions:                                         par
13     ...                                                if respirationMode = PCV then
14  main rule r_Main =                                     r_PCVStartExp[] endif
15  par                                                  if respirationMode = PSV then
16    if state = STARTUP then r_startup[] endif            par
17    if state = SELFTEST then r_selftest[] endif            state := PSV_STATE
18    if state = VENTILATIONOFF then                         r_PSVStartExp[]
19       r_ventilationoff[] endif                          endpar
20    if state = PCV_STATE then r_runPCV[] endif       endif endpar endif endpar
21    if state = PSV_STATE then r_runPSV[] endif     default init s0:
22  endpar                                            function state = STARTUP
```

Code 1. MVMController ASMETA model

iValve and oValve specify the interface IH between MVMController and HW components, and are the input for HW to set the valves status during ventilation.

Initially, the function state is initialized at the value STARTUP. At each step, depending on the current state value, a corresponding rule fires through by the r_Main execution. These call rules are specified in the definitions section.

In Code 1, on the right, we show, for example, the rule r_runPCV regulating the PCV mode. It in turn calls rules for the inspiration, r_runPCVInsp (line 6), and the expiration, r_runPCVExp (here missed). In PCV mode, the transition between inspiration and expiration is determined by the duration of each phase decided by the physician (when timers timerInspirationDurPCV, in case of inspiration, and timerExpirationDurPCV, in case of expiration, expire). When the inspiration time is passed (line 11), the controller goes to the PCV expiration phase (line 14). If the physician has required (by setting the value of the monitored function respirationMode) to move to PSV mode (line 15), the machine changes the state from PCV to PSV and executes the rule r_PSVStartExp (line 18). If a stop request (by the monitored function stopRequested, still part of the user input interface between the GUI and the MVMController) is received during the inspiration phase (line 8), it is stored (in stopVentilation) and will be executed in the expiration phase.

3.2 The ASMETA (ASM mETAmodeling) Toolset

ASMETA [1] provides the user with modeling notations, different analysis (V&V) techniques and automatic source code and test generators for ASMs to be applied at design-, development-, and operation- time [8]. In particular, a runtime simulation engine, AsmetaS@run.time [24], has been developed within ASMETA as extension of the offline simulator AsmetaS [17] to handle an ASM as a living model [10,26] to run in tandem with a real software system. AsmetaS@run.time supports simulation *as-a-service* features of AsmetaS and additional features such as model execution with timeout and model roll-back to the previous state after a failure step during model execution. All these features are accessible by UI dashboards (both in a graphical and in a command-line way). This runtime model simulation mechanism has been already used within an *enforcer* software tool [13] to sanitize input/output events for a running system or to prevent the execution of unsafe commands by the system.

4 Compositional Simulation of ASM Models

Consider the partitioning of a software system into distinct subsystems/components interacting for sharing resources in terms of input/output events. Since ASMs are a state-based formalism, i.e., outputs of the machine depend only on its current state and inputs, formal behavioral models of these subsystem/components may be specified in terms of a set of ASMs, each having its own input I (the monitored locations of the ASM), current state (the controlled locations of the ASM), and output O (the out locations of the ASM). We denote by *I/O ASM* each of these component models that are defined as follows:

Definition 1 (I/O ASM). *An* I/O ASM *is an ASM model m with a non-empty set I_m of input (or monitored) functions and a non-empty set O_m of out functions in its signature. We denote by (I_m, m, O_m) an I/O ASM, and by $curr\_state(m)$ the set of its locations values.*

We assume that I/O ASMs can interact in a black-box manner by binding input and out functions with the same symbol name and interpretation. Consider, for example, the typical cascade or sequence of two machines A and B, where the output of the machine A is the input of the machine B. We may view the cascade of these two machines as a single compound machine that reacts to an external input x by propagating instantaneously the effect of x through the cascade of A and B at each step (A reacts to x, then B reacts to the output of A). So we focus on ASMs having a well-defined I/O interface represented by (possibly parameterized) input and out ASM functions, which are, in fact, the interaction ports (or points) with the environment or other ASMs. We formally define the binding between I/O ASMs as follows:

Definition 2 (I/O ASM binding). *Given two* I/O ASMs, *m_i and m_j, an I/O binding exists from m_i to m_j iff $O_{m_i} \cap I_{m_j} \neq \emptyset$. We denote by $m_i \stackrel{B_{i,j}}{\Longrightarrow} m_j$ the I/O binding from m_i to m_j, where $B_{i,j} = O_{m_i} \cap I_{m_j}$ is the set of binding functions.*

We assume that if an I/O ASM binding exists between two models m_i and m_j, then at least one function symbol f must occur in both the two models' signatures and is used as out function for m_i and as input function for m_j, with the same domain and codomain and same interpretation.

The model `MVMcontroller` shown in Code 1 is an example of I/O ASM, having input functions $I = \{\mathsf{respirationMode}, \mathsf{stopRequest}\}$ and out functions $O = \{\mathsf{iValve}, \mathsf{oValve}\}$. The set O represents also the bindings of this model with the model of the hardware (not reported here), while the set I is the binding with the GUI component. Figure 1 provides a graphical view of the bindings among all the component models as wires labeled with the name of the UML interface representing the binding functions (the exchanged signals values).

Several I/O ASMs can execute and communicate over I/O bindings to form a whole ASM assembly.

Definition 3 (I/O ASM assembly). *An I/O ASM assembly is a set of I/O ASMs bound together by I/O ASM bindings.*

Figure 1 illustrates an I/O ASM assembly consisting of the I/O ASM models of the MVM system and the I/O ASM bindings among all the component models.

The execution of an assembly of I/O ASMs can be orchestrated (or coordinated) in accordance with a workflow expressible through different types of coordination constructs, defined below, with a specific semantics. Intuitively, a pipe connection $m_1 \mid m_2$ means that the output of m_1 is used as input to m_2, assuming a directional I/O binding exists between the two models, namely some input functions of m_2 are a subset of the out functions of m_1. Similarly, a bidirectional pipe $m_1 <\mid> m_2$ is like having two pipes where one is used for the reverse direction, i.e. the output (or a subset) from m_2 becomes the input of m_1 in addition to external input from the environment or other machines bound to m_1. In both these two series compositions, we assume a cascade synchrony in reacting to external input from the environment. In a parallel connection $m_1 \parallel m_2$, both the two models react to external input from the environment or from other machines separately. Such coordination constructs allow for the following (recursive) definition of a *composition formula* of I/O ASMs.

Definition 4 (I/O ASM composition formula). *A* composition *formula* c *over an ASM assembly* A *is a single I/O ASM* m *belonging to* A *or* $c_1 \mid c_2$ *or* $c_1 <\mid> c_2$ *or* $c_1 <\parallel> c_2$ *or* $c_1 \parallel c_2$, *where* c_1 *and* c_2 *are composition formulas and* $\{\mid, <\mid>, <\parallel>, \parallel\}$ *are composition operators.*

The composition formula $c = (m_1 <\mid> (m_2 \parallel m_3)) \mid m_4$ denotes, for example, the execution schema of four ASM models, where the output of the bidirectional pipe between *m1* and the parallel of *m2* and *m3*, is given as input to *m4* connected through a pipe. As another example, the I/O ASM assembly shown in Fig. 1 is executed using a composition formula made by two bidirectional pipes: $Hardware.asm <\mid> (MVMController.asm <\mid> Supervisor.asm)$.

Note that in case of a composition formula of the form $c_i \, op \, c_j$, we assume that there is no one model in common between c_i and c_j.

Before providing the operational semantics of the composition operators, we extend the definition of I/O ASM binding at the level of a composition formula, as follows:

Definition 5 (I/O ASM composition binding). *Given two I/O ASM composition formulas, c_1 and c_2, for a given I/O ASM assembly, an I/O composition binding $c_1 \stackrel{B_{1,2}}{\Longrightarrow} c_2$ exists from c_1 to c_2 iff there exists at least an I/O ASM m_{1i} occurring in c_1 and an I/O ASM m_{2j} occurring in c_2 such that an I/O binding $m_{1i} \stackrel{B_{1i,2j}}{\Longrightarrow} m_{2j}$ exists from m_{1i} to m_{2j}. $B_{1,2}$ is the union of all existing binding functions between models m_{1i} in c_1 and models m_{2j} in c_2:*

$$B_{1,2} = \bigcup_{m_{1i} \in c_1, m_{2j} \in c_2} B_{1i,2j}$$

As an example, consider the two composition formulas $c_1 = Hardware.asm$ and $c_2 = MVMController.asm <|> Supervisor.asm$ for the assembly shown in Fig. 1. c_1 and c_2 are bound by the binding $B_{1,2} = B_{HW,MVMController} \cup B_{HW,Supervisor}$ where $B_{HW,MVMController} = \{$IH functions$\}$ and $B_{HW,Supervisor} = \{$IHS functions$\}$.

Definition 6 (I/O ASM composition operators). *Let c be an I/O ASM composition formula. The operational semantics of c is defined as follows:*

Single model: $c = (I_m, m, O_m)$. *A step of c is an execution step of the ASM m on the inputs I_m provided by the I/O bindings and by the environment.*

(Simplex) pipe or sequence: $c = c_1 \mid c_2$. *We assume $c_1 \stackrel{B}{\Rightarrow} c_2$. First execute c_1 on inputs I_{c_1} provided by the I/O bindings and by the environment, and if c_1 succeeds, subsequently execute c_2 on the inputs I_{c_2} provided by the I/O binding B with c_1, and by the environment, and return the results as outputs.*

Half-duplex bidirectional pipe: $c = c_1 <|> c_2$. *We assume $c_1 \stackrel{B_{1,2}}{\Longrightarrow} c_2$ and $c_2 \stackrel{B_{2,1}}{\Longrightarrow} c_1$. First execute c_1 on the inputs I_{c_1} provided by its I/O bindings and by the environment, and if c_1 succeeds, subsequently execute c_2 on the inputs I_{c_2} provided by the I/O binding $B_{1,2}$ with c_1 and by the environment; then, return the outputs for the I/O binding $B_{2,1}$ with c_1.*

Full-duplex bidirectional pipe: $c = c_1 <\|> c_2$. *We assume $c_1 \stackrel{B_{1,2}}{\Longrightarrow} c_2$ and $c_2 \stackrel{B_{2,1}}{\Longrightarrow} c_1$. First execute both c_1 and c_2 simultaneously on their inputs I_{c_1} and I_{c_2} provided by their I/O bindings and by the environment; if both succeed, then return their outputs for their I/O bindings.*

Synchronous parallel split (or fork-join): $c = c_1 \parallel c_2$ *Execute both c_1 and c_2 separately on their inputs I_{c_1} and I_{c_2}, respectively, provided by their I/O bindings and by the environment, then return their outputs.*

In case an I/O ASM model fails, the composition expression fails and the faulty ASM model and all models already executed in the composition are rolled-back.

The simulation of an I/O ASM assembly is the result of the compositional simulation of its I/O ASM components according to the execution semantics of a precise composition formula. Concretely, it can be represented and managed in memory in terms of an expression tree as given by the following definition.

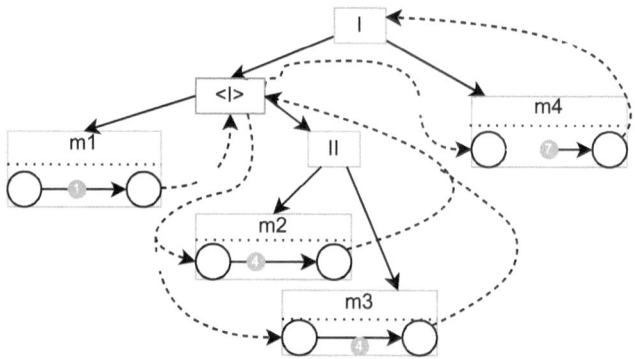

Fig. 2. Visit of a Compositional Simulation Tree for $(m_1 <|> (m_2 \parallel m_3)) \mid m_4$

Definition 7 (Compositional Simulation Tree of I/O ASMs). *Given a composition formula c of I/O ASMs m_i, $i = 1, ..., n$, a* Compositional Simulation Tree (CST) *of c is a binary tree $T_c = (V_c, E_c)$, where a leaf node in V_c is labelled by an ASM m_i together with its* curr_state(m_i), *and an internal node in V_c is labeled by a composition operator chosen from the set $\{\mid, <|>, <\parallel>, \parallel \}$.*

Intuitively, a step of an I/O ASM composition (i.e., a single compositional simulation step) is a recursive pre-order traversal of the corresponding CST to visit nodes (from the root to leaves) and evaluates them according to their type. The *execCST* in Algorithm 1 is the pseudocode of a simplified version of a CST traversal without considering the roll-back of models in case a failure occurs during model execution. Given a CST T_c for a composition c, the output of the algorithm is the set O_c of ASM out functions values in the (final) current state of the I/O ASM models (executed at each leaf node of T_c). The recursive traversal *exec* in Algorithm 2 is initially invoked (see line 2 of Algorithm 1) on the root node of a non-empty CST with an empty set $I_c{}^2$ of ASM input functions values for the I/O ASM models occurring in c. I_c will be populated during the tree traversal (when a model at a leaf node is executed) with function values provided externally by the environment or computed by other previously executed models in the tree. Algorithm 2 uses the subroutine $put(c_1, c_2, B_{1,2})$ to copy the values of the binding functions in $c_1 \xstack{B_{1,2}} c_2$ from $B_{1,2} \subseteq O_{c_1}$ to $B_{1,2} \subseteq I_{c_2}$. An intuitive graphical representation of the execution steps of the various composition operators is depicted on the left side of Algorithm 2.

Definition 8 (Runstep of an I/O ASM composition). *Given a composition formula c for an assembly A of I/O ASMs m_i, $i = 1, ..., n$, a* runstep *of the composition c is the depth (pre-order) traversal, given by Algorithm 1:* execCST, *of the compositional simulation tree T_c, which updates the current states of the component models m_i at the leaves of T_c.*

---

[2] We assume I_c is concretely realized as a map (or dictionary) that associates ASM function symbols (the keys) with their values.

Algorithm 1. $execCST$	**Algorithm 2.** $exec$

Algorithm 1. $execCST$

Input : T: a CST
1 **Function** execCST(T):
2   **if** $T.root \neq NULL$ **then**
3   exec(T.root,$\emptyset$);

Algorithm 2. $exec$

Input : node: a CST node
Input : I: object that maps ASM *monitored* functions to their values.
Output: O: an object that maps ASM *out* functions to their values.
1 **Function** exec($node$, I):
2   **if** $node.isLeaf()$ **then**
3   run a step of node.model on input I;
4   **return** node.model.getOutLocations();
5   **else**
6   c_1 = node.left;
7   c_2 = node.right;
8   **switch** $node.operator$ **do**
9   **case** $|$ **do**
10   O_1 = exec(c_1,I);
11   put(c_1,c_2,$B_{1,2}$);
12   O_2 = exec(c_2,I);
13   **return** $O_1 \cup O_2$;
14   **case** $< | >$ **do**
15   O_1 = exec(c_1,I);
16   put(c_1,c_2,$B_{1,2}$);
17   O_2 = exec(c_2,I);
18   put(c_2,c_1,$B_{2,1}$);
19   **return** $O_1 \cup O_2$;
20   **case** $< \| >$ **do**
21   O_1 = exec(c_1,I);
22   O_2 = exec(c_2,I);
23   put(c_1,c_2,$B_{1,2}$);
24   put(c_2,c_1,$B_{2,1}$);
25   **return** $O_1 \cup O_2$;
26   **case** $\|$ **do**
27   O_1 = exec(c_1,I);
28   O_2 = exec(c_2,I);
29   **return** $O_1 \cup O_2$;

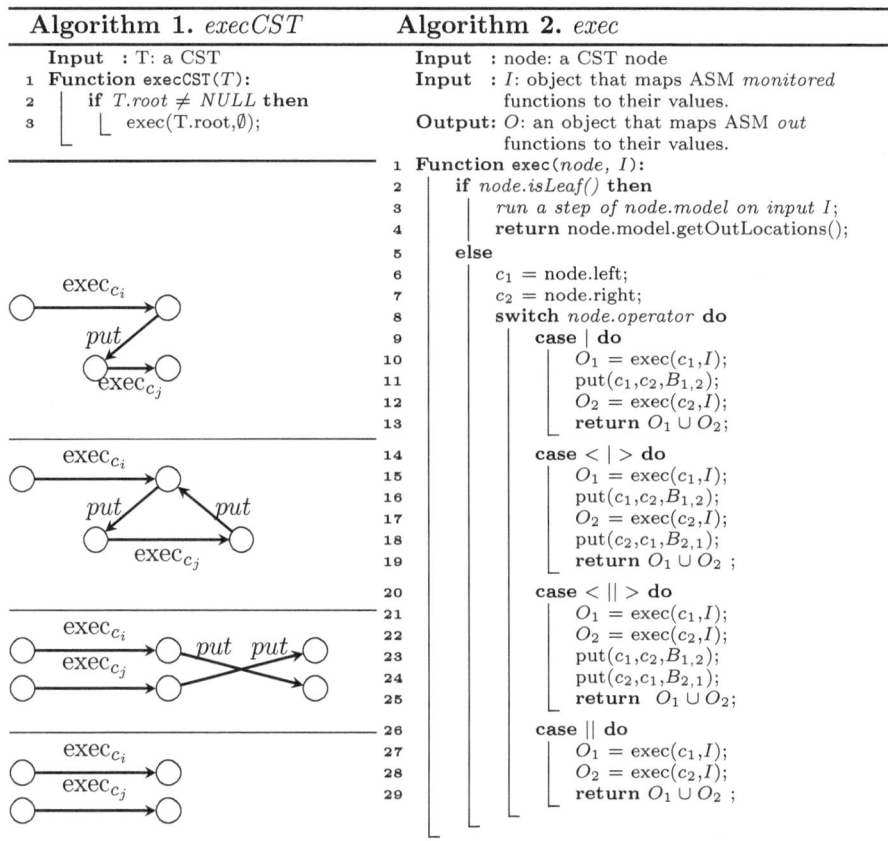

As an example, Fig. 2 depicts the depth visit of the CST of the composition $c = (m_1 < | > (m_2 \parallel m_3)) \mid m_4$: an execution step of the ASM models at leaf nodes of the tree updates the model's current state on the inputs provided by the I/O bindings (plus those provided by the environment but not shown in the picture). Labels 1, 4, and 7 denote the steps of the models to update their location values, while labels 2, 3, 5, 6, and 8 denote copying of the binding function values according to the composition operators.

5   Compositional Modeling and Simulation at Work

In order to support compositional I/O ASM simulation in practice, the tool AsmetaComp has been developed as part of the ASMETA tool-set, and it is based on the AsmetaS@run.time. We here show the compositional simulation tool on the MVM case study.

     AsmetaComp can be used through a basic graphical user interface (AsmetaComp GUI) or a command line (AsmetaComp shell). AsmetaComp GUI, after having specified the number of models involved in the composition, allows the user to define the

```
1    init −n 3
2    setup comp as Hardware.asm <|> (Controller.asm <|> Supervisor.asm)
3    run(comp, {adc_reply_m=RESPONSE;fan_working_m=true;pi_6_m=25;pi_6_reply_m=RESPONSE;
         temperature_m=25;state=STARTUP;insp_valve=CLOSED;exp_valve=OPEN;mCurrTimeSecs
         =1})
4    run(comp, {adc_reply_m=RESPONSE;fan_working_m=true;pi_6_m=25;pi_6_reply_m=RESPONSE;
         temperature_m=25;startupEnded=true;mCurrTimeSecs=2})
5    run(comp, {adc_reply_m=RESPONSE;fan_working_m=true;pi_6_m=25;pi_6_reply_m=RESPONSE;
         temperature_m=25;selfTestPassed=true;mCurrTimeSecs=3})
6    run(comp, {adc_reply_m=RESPONSE;fan_working_m=true;pi_6_m=25;pi_6_reply_m=RESPONSE;
         temperature_m=25;startVentilation=true;mCurrTimeSecs=4;run_command=false})
7    run(comp, {adc_reply_m=ERROR;fan_working_m=true;pi_6_m=25;pi_6_reply_m=RESPONSE;
         temperature_m=25;startVentilation=false;mCurrTimeSecs=5;run_command=false})...
8    run(comp, {adc_reply_m=ERROR;fan_working_m=true;pi_6_m=25;pi_6_reply_m=RESPONSE;
         temperature_m=25;startVentilation=false;mCurrTimeSecs=9;run_command=true})
```

Code 2. Composition script

composition formula and to execute the model composition. Using the `AsmetaComp` shell, instead, the user can set and run a composition of models interactively or via a script of commands. Such a script can be written using a textual composer language for the basic composition operators introduced in Sect. 4.

Applying `AsmetaComp` *to MVM.* Recall from Fig. 1 the four components of the MVM design. We have abstracted the GUI since the user interface is represented by the concept of ASM environment. Each of the other three components has been modeled as an I/O ASM[3], and validated individually by using the ASMETA tools. Then, we have composed the three models and executed their composition.

According to the MVM operation, the hardware sends values to both controller and supervisor and both of them return the configuration of hardware components based on the ventilation status (bidirectional pipe between hardware and controller/supervisor). Moreover, the controller sends information (i.e., its status and alarms raised) to the supervisor, which in turn returns its status to it (bidirectional pipe between controller and supervisor). Therefore, the MVM components interaction is captured by two nested bidirectional pipes: *Hardware.asm* <|> (*MVMController.asm* <|> *Supervisor.asm*).

Through the script reported in Code 2 for the `AsmetaComp` shell, we set this composition formula (by command `setup` at line 2) and express some compositional steps of the interacting MVM models, including inputs from the environment for the unbounded ASM monitored functions of models (by commands `run`). Specifically, we plan to simulate the startup (lines 3–4), selftest (line 5), start ventilation (line 6) and the intervention of the supervisor by setting the valves in safe mode (line 8) because there is an error in the power supply (line 7).

Code 3 reports the script execution trace. The components (hardware, controller and supervisor) perform the first compositional step where all the functions are initialized. In the second step the controller performs self test . In compositional step three the controller is in ventilation off waiting for starting ventilation, and the supervisor has concluded self-test and is waiting for the command from the controller when ventilation has begun. In the next compositional

---

[3] The models are available at https://github.com/asmeta/asmeta/tree/master/code/
experimental/asmeta.simulator%40run.time/examples/MVM/ConfModels.

Code 3. Simulation trace supervisor in FAILSAFE, the valves change state

step, the ventilator is ventilating and the supervisor detects it. In compositional step five, six and seven the power supply reports an error (adc_reply_m=ERROR) and in compositional step seven the supervisor is in FAILSAFE state. In the next compositional step, since the controller is in inspiration phase, the input valve is open and the output valve is closed. But in the last compositional step reported in Code 3, nevertheless the controller is in inspiration, the valves are in safe mode (input valve is closed and output valve is opened) because the supervisor has detected an error.

6   Discussion and Lesson Learned

Based on our modeling experience with the MVM system, we can conclude that splitting the system model into two or more sub-models and loosely coupling their simulation (co-simulation) provided us several advantages w.r.t. creating an entire system model. Since right from the very beginning, the sub-models were developed by different groups and for different target platforms (e.g., the MVM controller prototype on the Arduino board), this compositional modeling technique allowed us a high degree of flexibility in managing the separation of the modeling/analysis tasks in different modeling/development groups, each working in parallel at their own speed. As a result, subsystem models could be analyzed and validated/verified in isolation to prove their correctness according to the established I/O interfaces with the other subsystem models, so allowing us to speed up the overall formal development process from requirements to code. Moreover, this compositional modeling helped to clarify and make precise w.r.t. the documented system requirements, the communication protocol among MVM components. Defining the I/O bindings between the sub-models and capturing the computational causality between sub-models in terms of a compositional simulation formula is, however, not a trivial task and requires a clear understanding of how the involved subsystems react to I/O stimuli and of their communication and computation protocol.

In a wider context, with the proposed technique we intend to contribute in providing virtual models (e.g., an ASM model that is a virtual copy of a system controller) and leveraging model simulation at runtime. Models@run.time together with formal analysis and data analytics are the main enablers of the Digital Twin technology, which is a growing interest in any field. By model simulation, it is possible to interact with the digital twin of a real system (or a part of it) by simulating different *what-if* analysis scenarios [16] to identify the best actions to be then applied on the physical twin.

7   Related Work

Existing frameworks that inspired our approach to compositional model-based simulation are those related to workflow modeling and service orchestration (such as tools for the Business Process Model and Notation (BPMN) [4], and the Jolie language [3]), and to multi-state machine modeling (like Yakindu statecharts [6]). However, the proposed technique is oriented to a distributed model-based system simulation, to be used for example in practical contexts where model simulation is required at runtime and models have to be co-simulated along with real systems, such as runtime models that are part of the knowledge base of a self-adaptive and autonomous system [10] or of a *digital twin* plant [18].

In [21], a *choreography automaton* for the choreographic modeling of communicating systems is introduced as a system of communicating finite state machines whose transitions are labelled by synchronous/asynchronous interactions. Choreographies are suitable approaches to describe modern software architectures such as micro-services, but in this first compositional simulation mechanism we preferred to rely on a centralized synchronous communication semantics that is typical of IT service orchestration and automation platforms. We postpone as future work the definition and implementation of choreography constructs to deploy and enact a choreography-based execution of asynchronous I/O ASMs.

Within the context of component- and service- based architectures, ASMs have been used for service modeling and prototyping in the OASIS/OSOA standard Service Component Architecture (SCA) for heterogeneous service assembly.

In such a framework, abstract implementation (or prototype) of SCA components in ASM (SCA-ASM components) are co-executed *in place* with other component implementations [22]. In [20], a method for predicting service assembly reliability both at system-level and component-level is presented by combining a reliability model for an SCA assembly involving SCA-ASM components.

8   Conclusion and Future Directions

In this paper, we have formulated the concept of I/O ASM and introduced the compositional simulation technique of I/O ASMs. We have also presented, through the MVM case study, the scope and use of the compositional simulation of I/O ASMs as supported by the ASMETA tool `AsmetaComp`. `AsmetaComp` is intended to support distributed simulation of ASMs by allowing separate ASM system models to be connected and co-simulated together to form simulations of integrated systems, or a big ASM model to be divided into smaller sub-models that co-execute, possibly on separate local or remote processes/computers.

As future work, we want to conduct some experiments for evaluating the real benefits of a compositional simulation w.r.t. a conventional non-compositional simulation of one single monolithic model. These include the evaluation of the usability of the technique and, in particular, the understandability of the execution traces. We plan to investigate on the fault-detection capability of the proposed approach to guarantee that using composition does not reduce the ability to discover unacceptable behaviors.

We also want to support additional composition operators and patterns (such as I/O wrapping, conditional execution, iterated execution, alternate execution, attempt choice, non-deterministic choice, etc.), and alternative model roll-back semantics to allow more expressiveness in specifying how models interact in a compositional simulation. We also want to support choreography constructs for decentralized co-simulation of asynchronous ASMs as provided in the standard Business Process modeling Notation 2.0 (BPMN2) *Choreography Diagram* [4]. Moreover, we want to support the co-simulation of ASM models with other DES simulated/real subsystems into a federated, interoperable simulation environment, possibly in accordance with standards such as the IEEE 1516 High Level Architecture (HLA) and the Functional Mockup Interface (FMI) [2] for distributed co-simulation [19].

Acknowledgement. We thank the students Davide Santandrea and Michele Zenoni for their contribution in tool implementation and case study composition.

References

1. ASMETA (ASM mETAmodeling) toolset. https://asmeta.github.io/
2. Functional mock-up interface. https://fmi-standard.org/
3. Jolie. https://jolie-lang.org
4. Object management group business process model and notation. https://bpmn.org/

5. Straight Through Processing - STP, Investopedia, 18 October 2020. https://www. investopedia.com/terms/s/straightthroughprocessing.asp
6. YAKINDU Statechart Tools. https://itemis.com/en/yakindu/state-machine
7. Abba, A., et al.: The novel mechanical ventilator Milano for the COVID-19 pandemic. Phys. Fluids **33**(3), 037122 (2021)
8. Arcaini, P., Bombarda, A., Bonfanti, S., Gargantini, A., Riccobene, E., Scandurra, P.: The ASMETA approach to safety assurance of software systems. In: Raschke, A., Riccobene, E., Schewe, K.-D. (eds.) Logic, Computation and Rigorous Methods. LNCS, vol. 12750, pp. 215–238. Springer, Cham (2021). https://doi.org/10.1007/978-3-030-76020-5_13
9. Bañares, J.Á., Colom, J.M.: Model and simulation engines for distributed simulation of discrete event systems. In: Coppola, M., Carlini, E., D'Agostino, D., Altmann, J., Bañares, J.Á. (eds.) GECON 2018. LNCS, vol. 11113, pp. 77–91. Springer, Cham (2019). https://doi.org/10.1007/978-3-030-13342-9_7
10. Bencomo, N., Götz, S., Song, H.: Models@run.time: a guided tour of the state of the art and research challenges. Softw. Syst. Model. **18**(5), 3049–3082 (2019)
11. Bombarda, A., Bonfanti, S., Gargantini, A., Riccobene, E.: Developing a prototype of a mechanical ventilator controller from requirements to code with ASMETA. Electron. Proc. Theor. Comput. Sci. **349**, 13–29 (2021)
12. Bombino, M., Scandurra, P.: A model-driven co-simulation environment for heterogeneous systems. Int. J. Softw. Tools Technol. Transf. **15**(4), 363–374 (2013). https://doi.org/10.1007/s10009-012-0230-5
13. Bonfanti, S., Riccobene, E., Scandurra, P.: A runtime safety enforcement approach by monitoring and adaptation. In: Biffl, S., Navarro, E., Löwe, W., Sirjani, M., Mirandola, R., Weyns, D. (eds.) ECSA 2021. LNCS, vol. 12857, pp. 20–36. Springer, Cham (2021). https://doi.org/10.1007/978-3-030-86044-8_2
14. Börger, E., Raschke, A.: Modeling Companion for Software Practitioners. Springer, Heidelberg (2018). https://doi.org/10.1007/978-3-662-56641-1
15. Börger, E., Stärk, R.: Abstract State Machines: A Method for High-Level System Design and Analysis. Springer, Berlin (2003)
16. Fuller, A., Fan, Z., Day, C., Barlow, C.: Digital twin: enabling technologies, challenges and open research. IEEE Access **8**, 108952–108971 (2020)
17. Gargantini, A., Riccobene, E., Scandurra, P.: A metamodel-based language and a simulation engine for abstract state machines. J. UCS **14**(12), 1949–1983 (2008)
18. Grieves, M.: Origins of the Digital Twin Concept, August 2016
19. Huiskamp, W., van den Berg, T.: Federated simulations. In: Setola, R., Rosato, V., Kyriakides, E., Rome, E. (eds.) Managing the Complexity of Critical Infrastructures. SSDC, vol. 90, pp. 109–137. Springer, Cham (2016). https://doi.org/10.1007/978-3-319-51043-9_6
20. Mirandola, R., Potena, P., Riccobene, E., Scandurra, P.: A reliability model for service component architectures. J. Syst. Softw. **89**, 109–127 (2014)
21. Orlando, S., Pasquale, V.D., Barbanera, F., Lanese, I., Tuosto, E.: Corinne, a tool for choreography automata. In: Salaün, G., Wijs, A. (eds.) FACS 2021. LNCS, vol. 13077, pp. 82–92. Springer, Cham (2021). https://doi.org/10.1007/978-3-030-90636-8_5
22. Riccobene, E., Scandurra, P.: A formal framework for service modeling and prototyping. Formal Aspects Comput. **26**(6), 1077–1113 (2013). https://doi.org/10.1007/s00165-013-0289-0
23. Riccobene, E., Scandurra, P.: Model-based simulation at runtime with abstract state machines. In: Muccini, H., et al. (eds.) ECSA 2020. CCIS, vol. 1269, pp. 395–410. Springer, Cham (2020). https://doi.org/10.1007/978-3-030-59155-7_29

24. Riccobene, E., Scandurra, P.: Model-based simulation at runtime with abstract state machines. In: Software Architecture - 14th European Conference, ECSA 2020 Tracks and Workshops, Proceedings. Communications in Computer and Information Science, vol. 1269. Springer, Berlin (2020)
25. Talcott, C., et al.: Composition of languages, models, and analyses. In: Composing Model-Based Analysis Tools, pp. 45–70. Springer, Cham (2021). https://doi.org/10.1007/978-3-030-81915-6_4
26. Van Tendeloo, Y., Van Mierlo, S., Vangheluwe, H.: A multi-paradigm modelling approach to live modelling. Softw. Syst. Model. **18**(5), 2821–2842 (2018). https://doi.org/10.1007/s10270-018-0700-7
27. Weyns, D., Iftikhar, M.U.: Model-based simulation at runtime for self-adaptive systems. In: Kounev, S., Giese, H., Liu, J. (eds.) 2016 IEEE International Conference on Autonomic Computing, ICAC 2016. IEEE Computer Society (2016)

Specifying Source Code and Signal-based Behaviour of Cyber-Physical System Components

Joshua Heneage Dawes$^{(\boxtimes)}$ and Domenico Bianculli

University of Luxembourg, Luxembourg, Luxembourg
{joshua.dawes,domenico.bianculli}@uni.lu

Abstract. Specifying properties over the behaviour of components of Cyber-Physical Systems usually focuses on the behaviour of signals, i.e., the behaviour of the *physical* part of the system, leaving the behaviour of the cyber components implicit. There have been some attempts to provide specification languages that enable more explicit reference to the behaviour of cyber components, but it remains awkward to directly express the behaviour of both cyber and physical components in the same specification, using one formalism. In this paper, we introduce a new specification language, Source Code and Signal Logic (SCSL), that 1) provides syntax specific to both signals and events originating in source code; and 2) does not require source code events to be abstracted into signals. We introduce SCSL by giving its syntax and semantics, along with examples. We then provide a comparison between SCSL and existing specification languages, using an example property, to show the benefit of using SCSL to capture certain types of properties.

Keywords: Specification language · Temporal logic · Source code · Signals

1 Introduction

Analysing the behaviour of components of Cyber-Physical Systems (CPS) often involves analysing the behaviour of the signals generated by such components. Analysis can begin with capturing the expected behaviour of these signals using formal specifications. For example, one could write a specification to capture the *temporal property* that, if a given signal falls below a certain threshold then, no more than ten units of time later, the same signal has risen back to a safe value again. Ultimately, deciding whether a given system satisfies such a specification is the goal of Runtime Verification (RV) [8].

In the context of RV, a number of languages have been introduced for the CPS setting, one of the most notable examples being Signal Temporal Logic (STL) [24]. STL is based on Metric Temporal Logic (MTL) [23], which allows one to capture temporal properties over *atomic propositions*, that is, symbols

S. L. Tapia Tarifa and J. Proença (Eds.): FACS 2022, LNCS 13712, pp. 20–38, 2022.
https://doi.org/10.1007/978-3-031-20872-0_2

that can be associated with a Boolean value. STL extends these propositions to be predicates over real-valued signals. For example, one can capture the property that a signal x should always be strictly less than 10 by writing $\Box\ x < 10$. Here, the *always* temporal operator $\Box$ is inherited from MTL, while the ability to compare the value of a signal x with another quantity is a novelty of STL.

Another example of a specification language introduced in the context of CPS is the Hybrid Logic of Systems (HLS) [25]. This language allows explicit reference to the behaviour of *cyber* and *physical* components of a CPS by providing access to both timestamps and indices. Here, a cyber component is often a component that contains software, and a physical component is one that measures a physical process. While the timestamps provided by HLS allow one to refer, as usual, to the behaviour of some physical process, the indices allow one to capture properties of the behaviour of cyber components. The capture of the behaviour of cyber components assumes that behaviour like program variable changes has been abstracted into Boolean signals, which can then be accessed using the indices provided by HLS. Ultimately, describing the behaviour of cyber components is more natural when using HLS than when using STL.

In the domain of cyber components, in particular capturing properties over source code-level behaviour, recent contributions include Inter-procedural Control-Flow Temporal Logic (iCFTL) [16]. iCFTL allows one to write constraints over events such as program variable changes and function calls by providing specific syntax for doing so.

Ultimately, iCFTL and HLS are complementary: HLS allows one to capture behaviour of signals and cyber components of CPS, but ultimately assumes that everything is encoded in signals. iCFTL does not support signals, but provides syntax specific to source code-level behaviour.

Hence, in this paper we introduce *Source Code and Signal Logic* (SCSL), which is a combination of iCFTL and HLS. In particular, SCSL ,provides syntax specifically designed for dealing with signal and source code-level behaviour. We also introduce a semantics that deals with traces representing CPS runs that are on-going. In line with existing work, such as that by Bauer at al. [9,10], our semantics uses an extended truth domain in order to deal with situations in which information that is required by a given specification may not yet be available. Once SCSL is introduced, we then provide a comparison of SCSL with existing specification languages, by attempting to capture a property using those languages, along with SCSL.

Related Work. Since we provide a comparison of SCSL with existing specification languages, our description of related work does not go into more depth with respect to contributions from the RV community. Instead, we focus on a subset of contributions from Model Checking, since these languages served as a starting point for a lot of contributions from the RV community. An example of a contribution is that by Alur et al. [6], where *hybrid systems* are assumed to be represented by automata augmented with time [7]. Other contributions are numerous, ranging from providing specification languages for model checking of systems that contain both cyber and physical components [13] to falsification of specifications [5, 20].

In contrast with the RV community, contributions from model checking focus on static analysis of a model of the system under scrutiny. When a system involves continuous behaviour, such as signals, this can give rise to probabilistic model checking approaches, such as the COMPASS project [12]. Our work assumes that a trace has already been generated, either by the real system or by a simulation. Hence, we do not consider the probabilistic setting.

Paper Structure. Section 2 motivates our new language and defines its design goals. We describe a notion of trace for CPS in Sect. 3. In Sect. 4, we introduce the concept of hybrid traces. Section 5 presents the syntax of SCSL; Sect. 6 illustrates its semantics. We present a comparison of SCSLwith existing specification languages in Sect. 7. Finally, Sect. 8 describes on-going further work, and Sect. 9 concludes the paper.

2 Motivation and Language Design Goals

Our main goal is to introduce a language that enables engineers to capture properties concerning the behaviour of multiple components in the system under scrutiny. To this end, we assume that the expression of properties concerning both the *cyber* and *physical* components of a system can involve placing constraints over the behaviour of 1) signals and 2) source code components (that control the signals). Hence, a specification language that allows one to capture such properties must provide syntax (and a semantics) specific to both of these domains.

Considering the signal and source code domains separately, the properties that one could aim to express can be taken from HLS and iCFTL respectively. For example, over signals, one might express the property that the value of a signal temperature never exceeds 100. Over source code, one might express the property that one statement is reached from another within a certain amount of time.

Let us consider a property that requires us to talk about both signals and source code behaviour at the same time, $\mathcal{P}1$: "*if the value of the signal* temperature *exceeds 100, then the time taken to reach the next call of the function* fix *should not exceed one second*". Here, we can express the two components of this property in two separate specification languages, but it would be useful to express the property as a whole using a single language.

Consider a further example, $\mathcal{P}2$: "*if the value of the signal* temperature *exceeds 100, then the next change of the variable* adjustment *must leave it with a value that is proportional to the value of* temperature". For this property, there are no distinct components that we could express in HLS and iCFTL; we must be able to relate a quantity taken from source code directly to a quantity taken from a signal.

These two examples define the key design goals for our new specification language:

G1 One must be able to define constraints over signals and source code within the same specification.

G2 One must be able to define *hybrid* constraints, that is, single constraints that relate quantities from signals with quantities from source code.

Ultimately, RV serves as a supplement to existing testing approaches. Hence, while these design goals are met by many existing specification languages, the language features that allow them to be met introduce a lot of additional effort to the software verification and validation process. This is discussed in depth in Sect. 7, which highlights the extra effort required (for a representative set of specification languages) if one wants to capture properties like $\mathcal{P}1$ and $\mathcal{P}2$.

3   Background

In line with other approaches in RV, we refer to our representation of a system's execution as a *trace*. Given our focus on the behaviour of source code-level behaviour in control components, and signal behaviour, our notion of trace must contain information from source code-level events, and signals. Hence, we begin by describing the traces used by iCFTL and HLS.

3.1   Traces for Signals Used by HLS [25]

The traces used by HLS are intended to represent a set of signals with the assumption that a value of each signal in the system is available at each timestamp being considered. This assumption is encoded in the *records* used by HLS, which are tuples of the form $\langle ts, \mathsf{index}, s_1, \ldots, s_n \rangle$, for ts a real-numbered timestamp, index an integer index and each s_i representing the value of the signal s_i at the timestamp ts. A signal is a sequence of such records, with strictly increasing timestamps and consecutive indices.

3.2   Traces for Source Code Used by iCFTL [16]

The traces introduced for iCFTL are tightly coupled with the source code that generated them, in that a trace can be seen as a path through a program. To support this idea, iCFTL introduces *concrete states*, which are intuitively the *states* reached by a program execution after the execution of individual program statements.

Formally, a concrete state c is a triple $\langle ts, \mathsf{pPoint}, \mathsf{values} \rangle$, for ts a real-numbered timestamp, pPoint a *program point*, and values a map from program variables to values. A program point pPoint is the unique identifier of the program statement whose execution generated the concrete state. Hence, program points capture the intuition that concrete states represent instantaneous checkpoints, within a single procedure, that a program can reach.

For a program point pPoint, we define the predicate $\mathsf{changed}(\mathsf{pPoint}, \mathsf{x})$ to indicate whether the statement at the program point pPoint assigns a value to the program variable x. Similarly, we define the predicate called(pPoint, f) to indicate whether the statement involves a call of the function f^1.

Returning to concrete states, a concrete state $c = \langle ts, \mathsf{pPoint}, \mathsf{values} \rangle$ can be said to *be attained* at time ts, which we denote by $\text{TIME}(\langle ts, \mathsf{pPoint}, \mathsf{values}\rangle)$. Since a concrete state holds a map m from program variables to their values, we will denote $\mathsf{values}(x)$, for a program variable x, by $c(x)$.

When concrete states are arranged in a sequence (ordered by timestamps ascending), we call a pair of consecutive concrete states a *transition*, often denoted by tr, because one can consider the computation required to move from one concrete state, to the other. For a transition $tr = \langle ts, \mathsf{pPoint}, \mathsf{values} \rangle$, $\langle ts', \mathsf{pPoint}', \mathsf{values}' \rangle$, we denote by $\text{TIME}(tr)$ the timestamp ts. Since transitions represent the computation performed to move between concrete states, we can also talk about the *duration* of the transition tr, which we denote by $\text{DURATION}(tr)$ and define as $ts' - ts$.

Ultimately, a sequence of concrete states generated by a single procedure in code is referred to as a *dynamic run* $\mathcal{D}$. We remark that, since we consider dynamic runs as being generated by individual procedures, concrete states from the same procedure execution cannot share timestamps. Further, we can consider a system consisting of multiple procedures and group the dynamic runs (generated by each procedure) together into a triple $\langle \{\mathcal{D}_i\}, \mathsf{Procs}, \mathsf{runToProc} \rangle$, where $\{\mathcal{D}_i\}$ is a set of dynamic runs, Procs is a set of names of procedures in the program, and $\mathsf{runToProc}$ is a map that labels each dynamic run in $\{\mathcal{D}_i\}$ with the name of a procedure in Procs. We call this triple an *inter-procedural dynamic run*.

Inside an inter-procedural dynamic run, given a concrete state c from some dynamic run, we write $\mathsf{proc}(c)$ to mean $\mathsf{runToProc}(\mathcal{D})$ for $\mathcal{D}$ being the dynamic run in which c is found. Similarly, for a transition tr, we write $\mathsf{proc}(tr)$.

4   Hybrid Traces

Based on the notions of traces introduced by iCFTL and HLS, we must now combine them to yield a kind of trace that contains both source code events, and signal entries. Such a notion of trace will serve as the basis for SCSL. Further, we will assume that all traces are generated by a CPS whose execution is on-going.

Our first step in combining the two notions of trace is to assume a global clock from which all timestamps (whether they be attached to concrete states or records) can be taken. We highlight that our approach is so far being developed in the context of simulators, so the existence of a global clock is a reasonable assumption. With this global clock, we collect the system's inter-procedural dynamic run and its sequence of records into a tuple $\langle \mathsf{signalNames}, \mathsf{records}, \mathsf{sigID}, \{\mathcal{D}_i\}, \mathsf{Procs}, \mathsf{runToProc} \rangle$, whose elements are as such:

– signalNames is a set of names of signals in the CPS.

---

[1] These predicates can be computed via static analyses of source code.

Entry point

$\phi \quad \rightarrow \forall v \in P_Q : \phi$
$\quad | \; true \; | \; \phi \vee \phi \; | \; \neg \phi \; | \; A$

Atomic constraints

$cmp \rightarrow \; < | > | =$
$A \quad \rightarrow V \; cmp \; V$

Terms

$V \quad \rightarrow V_{Ts} \; | \; V_C \; | \; V_{Tr} \; | \; V_{\mathsf{m}} \; | \; n$
$V_{Ts} \; \rightarrow \mathsf{signal}.\mathrm{AT}(Ts) \; | \; f(V_{Ts})$
$\quad \quad | \; \mathrm{TIME}(C) \; | \; \mathrm{TIME}(Tr)$
$V_C \; \rightarrow C(x) \; | \; f(V_C)$
$V_{Tr} \; \rightarrow \mathrm{DURATION}(Tr) \; | \; f(V_{Tr})$
$V_{\mathsf{m}} \; \rightarrow \mathrm{TIMEBETWEEN}(Ts, Ts)$

Expressions

$Ts \quad \rightarrow ts \; | \; n_{\mathsf{pos}} \; | \; Ts + Ts \; | \; \mathrm{TIME}(C) \; | \; \mathrm{TIME}(Tr)$
$C \quad \rightarrow c \; | \; ts.\mathrm{NEXT}(P_C) \; | \; C.\mathrm{NEXT}(P_C)$
$\quad \quad | \; Tr.\mathrm{NEXT}(P_C) \; | \; \mathrm{BEFORE}(Tr) \; | \; \mathrm{AFTER}(Tr)$
$Tr \quad \rightarrow tr \; | \; ts.\mathrm{NEXT}(P_{Tr}) \; | \; C.\mathrm{NEXT}(P_{Tr})$
$\quad \quad | \; Tr.\mathrm{NEXT}(P_{Tr})$

Predicates

$P_Q \quad \rightarrow P_{QTs} \; | \; P_{QC} \; | \; P_{QTr}$
$P_{QTs} \rightarrow [Ts, Ts] \; | \; (Ts, Ts)$
$P_{QC} \rightarrow P_C \; | \; P_C.\mathsf{after}(Ts)$
$P_C \quad \rightarrow \mathsf{changes}(\mathsf{var}).\mathsf{during}(\mathsf{proc})$
$P_{QTr} \rightarrow P_{Tr} \; | \; P_{Tr}.\mathsf{after}(Ts)$
$P_{Tr} \quad \rightarrow \mathsf{calls}(\mathsf{proc}_1).\mathsf{during}(\mathsf{proc}_2)$

Fig. 1. The syntax of SCSL.

- records is a sequence of records, each containing signal entries for each of the signals represented in signalNames.
- sigID is a map that sends each signal name s in signalNames to a sequence of triples $\langle ts, \mathsf{index}, s \rangle$ derived from records. These triples are obtained by projecting the records in records with respect to the signal name s. Assuming that one has used sigID(s) to obtain a sequence of triples, we denote by sigID$(s)(ts)$ the signal value held in the triple whose timestamp is ts. This map is included in the tuple (i.e., it is part of the trace) so that the correspondence between signal names and sequences of triples is fixed for a given trace.
- $\{\mathcal{D}_i\}$, Procs and runToProc are as introduced earlier.

Ultimately, we refer to the tuple introduced above as a *trace*, which we will denote by $\mathcal{T}$. Using this notation, we will often write $\mathcal{T}(s)$ instead of sigID(s), with the understanding that we are implicitly referring to the map sigID held in $\mathcal{T}$. Hence, to refer to the value of a signal s at the timestamp ts in the trace $\mathcal{T}$, we write $\mathcal{T}(s)(ts)$.

Remark on Obtaining Traces. We highlight that, in practice, the parts of a trace that are dynamic runs can be obtained by instrumenting the relevant source code of a CPS [15]. Further, obtaining signal entries depends heavily on the use case (i.e., whether signals are generated by a simulator, or by physical sensors).

5   SCSL Syntax

We now introduce the syntax of SCSL, which allows one to construct specifications over traces, as defined in Sect. 4. We give the syntax as a context-free grammar in Fig. 1, in which all uses of non-terminal symbols are highlighted in blue. The rules given in the grammar are divided into groups that cover the key roles of certain parts of specifications. Further, n is a constant, and n_{pos} is a real number that is greater than zero.

Entry Point. A specification is constructed by first using the rule ϕ. This allows one to generate a quantifier, along with a subformula that will be subject to the quantification. The role of a quantifier is to capture events from a trace, including concrete states, transitions, and timestamps of signal entries. Hence, quantifiers consist of a *predicate* P_Q and a *variable* v; P captures values from a trace, which are each bound to v ready to be used elsewhere in the specification.

Aside from the form of quantifiers enforced by the grammar, we place two additional constraints: 1) if a quantifier is not the root quantifier in a specification, it must depend on its closest parent quantifier (i.e., it must use the timestamp of the event captured by its parent quantifier to capture events that occur after that event); and 2) a specification can have no free variables.

Predicates. The predicates that one can use are arranged in the grammar by what they capture: timestamps (P_{QTs}), concrete states (P_{QC}), or transitions (P_{QTr}).

Atomic Constraints. Once a quantifier has been used to assign values to a variable, the next step is to define constraints over those values. This is done using the rule A, which generates a comparison between two *terms*, that is, strings obtained using the rule V. These terms represent values that are obtained by extracting certain information from concrete states, transitions or timestamps. For example, from a concrete state c, one might wish to refer to the value of the program variable x in that state, so one would use V_C to generate $c(\mathbf{x})$.

Expressions. Using variables from quantifiers, one can either place a constraint on the immediate value held by those variables, or one can search forwards in time using a specific criterion. For example, given a concrete state c, one could denote the next transition after c in the trace by $c.\text{NEXT}(P_{Tr})$. A predicate could then be generated using P_{Tr} to reflect the desired criterion.

5.1   Examples

We now present two sets of example SCSL specifications, and show how the design goals introduced in Sect. 2 are met.

Artificial Examples. Our first set consists of artificial examples that have been constructed to showcase the expressive power of SCSL.

Example 1. *"Whenever the signal* temperature *drops below 100 during the first 10 min of a system run, the time until the variable* flag *in the procedure* monitor *is changed should be no more than 1 s."*

$$\forall ts \in [0, 60 * 10] : \textsf{temperature}.\text{AT}(ts) < 100 \rightarrow$$
$$\text{TIMEBETWEEN}(ts, \text{TIME}(ts.\text{NEXT}(\textsf{changes}(\textsf{flag}).\textsf{during}(\textsf{monitor})))) \leq 1 \qquad \text{(E1)}$$

In this specification, we refer to the value of the signal temperature at the times-tamp held in the variable *ts*, and we refer to the time taken to reach a specific variable change, from the time *ts*.

This specification shows that we have met design goal **G1**, because we can refer to both signals and source code events in the same specification.

Example 2. *"If the procedure* **adjust** *is called by the procedure* **control**, *then the signal* temperature *should be equal to 100 within 1 s."*

$$\forall tr \in \mathsf{calls}(\mathtt{adjust}).\mathsf{during}(\mathtt{control}) :$$
$$\text{DURATION}(tr) < 1 \wedge \exists ts \in [\text{TIME}(tr), \text{TIME}(tr) + 1] : \mathsf{temperature}.\text{AT}(ts) = 100 \quad \text{(E2)}$$

In this specification, we use the time at which a given function was called to select timestamps, and then refer to a signal value at each of the timestamps identified.

Example 3. *"Within the first ten seconds of a CPS execution, the value of the program variable* **x** *in the procedure* **p** *should reflect the value of the most recent value of the signal* signal"*

$$\forall ts \in [0, 10] : \mathsf{signal}.\text{AT}(ts) = ts.\text{NEXT}(\mathsf{changes}(\mathtt{x}).\mathsf{during}(\mathtt{p}))(\mathtt{x}), \quad \text{(E3)}$$

This specification involves the comparison of a value extracted from a signal, and a value extracted from a program variable.

This specification shows that we have met design goal **G2**, because we can write a single atomic constraint that uses information from a signal entry, and from a source code event.

The ArduPilot System. The ArduPilot [3] system acts as an autopilot for various types of vehicle both in the simulation setting and in the real-world setting. Here, we give examples derived by inspecting the source code found in their GitHub repository [4]. We have simplified the names of some program variables and procedures to save space.

Example 4. This property is derived from the code in the file **fence.cpp** [2]. In this code, the procedure **fence_check** checks for the copter leaving some *safe region*, called a *fence*. To this end, a program variable **new_breaches** holds each example of a breach of the fence that has been detected. If the copter strays more than 100 m outside the fence, its *mode* is set to *landing* by a call of the function **set_mode**. Hence, the property *"If a copter strays more than 100 m outside a fence, the mode should be changed within 1 unit of time"* can be expressed as follows:

$$\forall q \in \mathsf{changes}(\mathtt{new\_breaches}).\mathsf{during}(\mathtt{fence\_check}) :$$
$$(q(\mathtt{new\_breaches}) \neq \mathtt{null} \wedge \mathsf{distFence}.\text{AT}(\text{TIME}(q)) > 100)$$
$$\rightarrow \text{TIMEBETWEEN}(\text{TIME}(q), \quad \text{(E4)}$$
$$\text{TIME}(\text{BEFORE}(q.\text{NEXT}(\mathsf{calls}(\mathtt{set\_mode}).\mathsf{during}(\mathtt{fence\_check}))))) \leq 1$$

We have introduced the signal distFence, which we assume contains a value representing the distance of the copter from the fence in metres.

Example 5. This property is derived from the code in the file `crash-check.cpp` [1]. In this file, the procedure `thrust_loss_check` checks the behaviour of the copter's thrust. This involves checking the attitude, the throttle, and the vertical component of the velocity. If the checks reveal a problem, `set_thrust_boost` is called.

A property capturing this behaviour could be *"If (a) The attitude is less than or equal to an allowed deviation; (b) The throttle satisfies the predicate P; and (c) The vertical component of velocity is negative; then `set_thrust_boost` should be called within 1 unit of time"*, which could be captured by the specification

$$\forall ts \in [0, L] : (\text{att}.\text{AT}(ts) < \text{maxDev} \wedge P(\text{thr}.\text{AT}(ts)) \wedge \text{vel}_z.\text{AT}(ts) < 0) \rightarrow$$
$$\exists c \in \text{calls}(\texttt{set\_thrust\_boost}).\text{during}(\texttt{thrust\_loss\_check}).\text{after}(ts) : \quad \text{(E5)}$$
$$\text{TIMEBETWEEN}(ts, \text{TIME}(\text{BEFORE}(c))) \leq 1.$$

for L some positive real number. Here, we assume that att (attitude), thr (throttle), and vel_z are signals. In order to refer to their values over time, we quantify over the interval $[0, L]$ using the variable ts. We take maxDev to be a constant, and P to be some atomic constraint allowed by the syntax in Fig. 1 (both to be decided by the engineer).

6   Semantics

We now introduce a function that takes a trace and an SCSLspecification, and gives a *truth value* reflecting the status of the trace, with respect to the specification. For example, if the trace satisfies the specification (that is, holds the property captured by the specification), then our function should give a value that indicates as such. The situation should be similar if the trace does not satisfy the specification. However, if there is not enough information in the trace to decide whether it satisfies the specification, then our function should give a value reflecting this.

Our semantics function makes use of multiple components, including 1) a way to extract the information from a trace that is needed by a term (Sect. 6.1); and 2) a way to determine the truth value of an atomic constraint and, from there, the specification as a whole (Sect. 6.2).

6.1   Determining Values of Terms

We will support our introduction of the first components of our semantics for SCSLusing the specification in Example E2.

This example includes the atomic constraints $\text{DURATION}(tr) < 1$ and temperature.$\text{AT}(ts) = 100$, along with a quantifier whose predicate is $[\text{TIME}(tr), \text{TIME}(tr) + 1]$. In each case, there is an expression, or a term, whose value we must determine, given either the transition held in the variable tr, or the timestamp held in the variable ts. To this end, we introduce the eval and getVal functions. Leaving a complete definition to [17], we consider how these functions would be applied to the various components of our running example:

$\mathcal{T}, \beta, ts \vdash [Ts_1, Ts_2]$ iff $\mathsf{getVal}(\mathcal{T}, \beta, Ts_1) \leq ts \leq \mathsf{getVal}(\mathcal{T}, \beta, Ts_2)$

$\mathcal{T}, \beta, ts \vdash (Ts_1, Ts_2)$ iff $\mathsf{getVal}(\mathcal{T}, \beta, Ts_1) < ts < \mathsf{getVal}(\mathcal{T}, \beta, Ts_2)$

$\mathcal{T}, \beta, \langle ts, \mathsf{pPoint}, \mathsf{values} \rangle \vdash \mathsf{changes(x).during(func)}$
        iff $\mathsf{changed}(\mathsf{pPoint}, \mathsf{x})$ and $\mathsf{proc}(\langle ts, \mathsf{pPoint}, \mathsf{values} \rangle) = \mathsf{func}$

$\mathcal{T}, \beta, q \vdash \mathsf{changes(x).during(func).after}(Ts)$
        iff $\mathrm{TIME}(q) > \mathsf{eval}(\mathcal{T}, \beta, Ts)$ and $\mathcal{T}, \beta, q \vdash \mathsf{changes(x).during(func)}$

$\mathcal{T}, \beta, tr \vdash \mathsf{calls(f).during(func)}$ iff $\mathsf{called}(\mathsf{pPoint'}, \mathsf{f})$ and $\mathsf{proc}(tr) = \mathsf{func}$

$\mathcal{T}, \beta, tr \vdash \mathsf{calls(f).during(func).after}(Ts)$
        iff $\mathrm{TIME}(tr) > \mathsf{eval}(\mathcal{T}, \beta, Ts)$ and $\mathcal{T}, \beta, tr \vdash \mathsf{calls(f).during(func)}$

Fig. 2. The valuation relation for SCSL. Timestamps ts are assumed to be in some record. Transitions tr denote, as defined in Sect. 3.2, pairs of concrete states $\langle ts, \mathsf{pPoint}, \mathsf{values} \rangle, \langle ts', \mathsf{pPoint'}, \mathsf{values'} \rangle$.

First, consider the atomic constraint $\mathrm{DURATION}(tr) < 1$. Deciding a truth value for this atomic constraint requires us to determine 1) the transition held by the variable tr, and 2) the duration of that transition. In this case, the getVal function is responsible for deriving the final value of $\mathrm{DURATION}(tr)$, while the eval function is used to determine the transition held by tr. For temperature.$\mathrm{AT}(ts) = 100$, the getVal function must determine the value to which the term temperature.$\mathrm{AT}(ts)$ evaluates. This then requires the eval function to determine the timestamp held by the variable ts. For $[\mathrm{TIME}(tr), \mathrm{TIME}(tr) + 1]$, in order for the getVal function to determine the relevant values, the eval function is needed to determine the value stored in the variable tr, whose timestamp can then be extracted.

In all three cases, we need the eval function to determine the value held by a variable. Hence, we need some structure that quantifiers can use to communicate the values that they capture with the eval function. We call such a structure a *valuation*, which is a map that associates with each variable in a specification a concrete state, transition or timestamp.

Using this idea, we can say that the eval function will take a trace, a valuation, and an expression, and return a unique result. In addition, we say that the getVal function will take a trace, a valuation, and a term, and return a unique result. Hence, we will write $\mathsf{eval}(\mathcal{T}, \beta, expr)$ and $\mathsf{getVal}(\mathcal{T}, \beta, term)$.

This notation is used by Fig. 2, which gives a recursive definition of the *valuation relation*. This relation defines which concrete states, transitions, or timestamps are captured by a predicate, hence providing a way to construct valuations. The relation also makes use of the $\mathsf{changed}(\mathsf{pPoint}, \mathsf{x})$ and $\mathsf{called}(\mathsf{pPoint}, \mathsf{f})$ predicates defined in Sect. 3.2.

We conclude our description of the eval function and the valuation relation with two remarks.

Returning Null. We do not assume that traces will always contain the information that a given term references. For example, a specification might refer to $c.\mathrm{NEXT}(\mathsf{changes(x).during(p)})$, but we might work with a trace that does not

$$[\mathcal{T}, \beta, \forall v \in P : \varphi] = \bigsqcap_{v \vdash P} [\mathcal{T}, \beta \dagger [c \mapsto v], \varphi] \quad [\mathcal{T}, \beta, \varphi_1 \vee \varphi_2] = [\mathcal{T}, \beta, \varphi_1] \sqcup [\mathcal{T}, \beta, \varphi_2]$$

$$[\mathcal{T}, \beta, \neg\varphi] = \overline{[\mathcal{T}, \beta, \varphi]} \quad [\mathcal{T}, \beta, true] = true$$

$$[\mathcal{T}, \beta, V_1 \; cmp \; V_2] = \begin{cases} trueSoFar & \mathsf{getVal}(\mathcal{T}, \beta, V_1) \neq null \text{ and } \mathsf{getVal}(\mathcal{T}, \beta, V_2) \neq null \\ & \text{and } \mathsf{getVal}(\mathcal{T}, \beta, V_1) \; cmp \; \mathsf{getVal}(\mathcal{T}, \beta, V_2) \\ falseSoFar & \mathsf{getVal}(\mathcal{T}, \beta, V_1) \neq null \text{ and } \mathsf{getVal}(\mathcal{T}, \beta, V_2) \neq null \\ & \text{and } \neg(\mathsf{getVal}(\mathcal{T}, \beta, V_1) \; cmp \; \mathsf{getVal}(\mathcal{T}, \beta, V_2)) \\ inconclusive & \text{otherwise} \end{cases}$$

Fig. 3. The semantics function for SCSL.

contain the relevant concrete state. To deal with such cases, we allow the eval and getVal functions to return null.

Interpolation of Signals. Suppose that a specification contains the atomic constraint signal.AT(Ts) < 1, where Ts is some expression that yields a timestamp. Depending on the expression Ts, the signal signal is not certain to have a value at that timestamp. Hence, we must interpolate. Our strategy in this work is to find the closest timestamp *in the future* at which the signal has a value. Interpolation, a common practice in CPS monitoring [25], is required because, otherwise, one would have to know the precise timestamps of events for a given CPS run.

6.2   A Semantics Function

We next introduce a semantics function that takes a trace, along with an SCSL specification, and yields a truth value reflecting the status of that trace with respect to the specification. Our semantics function assumes that the trace given represents a CPS execution that is on-going. In particular, our semantics function holds the *impartiality* property [9], that is, it does not generate a definitive verdict, rather a *provisional* one, since processing further events can lead to a change in verdict. Specifically, our semantics function declares *falseSoFar*, *inconclusive*, or *trueSoFar*. These truth values have the total ordering *falseSoFar* $<$ *inconclusive* $<$ *trueSoFar*. We also have that $\overline{trueSoFar} \equiv falseSoFar$, and $\overline{inconclusive} \equiv inconclusive$. In order to generate truth values in either of these domains, our semantics function works as follows: for a given trace, an appropriate truth value is computed by recursing on the structure of a specification, computing a truth value for each subformula. Truth values come from atomic constraints, by deciding the truth value of each constraint with respect to the information available in the trace. These truth values are then propagated up through the specification by the recursion. In line with this intuition, a semantics function is presented in Fig. 3. The function takes a trace $\mathcal{T}$, a valuation β, and a subformula φ, and computes a truth value. We now give a brief description of the approach taken by each case in Fig. 3 to evaluate a given part of a specification.

- Computing $[T, \beta, \forall v \in P : \varphi]$ consists of computing the greatest-lower-bound of the set of truth values $[T, \beta \dagger [v \mapsto e], \varphi]$, for each e (whether it be a concrete state, a transition or a timestamp) that satisfies P, according to the relation defined in Fig. 2. The $\dagger$, or *map amend*, operator is used to extend valuations with new values. For example, $a \dagger [v \mapsto n]$ refers to the map that agrees with the map a on all values except v, which it sends to n. This operator is used to extend a valuation once a v that satisfies P has been found.
- Computing $[T, \beta, \varphi_1 \vee \varphi_2]$ consists of computing the truth values of the two disjuncts, and then computing their least-upper-bound.
- Computing $[T, \beta, \neg\varphi]$ consists of taking the complement of $[T, \beta, \varphi]$.
- Computing $[T, \beta, true]$ consists of deciding on a truth value for this case requires no further computation, other than taking the truth value already used in the subformula.
- Computing $[T, \beta, V_1 \; cmp \; V_2]$ involves the weight of the work performed by the semantics, and is responsible for generating truth values that are propagated up through the specification. Specifically, *provisional* truth values are generated, including *trueSoFar*, *inconclusive*, and *falseSoFar*, depending on whether 1) the information necessary was found in the trace; and 2) that information satisfies the constraint in question.

Table 1. Comparison of specification languages and their features.

	$\mathcal{S}_{ts}$	$\mathcal{S}_{index}$	$\mathcal{X}_{sig}$	$\mathcal{X}_{code}$	$\mathcal{X}_{ll}$	$\mathcal{X}_{het}$	$\mathcal{X}_{index}$	$\mathcal{X}_{ts}$	$\mathcal{X}_{metric}$
SCSL	✓	✗	✓	✓	✓	✓	✗	✓	✗
LTL	✗	✓	✗	✗	✗	✗	✗	✗	✗
MTL	✓	✗	✗	✗	✗	✗	✗	✗	✓
TLTL	✗	✓	✗	✗	✗	✗	✗	✓	✗
HyLTL	✗	✓	✓	✗	✗	✓	✗	✗	✗
STL	✓	✗	✓	✗	✓	✗	✗	✗	✓
STL*	✓	✗	✓	✗	✓	✗	✗	✗	✓
STL-MX	✓	✓	✓	✗	✓	✓	✗	✗	✓
HLS	✓	✓	✓	✗	✓	✓	✓	✓	✗
SB-TemPsy	✓	✗	✓	✗	✓	✗	✗	✓	✓

7 Language Comparison

We now present a comparison of SCSL with existing specification languages, in order to demonstrate the novelty of this new language. Table 1 presents an initial comparison by highlighting a number of key features, which are defined as follows:

$\mathcal{S}_{ts}$ Whether a specification language's semantics is defined using timestamps to refer to entries in the trace. The form of such *entries* differs according to the specification language. For example, STL's semantics considers the pair (s, t) of a signal s and a timestamp t, and uses the timestamp t to refer to the signal s at the given timestamp.

$\mathcal{S}_{index}$ Whether a specification language's semantics is defined using indices to refer to entries in the trace. For example, LTL's semantics considers the pair (ω, i) for a trace ω and index i.

$\mathcal{X}_{sig}$ Whether a specification language provides syntax specific to signals. For example, HLS provides the @t operator, which enables one to write $(s$ @t $t)$ to refer to the value of the signal s at time t.

$\mathcal{X}_{code}$ Whether a specification language provides syntax specific to events generated at the source code level. For example, SCSL enables one to easily measure the duration of a function call with DURATION(tr) (for tr holding a *transition*).

$\mathcal{X}_{ll}$ Whether a specification language's syntax is at a low level of abstraction with respect to the system being monitored. For example, LTL abstracts behaviour into *atomic propositions*, whereas SCSL assumes that traces contain explicit representations of key events, such as program variable value changes and function calls.

$\mathcal{X}_{het}$ Whether a specification language provides explicit support for heterogeneity (components of multiple types, such as sensors and source code-based control components). For example, with SCSL one can write the constraint that $q(\mathbf{x}) = s.\text{AT}(t)$, which involves measurements taken from both signal behaviour and source code execution. Hence, SCSL can be said to support heterogeneity.

$\mathcal{X}_{index}$ Whether a specification language allows reference to events in a trace by their index. For example, HLS enables one to get the event in a signal based on its index by writing $(s$ @i $i)$ for a signal s and an index i.

$\mathcal{X}_{ts}$ Whether a specification language allows explicit reference to timestamps. For example, TLTL provides the $\triangleleft$ operator, which gives the timestamp of the most recent occurrence of some atomic proposition.

$\mathcal{X}_{metric}$ Whether a specification language's temporal operators are augmented with metrics. For example, MTL attaches a metric to its temporal operator $\mathcal{U}$, yielding the operator $\mathcal{U}_{[a,b]}$.

Justifications of our classification of each language are presented in [17]. Ultimately, Table 1 demonstrates the key feature of SCSL: syntax specific to the domain of application, which is signal and source code-based behaviour. In particular, though displaying only a representative set of languages, the table illustrates that the languages introduced by or adapted for the RV community offer a high level of abstraction with respect to the system being analysed. When considering specifications that talk about the behaviour of cyber components, this leads to the need to define a correspondence between runtime events and symbols used in a specification. While this approach often enables a language to be highly expressive (with the addition of complex modal operators), the use of generalised syntax means that expressing simple properties (such as the time taken by a function call) requires effort beyond simply writing the specification.

We now demonstrate the usefulness of SCSLby attempting to express the property *"whenever the program variable x is changed, eventually there is a call of the function f that takes less than 1 unit of time"* in each of the languages previously discussed. We remark that we have opted to use a property that does not require reference to signals so that we can include a wider range of languages in our comparison. In addition, we will consistently make use of the following *atomic propositions*:

- changed$_x$ to represent whether the program variable x has been changed.
- called$_f$ to represent whether the function f has been called.
- returned$_f$ to represent whether the function f has returned.

Linear Temporal Logic (LTL) [26]. This language has a high level of abstraction and provides complex modal operators. Its semantics is over *untimed words*, that is, sequences of atomic propositions that encode discrete time.

While expressing the example property is possible in LTL, effort would be required to define the correspondence and ensure that the specification was properly written to capture the variable change and function call behaviour (such as the combination of passing control to a function, and control being returned to the caller). Such a correspondence would make use of changed$_x$, called$_f$ and returned$_f$, but would also include timeLessThan$_1$, representing whether the time that elapsed since the last call to f is less than 1 unit of time. We might then write the specification

$$\Box\,(\text{changed}_x \rightarrow \Diamond\,(\text{called}_f \rightarrow \Diamond\,(\text{returned}_f \wedge \text{timeLessThan}_1)))$$

with globally, $\Box$, and eventually, $\Diamond$, having the expected semantics.

One can see that much of the actual computation required for checking the property would be migrated to the definition of the correspondence between runtime events and atomic propositions.

Metric Temporal Logic (MTL) [23]. This language extends LTL by attaching metrics to modal operators, allowing time constraints to be placed on the occurrence of events. The semantics of MTL is defined over *timed words*, which are sequences of atomic propositions with timestamps attached. Using the correspondence defined at the beginning of this section, we could then write the specification

$$\Box\,(\text{changed}_x \rightarrow \Diamond\,(\text{called}_f \rightarrow (\Diamond_{[0,1]}\,\text{returned}_f))).$$

A new operator is $\Diamond_I$, which is the *metric eventually* operator. For example, $\Diamond_{[a,b]}\,p$ means that, *eventually*, after a number of units of time in the interval $[a, b]$, p will become true.

Timed Linear Temporal Logic (TLTL) [10]. This language extends LTL with clock variables, which take the form of additional syntax used to check the time since/until an event occurred/will occur. Using the correspondence defined at the beginning of this section, we could then write the specification

$$\Box\ (\text{changed}_x \rightarrow \Diamond\ (\text{called}_f \rightarrow \Diamond\ (\text{returned}_f \wedge \lhd_{\text{called}_f} < 1))).$$

Here, $\lhd_{\text{called}_f}$ refers to the time at which the atomic proposition called_f was most recently true.

Hybrid Linear Temporal Logic (HyLTL) [13]. This language supports *hybrid behaviour*, meaning a combination of discrete and continuous behaviour, by extending LTL. Expressing the example property would be similar to LTL.

Signal Temporal Logic (STL) and variants [14,21,24]. Signal Temporal Logic [24], Signal Temporal Logic with a freeze quantifier (STL*) [14], and Mixed Time Signal Temporal Logic (STL-MX) [21] are all temporal logics whose semantics are defined over real-valued functions. While STL-MX is aimed at the heterogeneous setting (supporting both dense and discrete time), STL and STL* do not provide direct support for heterogeneity.

Since the behaviour of heterogeneous systems could be supported via instrumentation, one could capture the example property by abstracting the relevant system behaviour into signals, and using STL or its variants to express properties over that abstraction. However, this approach would require effort to 1) abstract complex behaviour into signals; and 2) correctly capture properties over such behaviour as properties over signals.

The Hybrid Logic of Systems (HLS) [25]. This language is a linear time, temporal logic whose semantics is defined over sequences of *records*, which are tuples $\langle t, i, v_1, \ldots, v_n \rangle$ for t a timestamp, i an index, and v_i signal values. Expressing the example property would require abstraction of the variable change and function call/return behaviour required by the property into Boolean signals. It would then be possible to use timestamp and index-based quantifiers to imitate the semantics of the modal operators provided by the other temporal logics considered so far. Hence, the atomic propositions used in previous examples would be interpreted as Boolean signals (to use the terminology in Sect. 4, sequences of *records* that associate timestamps with truth values). One further signal, timeSinceCall_f, would be necessary, to capture the amount of time since the most recent call of the function f. We assume that this signal would be computed given the other three signals.

We must then translate the modal operator $\Box$ into HLS, which can be expressed by $\forall t \in [0, L]$, for L the length of the trace being considered. Further, we can translate $\Diamond$ into HLS by writing $\exists t \in [t', L]$, for some *starting timestamp* t' and L again the length of the trace. Using this translation, the example property can be expressed as

$$\forall t \in [0, L] : ((\mathsf{changed}_x \; @t \; t) = 1 \rightarrow \exists t' \in [t, L] :$$
$$((\mathsf{called}_f \; @t \; t') = 1 \rightarrow \exists t'' \in [t', L] :$$
$$((\mathsf{returned}_f \; @t \; t'') = 1 \wedge (\mathsf{timeSinceCall}_f \; @t \; t'') < 1)$$
$$)$$
$$).$$

SB-TemPsy-DSL [11]. This language is a domain-specific, pattern-based language designed for expressing properties such as spiking, oscillation, undershoot and overshoot of signals. Its syntax follows the "scope" and "pattern" structure proposed by Dwyer et al. [19]. Its semantics is defined over traces which are assumed to be functions from timestamps to valuations of all signals being considered.

While runtime events can be extracted into signals, it would be non-trivial to express the property under consideration in SB-TemPsy-DSL, since the syntax focuses on a specific set of behaviours that a continuous signal could demonstrate.

Source Code and Signal Logic. Given our classification of SCSL, we highlight that:

– The lack of explicit referencing of indices ($\mathcal{X}_{\mathsf{index}}$) is not a disadvantage because SCSLprovides syntax specific to certain behaviour of cyber components of systems (namely source code level behaviour).
– The lack of metrics ($\mathcal{X}_{\mathsf{metric}}$) does not pose a problem because one can make explicit reference to timestamps.

The example property could be expressed as

$$\forall q \in \mathsf{changes}(\mathsf{x}).\mathsf{during}(\mathsf{p}) : \exists ts \in [\text{TIME}(q), L] :$$
$$\text{DURATION}(ts.\text{NEXT}(\mathsf{calls}(\mathsf{f}).\mathsf{during}(\mathsf{p}))) < 1.$$

where p is a procedure in the source code of the CPS under scrutiny, and L is the length of the trace.

Importantly, here there is no need for definition of a correspondence between runtime events and symbols used in the specification. We acknowledge that this specialisation of the syntax means that SCSL can only be used to express properties concerning the behaviour for which it was specifically designed. However, we argue that this enables a more intuitive language to be developed.

Ultimately, SCSL is a language with which one can capture source code and signal-based properties by referring directly to the events with which the properties are concerned.

7.1 Implications for Software Verification and Validation Processes

Throughout this section, we have seen that, while many existing languages allow one to capture the types of specifications with which this work is concerned, considerable additional work is usually required.

Taking TLTL as an example, if events that take place during an execution of a CPS are correctly encoded as atomic propositions, it is indeed possible to capture properties that concern both signals and source code-level events. However, this places considerable pressure on engineers to correctly define this correspondence. SCSL, on the other hand, is designed specifically for the signal and source code-level of granularity, meaning that there is no effort in the software verification and validation process beyond writing (and maintaining) the specification.

Similarly to the argument used in the initial introduction of CFTL [18] (the language that inspired iCFTL), we observe that, in some cases, the requirement to define a correspondence between runtime events and atomic propositions in a specification can be beneficial. In fact, such an approach can lead to a specification language that can be used to capture properties across a wide range of behaviours. This is indeed the case for the JAVA-MAC framework [22], in which one must first construct a specification, and then use a separate language to define the correspondence between runtime events and atomic propositions in the specification. However, as we have seen in this section, for a specific domain of application, it can be beneficial to use a language with specific features.

8 Ongoing Work

Our current work involves evaluating monitoring algorithms that we have developed for SCSL, based on the semantics given in Sect. 6. The evaluation has two objectives: investigating the performance of our monitoring algorithms in various situations (i.e., for various specifications and traces); and investigating the expressiveness of SCSL. We will test the expressiveness by selecting open source projects and attempting to capture representative requirements from those projects using SCSL.

Preliminary evaluations have given promising results, showing that it is feasible to construct algorithms for monitoring for SCSLspecifications in settings where 1) the trace is still being observed, as the system under scrutiny continues executing; and 2) the entire trace has already been observed. When the trace is still being observed, our preliminary results show that our online monitoring algorithm that can *keep up* with high event rates generated by systems under scrutiny. Alternatively, when the entire trace has already been observed, our results show that our offline monitoring algorithm scales approximately linearly with the trace length, in terms of time taken and memory consumed. Ultimately, due to space restrictions, we cannot include descriptions of these algorithms or preliminary results in this paper.

9 Conclusion

In this paper, we have introduced the new specification language SCSL, which allows engineers to explicitly specify the behaviour of source code-based components and signal-generating components of CPS. Our introduction of this new language has included a syntax, a semantics (suitable for the online and offline settings), and a comparison with existing specification languages using an example property. This comparison highlighted the benefits of SCSL: a syntax specialised to source code and signal-level behaviour, along with a semantics that assumes traces that contain information specific to signals and source code-level events.

As part of future work, we plan to investigate characteristics of SCSL, such as monitorability and satisfiability of specifications, along with diagnostics of specification violations. We also plan to carry out an extensive evaluation of the expressiveness of SCSLin the CPS domain.

Acknowledgments. The research described has been carried out as part of the COS-MOS Project, which has received funding from the European Union's Horizon 2020 Research and Innovation Programme under grant agreement No. 957254.

References

1. `Copter::crash_check` function - ArduPilot. https://github.com/ArduPilot/ardupilot/blob/a40e0208135c73b9f2204d5ddc4a5f281000f3f1/ArduCopter/crash_check.cpp#L100, accessed: 2022-04-13
2. `Copter::fence_check` function - ArduPilot. https://github.com/ArduPilot/ardupilot/blob/36f3fb316acf71844be80e0337fdc66515b4cf50/ArduCopter/fence.cpp#L9. Accessed 13 Apr 2022
3. The ArduPilot autopilot. https://ardupilot.org. Accessed 13 Apr 2022
4. The ArduPilot GitHub repository. https://github.com/ArduPilot/ardupilot. Accessed 13 Apr 2022
5. Abbas, H., Fainekos, G., Sankaranarayanan, S., Ivancic, F., Gupta, A.: Probabilistic temporal logic falsification of cyber-physical systems. ACM Trans. Embed. Comput. Syst. **12**(2s), 95:1–95:30 (2013). https://doi.org/10.1145/2465787.2465797
6. Alur, R., et al.: The Algorithmic analysis of hybrid systems. Theor. Comput. Sci. **138**(1), 3–34 (1995). https://doi.org/10.1016/0304-3975(94)00202-T
7. Alur, R., Dill, D.L.: A theory of timed automata. Theor. Comput. Sci. **126**(2), 183–235 (1994). https://doi.org/10.1016/0304-3975(94)90010-8
8. Bartocci, E., Falcone, Y., Francalanza, A., Reger, G.: Introduction to runtime verification. In: Bartocci, E., Falcone, Y. (eds.) Lectures on Runtime Verification. LNCS, vol. 10457, pp. 1–33. Springer, Cham (2018). https://doi.org/10.1007/978-3-319-75632-5_1
9. Bauer, A., Leucker, M., Schallhart, C.: Comparing LTL semantics for runtime verification. J. Logic Comput. **20**(3), 651–674 (2010). https://doi.org/10.1093/logcom/exn075
10. Bauer, A., Leucker, M., Schallhart, C.: Runtime verification for LTL and TLTL. ACM Trans. Softw. Eng. Methodol. **20**(4), 1-64 (2011). https://doi.org/10.1145/2000799.2000800
11. Boufaied, C., Menghi, C., Bianculli, D., Briand, L., Parache, Y.I.: Trace-checking signal-based temporal properties: a model-driven approach. In: Proceedings of the 35th IEEE/ACM International Conference on Automated Software Engineering, pp. 1004–1015. ASE 2020, Association for Computing Machinery, New York, NY, USA (2020). https://doi.org/10.1145/3324884.3416631

12. Bozzano, M., Bruintjes, H., Cimatti, A., Katoen, J.-P., Noll, T., Tonetta, S.: COM-PASS 3.0. In: Vojnar, T., Zhang, L. (eds.) TACAS 2019. LNCS, vol. 11427, pp. 379–385. Springer, Cham (2019). https://doi.org/10.1007/978-3-030-17462-0_25
13. Bresolin, D.: HyLTL: a temporal logic for model checking hybrid systems. Electron. Proc. Theor. Comput. Sci. **124**, 73–84 (2013). https://doi.org/10.4204/eptcs.124.8
14. Brim, L., Dluhos, P., Safránek, D., Vejpustek, T.: STL*: extending signal temporal logic with signal-value freezing operator. Inf. Comput. **236**, 52–67 (2014). https://doi.org/10.1016/j.ic.2014.01.012
15. Dawes, J.H.: Towards Automated Performance Analysis of Programs by Runtime Verification (2021). https://cds.cern.ch/record/2766727
16. Dawes, J.H., Bianculli, D.: Specifying properties over inter-procedural, source code level behaviour of programs. In: Feng, L., Fisman, D. (eds.) RV 2021. LNCS, vol. 12974, pp. 23–41. Springer, Cham (2021). https://doi.org/10.1007/978-3-030-88494-9_2
17. Dawes, J.H., Bianculli, D.: Specifying Properties over Inter-procedural, Source Code Level Behaviour of Programs (2022). http://hdl.handle.net/10993/52185, extended version
18. Dawes, J.H., Reger, G.: Specification of temporal properties of functions for runtime verification. In: Proceedings of the 34th ACM/SIGAPP Symposium on Applied Computing, pp. 2206–2214. SAC 2019, Association for Computing Machinery, New York, NY, USA (2019). https://doi.org/10.1145/3297280.3297497
19. Dwyer, M.B., Avrunin, G.S., Corbett, J.C.: Patterns in property specifications for finite-state verification. In: Proceedings of the 21st International Conference on Software Engineering. p. 411–420. ICSE '99, Association for Computing Machinery, New York, NY, USA (1999). https://doi.org/10.1145/302405.302672
20. Fainekos, G., Hoxha, B., Sankaranarayanan, S.: Robustness of specifications and its applications to falsification, parameter mining, and runtime monitoring with S-TaLiRo. In: Finkbeiner, B., Mariani, L. (eds.) RV 2019. LNCS, vol. 11757, pp. 27–47. Springer, Cham (2019). https://doi.org/10.1007/978-3-030-32079-9_3
21. Ferrère, T., Maler, O., Ničković, D.: Mixed-time signal temporal logic. In: André, É., Stoelinga, M. (eds.) FORMATS 2019. LNCS, vol. 11750, pp. 59–75. Springer, Cham (2019). https://doi.org/10.1007/978-3-030-29662-9_4
22. Kim, M., Viswanathan, M., Kannan, S., Lee, I., Sokolsky, O.: Java-MaC: a runtime assurance approach for java programs. Formal Meth. Syst. Des. **24**, 129–155 (2004). https://doi.org/10.1023/B:FORM.0000017719.43755.7c
23. Koymans, R.: Specifying real-time properties with metric temporal logic. Real-Time Syst. **2**(4), 255–299 (1990). https://doi.org/10.1007/BF01995674
24. Maler, O., Nickovic, D.: Monitoring temporal properties of continuous signals. In: Lakhnech, Y., Yovine, S. (eds.) FORMATS/FTRTFT -2004. LNCS, vol. 3253, pp. 152–166. Springer, Heidelberg (2004). https://doi.org/10.1007/978-3-540-30206-3_12
25. Menghi, C., Viganò, E., Bianculli, D., Briand, L.: Trace-checking CPS properties: bridging the cyber-physical gap. In: Proceedings of the 43rd International Conference on Software Engineering (ICSE'21), 23–29 May 2021, Virtual Event, Spain, pp. 847–859. IEEE, Los Alamitos, CA, USA (2021)
26. Pnueli, A.: The temporal logic of programs. In: 2013 IEEE 54th Annual Symposium on Foundations of Computer Science, pp. 46–57. IEEE Computer Society, Los Alamitos, CA, USA (oct 1977). https://doi.org/10.1109/SFCS.1977.32, https://doi.ieeecomputersociety.org/10.1109/SFCS.1977.32

Formally Characterizing the Effect of Model Transformations on System Properties

Rikayan Chaki[1] and Anton Wijs[2]([✉])

[1] Free University of Bozen-Bolzano, Bolzano, BZ 39100, Italy
`rikayan.chaki@stud-inf.unibz.it`
[2] Eindhoven University of Technology, Eindhoven, AZ 5612, The Netherlands
`A.J.Wijs@tue.nl`

Abstract. In Model-Driven Software Development, models and model transformations are the primary artefacts to develop software in a structured way. Models have been subjected to formal verification for a long time, but the field of formal model transformation verification is relatively young. Existing techniques, when they focus on the effect transformations have on the system components they are applied on, limit their analysis to checking for the preservation of semantics or particular properties, but it is not always the intention of a transformation to preserve these. We propose an approach to characterize the effect of applying a (formal description of a) model transformation when applied on a component that satisfies a given functional property. The given functional property is formalized in Action-based LTL, and our characterization is captured by a system of modal μ-calculus equations.

Keywords: Labeled transition systems · Model transformations · Linear temporal logic · μ-calculus · Büchi automata, system evolution

1 Introduction

Model-driven Software Engineering (MDE) [19] is an approach to software development in which models are the primary artefact to reason about the software under development. To support the workflow from initial model to concrete software, *model transformations* are applied, to automatically transform models to, for instance, refactor them, add information, or generate source code.

How to (formally) verify models has been studied for decades, but the verification of model transformations has only relatively recently started to gain attention. Several surveys have been published in which this field is explored, ranging from informal to formal approaches, and from syntactical checks to the analysis of the semantics of transformations [3,29].

In this paper, we focus on the formal verification of model-to-model transformations, considering code generation as being out of scope, since that tends

S. L. Tapia Tarifa and J. Proença (Eds.): FACS 2022, LNCS 13712, pp. 39–58, 2022.
https://doi.org/10.1007/978-3-031-20872-0_3

to require different verification approaches [29]. The aim is to develop a theory that can be used in combination with formal verification approaches for model analysis. The ultimate goal is to be able to efficiently apply formal verification during the entire MDE workflow.

Most of the work done so far on model transformation verification is focussed on syntactical checks, for instance checking whether the transformation is guaranteed to output models or code that is syntactically correct. Whenever semantical checks are performed, in the majority of cases, it is checked whether a transformation guarantees the *preservation* either of the complete semantics of the input model, or of particular properties, often formalized using temporal logic [5,6,10,14,15,17,25,28,32]. While these guarantees are very useful, sometimes semantics and/or properties are not intended to be preserved, and in those cases, very few techniques exist to reason about the applied changes.

The current paper addresses how to formally reason about model-to-model transformations, even when they do not preserve particular properties of input models, *independent* of the models on which they are applied, i.e., no concrete input models are involved in the analysis. In our setting, we focus on systems that consist of multiple *components*, and model transformations being defined as sets of model transformation rules. Each individual model transformation rule addresses the transformation of a single component. In this setting, we wish to reason about the effect a model transformation rule may have on a component on which it can be applied. This effect is expressed w.r.t. a particular functional property already satisfied by the component before it is transformed. In other words, the question we try to answer is: *Which functional property is a component C guaranteed to satisfy, after it has been transformed using a model transformation rule R, given the fact that C satisfies a given functional property φ before it is transformed?* Of course, we want this new property to be as insightful as possible, and not just expressing deadlock-freedom, for instance. We want this new property to be derived from φ, thereby expressing how R affects this property.

To do this formally, we use *Labeled Transition Systems* (LTSs) to define the behaviour of components, and *LTS transformation rules*, based on graph transformation rules, to formalize model transformations. Given a particular Action-based Linear Temporal Logic (ALTL) formula φ [13,26], we derive a temporal logic formula that expresses what the guaranteed effect is of the transformation rule on φ, when applied on an input LTS satisfying φ. In that sense, we formalize the effect the transformation has on the given input property. To express the formula characterizing this effect, we use the modal μ-calculus [20]. The construction of this formula is achieved by translating a Büchi automaton to a system of modal μ-calculus equations, similar to how this is defined in [11,18,22].

The structure of the paper is as follows. In Sect. 2, we discuss related work. Section 3 introduces the basic concepts used in the paper. Sect. 4 explains the formalization approach to model transformations. Here, we build on the theory developed by Wijs *et al.* in a number of papers [27,28,33–35,37]. In Sect. 5, we propose our technique to characterize the effect of model transformations, and

Sect. 6 focusses on a second example, on which we have applied an implementation of our technique. Finally, Sect. 7 provides our conclusions and a discussion on possible future work.

2   Related Work

Refinement checking [1,21], supported by tools such as RODIN [2], FDR3[1] and CSP-CASL-PROVER [16], checks if one model refines another. In contrast to our approach, model transformations are not represented as artefacts independent of the models they can be applied on. Instead, in refinement checking, two concrete models are compared, making it not possible to reason about the general effect of a model transformation.

The BART tool[2] involves refinement rules to transform B specifications, and checks whether the resulting specification is correct. Other work focussed on B addresses strictly refining functionality [23].

There are multiple approaches to prove semantics preservation of model transformations by proving that a transformed model is (weakly) bisimilar to the original model [6,10,14,15,17,25]. One of the authors of the current paper has proposed a technique to verify property preservation by means of bisimulation checks [27,28,33–35,37]. In [8], a technique is proposed to reason about the effect of updating a system component. Property preservation is also the main focus there.

In [7], a technique is proposed to update CTL formulae as functional behaviour is added to a model. Replacing or removing existing behaviour is not allowed. Work on *incremental model checking* [30,31] is more flexible regarding system changes, but, similar to refinement checking, does not focus on model transformations as separate artefacts to be subjected to verification. Instead, concrete transformed models are compared to the original ones, as is done in some model transformation verification approaches [17,25,32]. This is less general, requiring the verification to be applied every time the transformation is performed.

Our translation scheme from Büchi automata to systems of modal μ-calculus equations is similar to the ones proposed in earlier work [11,18,22], except that our scheme addresses action-based Büchi automata that correspond with Action-based LTL formulae [13].

3   Background

There are some concepts that must be understood to appreciate the paper. In this section, these are explained. The semantics of system components are represented by *Labeled Transition Systems* (LTSs):

---

[1] http://www.fsel.com/fdr3.html.

[2] http://www.tools.clearsy.com/tools/bart.

Definition 1 (Labeled Transition System). *An LTS M is defined as a tuple $M = (Q, A, \delta)$, where*

- *Q is a (finite) set of states;*
- *A is a set of actions;*
- *$\delta \subseteq Q \times A \times Q$ is a transition relation.*

Sometimes, a non-empty, finite number of states of an LTS is marked as being *initial.* We refer to this set of states with $Q_0 \subseteq Q$. We use a subscript-based notation to refer to the attributes of LTSs. As an example of this notation, $\mathcal{G}_Q$, $\mathcal{G}_{Q_0}$, $\mathcal{G}_A$, and $\mathcal{G}_\delta$ refer to the set of states, initial states, action set, and transition relation of the LTS $\mathcal{G}$.

With $q \xrightarrow{a}_M q'$, where $q, q' \in M_Q$ and $a \in M_A$ of some LTS M, we express that $(q, a, q') \in M_\delta$.

With $Act_M(q)$, we refer to the set of actions that are associated with the outgoing transitions of state q, i.e., $Act_M(q) = \{a \in M_A \mid \exists q' \in M_Q.q \xrightarrow{a}_M q'\}$.

In this paper, we only consider LTSs that are *weakly connected*, i.e., the undirected version of an LTS is a single connected component (from each state, there is a path to each other state). When there are initial states, we furthermore assume that every state in an LTS is reachable from at least one initial state.

An (infinite) *execution* of a system component is represented by a *trace* in the LTS expressing the potential behaviour of that component. A trace of an LTS M is formally defined as an *infinite* sequence $w = \langle w_0, w_1, \ldots \rangle$ that satisfies the following two conditions:

1. $\forall i \in \mathbb{N}.w_i \in M_A$;
2. $\exists q_0 \in M_{Q_0} \wedge \forall i \in \mathbb{N}.\exists q_{i+1} \in M_Q.q_i \xrightarrow{w_i}_M q_{i+1}$.

We use w_i to denote the ith element of the trace w, and w^i to denote the trace starting in w_i, i.e., $w^i = \langle w_i, w_{i+1}, \ldots \rangle$.

The *language* of an LTS M, denoted by $L(M)$, is defined as consisting of exactly all traces of M.

To express functional properties of components, we use an action-based version of LTL [26], usually referred to as ALTL [13].

Definition 2 (Action-based Linear Temporal Logic (ALTL)). *The syntax of an ALTL formula φ is recursively defined as*

$$\varphi = a \mid \neg\varphi_1 \mid \varphi_1 \wedge \varphi_2 \mid X\,\varphi_1 \mid \varphi_1 \,U\, \varphi_2$$

where φ_1 and φ_2 are ALTL formulae and a is an action belonging to the LTS on which the property is validated.

The operators X and U are the next and until operators, respectively. The semantics of an ALTL formula φ is defined by the relation $M, w \models \varphi$, where M is an LTS and w is a trace of this LTS, as follows.

$$M, w \models a \iff w_1 = a$$
$$M, w \models \neg\varphi_1 \iff M, w \not\models \varphi_1$$
$$M, w \models \varphi_1 \wedge \varphi_2 \iff M, w \models \varphi_1 \wedge M, w \models \varphi_2$$
$$M, w \models \mathsf{X}\varphi_1 \iff M, w^1 \models \varphi_1$$
$$M, w \models \varphi_1 \mathsf{U} \varphi_2 \iff \exists j . M, w^j \models \varphi_2 \wedge \forall k . k < j \implies M, w^k \models \varphi_1$$

An ALTL formula φ is said to be satisfied by an LTS M iff all the traces in $L(M)$ satisfy φ.

A shorthand for $\bigvee_{a \in M_A} a$ is TRUE, while FALSE represents the proposition that cannot be matched by any of the actions in M_A. Additional syntax can be defined using the existing syntax. Typical additional syntax is $\varphi_1 \vee \varphi_2 = \neg(\neg\varphi_1 \wedge \neg\varphi_2)$, $\varphi_1 \Rightarrow \varphi_2 = \neg\varphi_1 \vee \varphi_2$, $\mathsf{F}\varphi_1 = \text{TRUE} \mathsf{U} \varphi_1$ and $\mathsf{G}\varphi_1 = \neg\mathsf{F}\neg\varphi_1$.

An ALTL formula φ can be translated to an action-based *Büchi automaton* $\mathcal{B}^\varphi$ [9,13]. Typically, this is done for the automata-based way of verifying that an LTS satisfies an ALTL formula.

Definition 3 (Büchi automaton). *A Büchi automaton $\mathcal{B}$ is defined as a tuple $(Q, Q_0, \Omega, \delta, F)$, where*

- *Q is a (finite) set of states;*
- *$Q_0 \subseteq Q$ is a (non-empty) set of initial states;*
- *Ω is a set of logical action formulae over actions;*
- *$\delta \subseteq Q \times \Omega \times Q$ is a transition relation;*
- *$F \subseteq Q$ is a set of accepting states.*

Similarly as for LTSs, we use a subscript-based notation to refer to attributes of Büchi automata. For instance, $\mathcal{B}_Q$ refers to the states of Büchi automaton $\mathcal{B}$. The semantics of a Büchi automaton mandates that all of its valid traces must encounter at least one of the states in F infinitely often. In other words, for a Büchi automaton $\mathcal{B}$, $L(\mathcal{B})$ contains all traces with an infinite number of accepting states.

The labels in action-based Büchi automata typically are logical action formulae ω containing actions, combined with $\wedge$ and $\neg$. These formulae express requirements for the actions occurring in LTSs. For instance, the Büchi automaton label $\neg a \wedge \neg b$ can be matched with any LTS action other than a or b.

Finally, to characterize model transformations, we need a logic that is more expressive than ALTL. For this, we use the modal μ-calculus [20]:

Definition 4 (Modal μ-calculus). *A μ-calculus formula Φ is defined as*

$$\Phi = \text{FALSE} \mid \neg\Phi_1 \mid \Phi_1 \wedge \Phi_2 \mid [a]\Phi_1 \mid \mu X . \Phi_1 \mid \nu X . \Phi_1 \mid X$$

where X is a variable, Φ_1 and Φ_2 are μ-calculus formulae, and a is an action belonging to the model on which the property is validated.

The semantics of Φ produces a set of states with respect to a given LTS M. The interpretation $[\![\Phi]\!]_{M,\eta}$ defines this set for a μ-calculus formula Φ, w.r.t.

LTS M and propositional context $\eta : \mathcal{X} \rightarrow 2^{M_Q}$, which is a partial function mapping propositional variables (taken from the set of all such variables $\mathcal{X}$) to sets of states. The notation $\eta \oslash [Q'/X]$ represents a propositional context equal to η except for the fact that variable X is mapped to the state set Q'. The interpretation is now defined as follows.

$$[\![\text{FALSE}]\!]_{M,\eta} = \emptyset$$
$$[\![\neg\Phi_1]\!]_{M,\eta} = M_Q \setminus [\![\Phi_1]\!]_{M,\eta}$$
$$[\![\Phi_1 \wedge \Phi_2]\!]_{M,\eta} = [\![\Phi_1]\!]_{M,\eta} \cap [\![\Phi_2]\!]_{M,\eta}$$
$$[\![[a]\Phi_1]\!]_{M,\eta} = \{q \in M_Q \mid \forall q' \in M_Q . q \xrightarrow{a}_{M_\delta} q' \Rightarrow q' \in [\![\Phi_1]\!]_{M,\eta}\}$$
$$[\![\mu X.\Phi_1]\!]_{M,\eta} = \bigcap\{Q' \subseteq M_Q \mid Q' \supseteq [\![\Phi_1]\!]_{M,\eta \oslash [Q'/X]}\}$$
$$[\![\nu X.\Phi_1]\!]_{M,\eta} = \bigcup\{Q' \subseteq M_Q \mid Q' \subseteq [\![\Phi_1]\!]_{M,\eta \oslash [Q'/X]}\}$$
$$[\![X]\!]_{M,\eta} = \eta(X)$$

Additional syntax can be defined using the existing syntax. Typical additional syntax is $\text{TRUE} = \neg\text{FALSE}$, $\Phi_1 \vee \Phi_2 = \neg(\neg\Phi_1 \wedge \neg\Phi_2)$ and $\langle a \rangle \Phi_1 = \neg[a]\Phi_1$.

A modal μ-calculus formula can be written as a *system of modal μ-calculus equations* using the *equational modal μ-calculus*.

Definition 5 (Equational modal μ-calculus). *A system of modal μ-calculus equations E is a finite list of equations, each of the form $\nu X_i = \Phi_i$ or $\mu X_i = \Phi_1$. In such a system E, the X_i are all distinct, and the Φ_i are basic μ-calculus formulae, i.e., do not contain μ or ν. In a system E with n equations, Φ_i can use any of the variables $X_0, \ldots, X_{i-1}, X_{i+1}, \ldots, X_n$ as unbounded variables. All occurrences of X_i in Φ_i must be bounded.*

An LTS M is said to satisfy a system E iff all of its initial states are contained in the interpretation of the variable of the first equation, i.e., $M_{Q_0} \subseteq [\![X_0]\!]_{M,\eta}$.

4   Formalized Model Transformations

A model transformation specification defines when and how a given model should be transformed. The approach used to formalize such specifications, in a setting in which the individual components of a model are represented by LTSs, is adapted from earlier work of one of the authors [27,28,33–35,37]. In this approach, a rule R, defined as a pair of *pattern LTSs* $(\mathcal{L}, \mathcal{R})$, is used to define a model transformation specification.

Definition 6 (Transformation rule). *A transformation rule $R = (\mathcal{L}, \mathcal{R})$ consists of LTSs $\mathcal{L}$ and $\mathcal{R}$, with $\mathcal{L}_Q \cap \mathcal{R}_Q \neq \emptyset$.*

The states common to $\mathcal{L}$ and $\mathcal{R}$, i.e., $\mathcal{L}_Q \cap \mathcal{R}_Q$, are known as *glue states*. These serve as anchors when transforming an LTS using R.

To reason about model transformations, we first define *LTS morphism*.

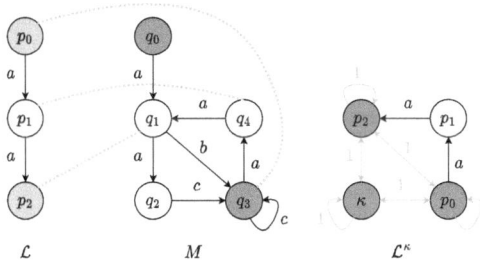

Fig. 1. A pattern LTS $\mathcal{L}$, an LTS M on which $\mathcal{L}$ can be matched, and κ-extended $\mathcal{L}$.

Definition 7 (LTS morphism). *An LTS morphism $f : M^0 \to M^1$ between LTSs M^0, M^1 is a pair of functions $f = (f_Q : M_Q^0 \to M_Q^1, f_\delta : M_\delta^0 \to M_\delta^1)$ such that for all $q \xrightarrow{a}_{M^0} q'$ it holds that $f_\delta(q \xrightarrow{a}_{M^0} q') = f_Q(q) \xrightarrow{a}_{M^1} f_Q(q')$.*

For LTSs, it is meaningless to have multiple transitions with the same source state, target state, and action label. When all transitions in an LTS are unique in this sense, a state function f_Q implies a transition function f_δ. Therefore, we reason in this paper about LTS morphisms as mappings between LTS states.

A rule R is said to be *applicable* on a given LTS M if and only if the set of states of $\mathcal{L}$ can be mapped onto a set of states of M such that the transitions between the states of $\mathcal{L}$ can also be mapped on transitions of M, using the state mapping.

Definition 8 (Match). *Given a transformation rule $R = (\mathcal{L}, \mathcal{R})$, a pattern LTS $\mathcal{P} \in \{\mathcal{L}, \mathcal{R}\}$ of R has a match $m : \mathcal{P}_Q \to M_Q$ on an LTS M iff m is an injective LTS morphism and for all $p \in \mathcal{P}_Q, q \in M_Q$:*

1. *$m(p) = q \wedge q \in M_{Q_0} \implies p \in \mathcal{L}_Q \cap \mathcal{R}_Q$;*
2. *$q \xrightarrow{a}_M m(p) \wedge (\neg \exists p' \in \mathcal{P}_Q . p' \xrightarrow{a}_{\mathcal{P}} p \wedge m(p') = q) \implies p \in \mathcal{L}_Q \cap \mathcal{R}_Q$;*
3. *$m(p) \xrightarrow{a}_M s \wedge (\neg \exists p' \in \mathcal{P}_Q . p \xrightarrow{a}_{\mathcal{P}} p' \wedge m(p') = q) \implies p \in \mathcal{L}_Q \cap \mathcal{R}_Q$.*

A match is a behaviour preserving morphism of a pattern LTS in an LTS. Condition 1 of Definition 8 states that if a pattern LTS state is mapped on an initial state, then the former must be a glue state. A consequence of this is that initial states are never removed when applying a transformation. The other two conditions ensure that no transitions can be removed, when transforming, that are not matched by the pattern LTS. This ensures that no side-conditions occur.

Figure 1 shows a pattern LTS $\mathcal{L}$ on the left, with the glue states colored red. In the middle an LTS M is shown, with initial states colored blue. Note that the only possible match m for $\mathcal{L}_Q$ on M_Q is defined as $m(p_0) = q_3, m(p_1) = q_4, m(p_2) = q_1$. In particular, the mapping m' defined as $m'(p_0) = q_0, m'(p_1) = q_1, m'(p_2) = q_2$ is *not* a match, as it violates conditions 2 and 3 of Definition 8: the state q_1 of M has an incoming transition $q_4 \xrightarrow{a}_M q_1$ and an outgoing transition $q_1 \xrightarrow{b}_M q_3$ that are both involved in the mapping, but state p_1, which is matched on state q_1, is not a glue state.

An LTS M can be transformed if R is applicable. Transformation proceeds by using the glue states as anchors and replacing the states of M matched on non-glue states of $\mathcal{L}$ by new states matched on non-glue states of $\mathcal{R}$, in a manner that preserves the relationship between the glue and non-glue states. The transformation induced by a rule R, as described above, can be obtained by transforming M iteratively over every match m found with respect to R on M. In the following, note that with $m(\mathcal{P}_Q)$, we refer to the set of all states to which the states in $\mathcal{P}_Q$ are mapped by m.

Definition 9 (LTS Transformation). *Let M be an LTS and $R = (\mathcal{L}, \mathcal{R})$ a transformation rule with match $m : \mathcal{L}_Q \to M_Q$. Moreover, consider match $\hat{m} : \mathcal{R}_Q \to T(M)_Q$, with $T(M)$ the transformed M, and for all $q \in \mathcal{L}_Q \cap \mathcal{R}_Q$, we have $\hat{m}(q) = m(q)$, and for all $q \in \mathcal{R}_Q \setminus \mathcal{L}_Q$, we have $\hat{m}(q) \notin M_Q$ (by which $\hat{m}$ introduces new states in $T(M)$). The* transformation of M, via R with matches $m, \hat{m}$ *is defined as $T(M)$, with*

- $T(M)_Q = (M_Q \setminus m(\mathcal{L}_Q)) \cup \hat{m}(\mathcal{R}_Q)$;
- $T(M)_\delta = (M_\delta \setminus \{m(p) \xrightarrow{a}_M m(p') \mid p \xrightarrow{a}_{\mathcal{L}} p'\}) \cup \{\hat{m}(p) \xrightarrow{a} \hat{m}(p') \mid p \xrightarrow{a}_{\mathcal{R}} p'\}$;
- $T(M)_A = \{a \mid \exists q \xrightarrow{a}_{T(M)} q'\}$.

It may be the case that $\mathcal{L}$ can be matched more than once on M, and these matches may overlap. In this paper, we assume that transforming an LTS M to $T(M)$ by applying a rule R on all its matches on M is *confluent*, i.e., leads to one unique result, independent of the order in which the matches are processed. This depends on the pattern LTSs of R. Techniques exist to determine whether a transformation rule is confluent [36].

In this work, the aim is to reason about the model transformations without involving any concrete components on which the transformations are applicable. Thus, a representation is needed for all components on which a given transformation rule can be applied. This representation must be able to *simulate* all behaviour of all those components. In the following definition, we introduce the notion of LTS simulation, with 1 being a special action that can simulate any other action.

Definition 10 (LTS simulation). *Given LTSs M, M' with the special action $1 \in M_A$, $1 \notin M'_A$. We say that M can simulate M' iff $M'_A \subset M_A$ and there exists a binary relation $D \subseteq M_Q \times M'_Q$ such that*

1. *for every $q' \in M'_{Q_0}$, there exists a $q \in M_{Q_0}$ such that $q \, D \, q'$, and*
2. *for all $q \in M_Q$, $q' \in M'_Q$, $q \, D \, q'$ implies that if $q' \xrightarrow{a}_{M'} \hat{q}'$ for some $\hat{q}' \in M'_Q$, $a \in M'_A$, then either $q \xrightarrow{a}_M \hat{q}$ or $q \xrightarrow{1}_M \hat{q}$ for some $\hat{q} \in M_Q$, and $\hat{q} \, D \, \hat{q}'$.*

Given a left pattern LTS $\mathcal{L}$ of a rule R, we define a suitable representation for all components on which R is applicable as the κ-*extended version* of $\mathcal{L}$, called $\mathcal{L}^\kappa$. We present a simplified version of its original definition [28], suitable for our purpose.

Definition 11 (κ-extended $\mathcal{L}$). *Given a transformation rule $\mathcal{P} = (\mathcal{L}, \mathcal{R})$, the κ-extension of $\mathcal{L}$, denoted by $\mathcal{L}^{\kappa}$, is defined as:*

- $\mathcal{L}_Q^{\kappa} = \mathcal{L}_Q \cup \{\kappa\}$;
- $\mathcal{L}_A^{\kappa} = \mathcal{L}_A \cup \{1\}$;
- $\mathcal{L}_{\delta}^{\kappa} = \mathcal{L}_{\delta} \cup \{\kappa \xrightarrow{1} \kappa\} \cup \{\kappa \xrightarrow{1} p, p \xrightarrow{1} \kappa, p \xrightarrow{1} p' \mid p, p' \in \mathcal{L}_Q \cap \mathcal{R}_Q\}$;

Furthermore, we define $\mathcal{L}_{Q_0}^{\kappa} = \{\kappa\} \cup (\mathcal{L}_Q \cap \mathcal{R}_Q)$. The state κ is a new state, and 1 a unique label, representing behaviour to, from and between glue states and κ.

If one considers an LTS M on which $\mathcal{L}$ is applicable, there must exist at least one match m associating the states and transitions of $\mathcal{L}$ with (a subset of) the states and transitions of M. In such a scenario, one can consider the state κ to be an abstraction of all the states of M that are not matched in m, and the 1-transitions to represent the transitions not matched in m. The action 1 represents therefore all the actions of transitions that are not matched. Note that in the κ-extended $\mathcal{L}$, all the glue states and κ are considered initial. For the glue states, we do this since for an arbitrary LTS, glue states can be matched on its initial states, and likewise, κ may also represent initial states.

On the right of Fig. 1, the κ-extension is given of the pattern LTS $\mathcal{L}$ shown on the left. The 1-transitions have been colored gray, to highlight the original pattern. For convenience, two 1-transitions between two states in both directions are represented by a single bidirectional edge, and the states of $\mathcal{L}^{\kappa}$ have been organised to resemble M w.r.t. match m. $\mathcal{L}^{\kappa}$ can simulate all behaviour of M. For instance, state q_0 is represented by κ, and the transition $q_0 \xrightarrow{a}_M q_1$ is represented by $\kappa \xrightarrow{1}_{\mathcal{L}^{\kappa}} p_2$. Likewise, state q_2 is also represented by κ, and transition $q_2 \xrightarrow{c}_M q_3$ can be simulated by $\kappa \xrightarrow{1}_{\mathcal{L}^{\kappa}} p_0$. Finally, transition $q_1 \xrightarrow{b}_M q_3$ is simulated by $p_2 \xrightarrow{1}_{\mathcal{L}^{\kappa}} p_0$, and $q_3 \xrightarrow{c}_M q_3$ is simulated by $p_0 \xrightarrow{1}_{\mathcal{L}^{\kappa}} p_0$.

With the following lemma, we prove that an LTS on which a rule $R = (\mathcal{L}, \mathcal{R})$ is applicable can be simulated by $\mathcal{L}^{\kappa}$.

Lemma 1. *Any LTS M on which a rule $R = (\mathcal{L}, \mathcal{R})$ is applicable can be simulated by $\mathcal{L}^{\kappa}$.*

Proof. We prove this for a single match $m : \mathcal{L}_Q \to M_Q$. The case that there are multiple matches of $\mathcal{L}_Q$ on M_Q can be proven in a similar way. We prove the lemma by defining a simulation relation D that adheres to the match m, i.e., for all $p \in \mathcal{L}_Q$, $q \in M_Q$ with $m(p) = q$, we have $p \, D \, q$. For all other states $q' \in M_Q$, not matched on by a state in $\mathcal{L}_Q$, we define $\kappa \, D \, q'$. That this defines a simulation relation can be seen as follows. For every state $q \in M_Q$, we distinguish two cases:

1. There exists a state $p \in \mathcal{L}_Q$ with $m(p) = q$. In this case, for every transition $q \xrightarrow{a}_M q'$, we distinguish two cases:
   (a) There exists a state $p' \in \mathcal{L}_Q$ with $m(p') = q'$. In this case, either $p \xrightarrow{a}_{\mathcal{L}^{\kappa}} p'$ and $p' \, D \, q'$, if this a-transition is part of $\mathcal{L}$ between p and p', or $p \xrightarrow{1}_{\mathcal{L}^{\kappa}} p'$ and $p' \, D \, q'$, if both p and p' are glue states.

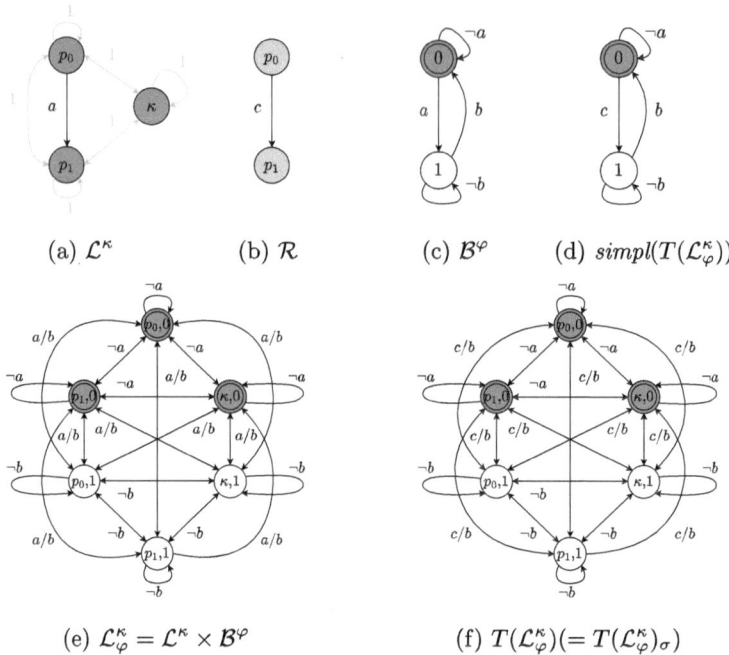

(a) $\mathcal{L}^{\kappa}$ (b) $\mathcal{R}$ (c) $\mathcal{B}^{\varphi}$ (d) $simpl(T(\mathcal{L}^{\kappa}_{\varphi}))$

(e) $\mathcal{L}^{\kappa}_{\varphi} = \mathcal{L}^{\kappa} \times \mathcal{B}^{\varphi}$ (f) $T(\mathcal{L}^{\kappa}_{\varphi})(= T(\mathcal{L}^{\kappa}_{\varphi})_{\sigma})$

Fig. 2. An example rule and its effect on systems satisfying $\mathsf{G}(a \Rightarrow \mathsf{F}\,b)$.

   (b) There does not exist a state $p' \in \mathcal{L}_Q$ with $m(p') = q'$. In that case, we
      have $p \xrightarrow{1}_{\mathcal{L}^{\kappa}} \kappa$ with $\kappa \, D \, q'$.

2. There does not exist a state $p \in \mathcal{L}_Q$ with $m(p) = q$. In this case, we have
   $\kappa \, D \, q$. For every transition $q \xrightarrow{a}_M q'$, we distinguish two cases:

   (a) There exists a state $p' \in \mathcal{L}_Q$ with $m(p') = q'$. In this case, $\kappa \xrightarrow{a}_{\mathcal{L}^{\kappa}} q'$ and
      $p' \, D \, q'$.

   (b) There does not exist a state $p' \in \mathcal{L}_Q$ with $m(p') = q'$. In that case, we
      have $\kappa \xrightarrow{1}_{\mathcal{L}^{\kappa}} \kappa$ with $\kappa \, D \, q'$.   □

5 Transformation Characterization

This section presents the technique we propose to characterize the effect of (formalized) model transformations. Suppose that we want to reason about the
model transformation illustrated by the rule R in Fig. 2, Subfigs. 2a and 2b,
which effectively replaces a-transitions by c-transitions. We wish to apply this
rule on systems satisfying the ALTL formula $\varphi = \mathsf{G}(a \Rightarrow \mathsf{F}\,b)$, i.e., globally, when
a occurs, eventually b occurs. We use this property and the model transforma-
tion specification of R in the remainder of this section to illustrate our developed
approach, which consists of four steps.

5.1   Merging the ALTL Formula φ and the κ-extended LTS $\mathcal{L}^\kappa$

The first step of our approach is to construct $\mathcal{L}^\kappa$, and to *refine* $\mathcal{L}^\kappa$ to make explicit that any LTS on which R is applied satisfies φ. This adds information to $\mathcal{L}^\kappa$ that will help us to formalize how R, when applied, affects an LTS satisfying φ. Inspired by the standard automata-based method to verify an LTL property [4], we first convert the property φ into a Büchi automaton $\mathcal{B}^\varphi$. For our example, Subfig. 2c shows the Büchi automaton corresponding to φ. State 0 is the only accepting state (indicated by its double border): note that any trace in which the occurrence of an a is eventually followed by a b is accepted by this automaton.

In the automata-based method of LTL verification, the cross-product of a Kripke structure and the Büchi automaton must be constructed. For our action-based setting, we define the cross-product of a κ-extended $\mathcal{L}$ of a rule $R = (\mathcal{L}, \mathcal{R})$ and an (action-based) Büchi automaton, denoted by $\mathcal{L}^\kappa \times \mathcal{B}^\varphi$. Since our κ-extended LTSs have 1-transitions, which should be interpreted as in Büchi automata, we interpret all actions in $\mathcal{L}^\kappa$ as logical action formulae, where an action a ($\neq 1$) can only be matched by a.

Before we can define the cross-product, we must define the function γ, and the interpretation $\langle\!\langle \omega \rangle\!\rangle_{\mathcal{L}^\kappa}$ of the logical action formula ω. The latter defines the set of actions in $\mathcal{L}^\kappa_A$ that can match ω. We define $\langle\!\langle \omega \rangle\!\rangle_{\mathcal{L}^\kappa}$ as follows, with ω_1, ω_2 logical action formulae and a an action:

$$\langle\!\langle \mathrm{TRUE} \rangle\!\rangle_{\mathcal{L}^\kappa} = \mathcal{L}^\kappa_A$$
$$\langle\!\langle a \rangle\!\rangle_{\mathcal{L}^\kappa} = \mathcal{L}^\kappa_A \cap \{a\}$$
$$\langle\!\langle \neg\omega \rangle\!\rangle_{\mathcal{L}^\kappa} = \mathcal{L}^\kappa_A \setminus \langle\!\langle \omega \rangle\!\rangle_{\mathcal{L}^\kappa}$$
$$\langle\!\langle \omega_1 \vee \omega_2 \rangle\!\rangle_{\mathcal{L}^\kappa} = \langle\!\langle \omega_1 \rangle\!\rangle_{\mathcal{L}^\kappa} \cup \langle\!\langle \omega_2 \rangle\!\rangle_{\mathcal{L}^\kappa}$$
$$\langle\!\langle \omega_1 \wedge \omega_2 \rangle\!\rangle_{\mathcal{L}^\kappa} = \langle\!\langle \omega_1 \rangle\!\rangle_{\mathcal{L}^\kappa} \cap \langle\!\langle \omega_2 \rangle\!\rangle_{\mathcal{L}^\kappa}$$

The function γ is given two logical action formulae ω, ω'. It first identifies, for these two logical action formulae, the set $\langle\!\langle \omega \rangle\!\rangle_{\mathcal{L}^\kappa} \cap \langle\!\langle \omega' \rangle\!\rangle_{\mathcal{L}^\kappa}$, and then produces a logical action formula corresponding to that set.

Definition 12 (Cross-product $\mathcal{L}^\kappa \times \mathcal{B}$). *Given a κ-extended $\mathcal{L}$ and Büchi automaton $\mathcal{B}$. The product $\mathcal{L}^\kappa \times \mathcal{B}$ is then a Büchi automaton, consisting of the following, where ω, ω' are logical action formulae.*

- $(\mathcal{L}^\kappa \times \mathcal{B})_Q = \mathcal{L}^\kappa_Q \times \mathcal{B}_Q;$
- $(\mathcal{L}^\kappa \times \mathcal{B})_{Q_0} = \mathcal{L}^\kappa_{Q_0} \times \mathcal{B}_{Q_0};$
- $\langle q, s \rangle \xrightarrow{\gamma(\omega,\omega')}_{\mathcal{L}^\kappa \times \mathcal{B}} \langle q', s' \rangle \iff q \xrightarrow{\omega}_{\mathcal{L}^\kappa} q' \wedge s \xrightarrow{\omega'}_{\mathcal{B}} s' \wedge \langle\!\langle \omega \rangle\!\rangle_{\mathcal{L}^\kappa} \cap \langle\!\langle \omega' \rangle\!\rangle_{\mathcal{L}^\kappa} \neq \emptyset;$
- $(\mathcal{L}^\kappa \times \mathcal{B})_A = \{\omega \mid \exists \langle q, s \rangle, \langle q', s' \rangle . \langle q, s \rangle \xrightarrow{\omega}_{\mathcal{L}^\kappa \times \mathcal{B}} \langle q', s' \rangle\};$
- $(\mathcal{L}^\kappa \times \mathcal{B})_F = \{\langle q, s \rangle \in (\mathcal{L}^\kappa \times \mathcal{B})_Q \mid s \in \mathcal{B}_F\}.$

In the following, we refer with $\mathcal{L}^\kappa_\varphi$ to the cross-product $\mathcal{L}^\kappa \times \mathcal{B}^\varphi$. Note that as $\mathcal{L}^\kappa$ can simulate all behaviour of systems on which R is applicable, and as $\mathcal{B}^\varphi$ accepts all traces satisfying φ, the cross-product $\mathcal{L}^\kappa_\varphi$ is a Büchi automaton accepting all potential traces of LTSs on which R is applicable that satisfy φ.

For our example, Subfig. 2e presents the cross-product $\mathcal{L}_\varphi^\kappa$. For illustrative purposes, two transitions between two states in both directions are combined into a single bidirectional edge, and if those transitions have two different labels a and b, the label a/b is shown, with the first label being applicable for the transition in the direction in which $\mathcal{B}^\varphi$ moves to a state with a higher index. For instance, the edge between states $\langle s_1, 0 \rangle$ and $\langle s_0, 1 \rangle$ with label a/b represents an a-transition from $\langle s_1, 0 \rangle$ to $\langle s_0, 1 \rangle$ and a b-transition from $\langle s_0, 1 \rangle$ to $\langle s_1, 0 \rangle$.

5.2   Transforming $\mathcal{L}_\varphi^\kappa$

The aim of step 2 in our process is to transform the Büchi automaton that resulted from step 1, by identifying all possible matches of $\mathcal{L}$ on $\mathcal{L}_\varphi^\kappa$ and applying LTS transformation for all those matches. Some adjustments are needed for this, as Definitions 7, 8 and 9 address how to match on and transform LTSs, as opposed to Büchi automata. Essentially, we can interpret Büchi automata as LTSs, if we ignore the acceptance status of the states, and if we interpret logical action formulae as the actions.

Since the transitions of $\mathcal{L}_\varphi^\kappa$ are associated with logical action formulae, we are restricted to matching the transitions of $\mathcal{L}$ on transitions of $\mathcal{L}_\varphi^\kappa$ with logical action formulae that syntactically correspond with single action labels, e.g., labels a and b as opposed to $\neg a$, $\neg b$, $a \vee b$, etc. Extending Definitions 7 and 8 to support matches on logical action formulae in general would not be a sound approach. For instance, consider the logical action formula $\neg b$, when trying to match an a-transition. We could define that an a-transition of $\mathcal{L}$ can be matched on a $\neg b$-transition of $\mathcal{L}_\varphi^\kappa$, as the latter transition *might* map on an a-transition of some LTS represented by $\mathcal{L}^\kappa$. However, $\mathcal{L}^\kappa$ likewise also represents LTSs for which this $\neg b$-transition is not mapped to an a-transition, and for those cases, this match would not be possible. Hence, if we want to reason about the guaranteed effect of a model transformation, we should be conservative in our matching.

For our running example, Subfig. 2f presents the transformed cross-product, denoted by $T(\mathcal{L}_\varphi^\kappa)$. Note that in this example, all the a-transitions have been transformed to c-transitions, but a has not been completely removed from the logical action formulae. In particular, the formula $\neg a$ cannot be transformed, for the reason given above. In this case, however, it is also obvious that matches on $\neg a$-transitions are not possible: $\neg a$ refers to all actions except a.

5.3   Detecting and Removing Non-Accepting Cycles in $T(\mathcal{L}_\varphi^\kappa)$

In the context we consider, we have $L(M) \subseteq L(\mathcal{L}^\kappa) = L(\mathcal{L}_\varphi^\kappa) \subseteq L(\mathcal{B}^\varphi)$. By Lemma 1, we have $L(M) \subseteq L(\mathcal{L}^\kappa)$, and $L(\mathcal{L}^\kappa) = L(\mathcal{L}_\varphi^\kappa)$ follows from the fact that $\mathcal{L}^\kappa$ satisfies φ. We wish to achieve that $T(\mathcal{L}_\varphi^\kappa)$ represents the property characterizing the effect of applying R on an LTS M satisfying φ, i.e., we want $L(T(\mathcal{L}_\varphi^\kappa))$ to express what $L(T(\mathcal{L}^\kappa))$ satisfies. However, we do not necessarily have $L(T(\mathcal{L}^\kappa)) \subseteq L(T(\mathcal{L}_\varphi^\kappa))$. In particular, if R introduces non-accepting cycles in $T(\mathcal{L}_\varphi^\kappa)$, then any trace w leading to such a cycle and traversing it infinitely often is not accepted by $T(\mathcal{L}_\varphi^\kappa)$, but it does correspond to a trace in $T(\mathcal{L}^\kappa)$.

Our characterizing property should reflect that newly introduced cycles can be present. To do this, we must identify when a transformation rule *may* introduce new cycles, and ensure that at least one state in those cycles is accepting.

A transformation rule R may introduce a new cycle in an LTS M when it adds new transitions to M, ignoring transition labels. Let $conn(M) = \{(q_0, q_1) \mid \exists a \in M_A.q_0 \xrightarrow{a}_M q_1\}$ be the set of all pairs of states in the LTS M between which a transition exists in M_δ. Then, formally, R may introduce new cycles iff $conn(T(\mathcal{L}_\varphi^\kappa)) \setminus conn(\mathcal{L}_\varphi^\kappa) \neq \emptyset$. In particular, if $conn(T(\mathcal{L}_\varphi^\kappa)) \subseteq conn(\mathcal{L}_\varphi^\kappa)$, then no new connections are made between states in $\mathcal{L}_\varphi^\kappa$, and since $\mathcal{L}_\varphi^\kappa$ represents $M \times \mathcal{B}^\varphi$ for all M satisfying φ, i.e., it does not contain non-accepting cycles, neither does $T(\mathcal{L}_\varphi^\kappa)$.

Based on these observations, we propose the following procedure: for every transition $\langle p, s \rangle \xrightarrow{a}_{T(\mathcal{L}_\varphi^\kappa)} \langle p', s' \rangle$ with $(\langle p, s \rangle, \langle p', s' \rangle) \in conn(T(\mathcal{L}_\varphi^\kappa)) \setminus conn(\mathcal{L}_\varphi^\kappa)$, check whether this transition is in a non-accepting cycle in $T(\mathcal{L}_\varphi^\kappa)$ using a Depth-First Search. If it is, mark $\langle p, s \rangle$ as an accepting state. The worst-case time complexity of this procedure is $\mathcal{O}(|T(\mathcal{L}_\varphi^\kappa)_\delta|^2)$, but pattern LTSs are typically very small, so this is reasonable. We refer to the resulting Büchi automaton as $T(\mathcal{L}_\varphi^\kappa)_\sigma$.

5.4   Constructing a Characteristic Formula for $T(\mathcal{L}_\varphi^\kappa)_\sigma$

In the final stage, the goal is to obtain a temporal logic formula that characterizes the effect of the model transformation on the ALTL formula that the original component satisfies. Since ALTL is strictly less expressive than Büchi automata, ALTL is not suitable to translate $T(\mathcal{L}_\varphi^\kappa)_\sigma$ to a temporal logic formula. Instead, we propose the following scheme to translate an action-based Büchi automaton $\mathcal{B}$ to a system of modal μ-calculus equations E.

For every state $q \in \mathcal{B}_Q$, a variable X_q and a formula Ψ_q is defined such that

$$X_q = \Psi_q \overset{\text{def}}{=} \bigvee \{ \langle \omega \rangle \text{TRUE} \wedge [\omega] \bigvee \{X_{q'} \mid q \xrightarrow{\omega}_\mathcal{B} q'\} \mid \omega \in Act_\mathcal{B}(q)\}$$

Subsequently, we define an additional equation $X = \Psi' \overset{\text{def}}{=} \bigvee \{X_q \mid q \in \mathcal{B}_{Q_0}\}$. All these equations are finally combined into the following set of equations E_S:

$$E_S = \{\mu X = \Psi'\} \cup \{\mu X_q = \Psi_q \mid q \in \mathcal{B}_Q \setminus \mathcal{B}_F\} \cup \{\nu X_q = \Psi_q \mid q \in \mathcal{B}_F\}$$

The intuition behind the use of the fixed point operators μ and ν is as follows: the minimal fixed point operator μ refers to behaviour that can only be exhibited a finite number of times, while the maximal fixed point operator ν refers to behaviour that must be exhibited an infinite number of times. Hence, ν is placed before equations relating to Büchi accepting states, while μ is used for equations relating to Büchi non-accepting states.

With a function f, we finally order the equations in E_S to obtain a list, i.e., each equation X' is assigned to a unique index $0 \leq i < |E_S|$, resulting

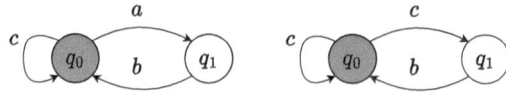

Fig. 3. $G(a \Rightarrow F\,b)$ is satisfied on the left, $G(c \Rightarrow F\,b)$ is not satisfied on the right.

in $f(i) = X'$. The order can be chosen arbitrarily, except that we must have $f(0) = X$, since X defines the starting point for verification.

Our translation scheme closely resembles previous schemes [11,18,22], with the key difference being that we involve logical action formulae instead of propositional formulae. In particular, [18] translates a state change from a state s to a state s' under the condition that the propositional formula p is **true** to the equation $X_s = p \wedge [\text{TRUE}]X_{s'}$.[3] In our case, we express with $\langle a \rangle \text{TRUE}$ that one can change state by following an a-transition, hence $\langle a \rangle \text{TRUE}$ can be seen as the condition for the state change expressed by $X_s = \langle a \rangle \text{TRUE} \wedge [a]X_{s'}$.

We illustrate our translation scheme using the Büchi automaton in Subfig. 2c. The following equation system encodes this Büchi automaton:

$$\mu X = X_0$$
$$\nu X_0 = (\langle \neg a \rangle \text{TRUE} \wedge [\neg a]X_0) \vee (\langle a \rangle \text{TRUE} \wedge [a]X_1)$$
$$\mu X_1 = (\langle \neg b \rangle \text{TRUE} \wedge [\neg b]X_1) \vee (\langle b \rangle \text{TRUE} \wedge [b]X_0)$$

In the final step of our technique, to obtain a new property formalized as a temporal logic formula, we first simplify the Büchi automaton $T(\mathcal{L}_\varphi^\kappa)_\sigma$, to minimize the number of states in it. For instance, the techniques implemented in the SPOT framework can be used for this [12]. We refer with the function *simpl* to this operation. In our running example, the Büchi automaton in Subfig. 2f can be minimized, using SPOT, to the Büchi automaton shown in Subfig. 2d.

Finally, using our translation scheme, we construct a system of modal μ-calculus equations. For our example, we obtain the following system E_φ:

$$\mu X = X_0$$
$$\nu X_0 = (\langle \neg a \rangle \text{TRUE} \wedge [\neg a]X_0) \vee (\langle c \rangle \text{TRUE} \wedge [c]X_1)$$
$$\mu X_1 = (\langle \neg b \rangle \text{TRUE} \wedge [\neg b]X_1) \vee (\langle b \rangle \text{TRUE} \wedge [b]X_0)$$

It is interesting to note that this new formula *almost* describes what we perhaps would expect from applying rule R in general, namely that after the a-transitions have been transformed to c-transitions, the property $G(c \Rightarrow F\,b)$ would be satisfied. However, consider an LTS satisfying $G(a \Rightarrow F\,b)$. This LTS may already contain c-transitions. An example of this is shown in Fig. 3 on the left. For instance, the trace $\langle c, c, \ldots \rangle$ satisfies $G(a \Rightarrow F\,b)$. On the right in Figure 3, we have an LTS that would result from applying the transformation rule R of Fig. 2 on the LTS on the left in Fig. 3. Note that the LTS on the right

---

[3] In the μ-calculus as defined in [18], this is expressed as $X_s = p \wedge [\cdot]X_{s'}$.

does not satisfy $G(c \Rightarrow F b)$, as the trace $\langle c, c, \ldots \rangle$ does not satisfy this property. However, it does satisfy the property expressed by E_φ.

In the following Theorem, we address the correctness of our technique.

Theorem 1. *Given an LTS M satisfying an ALTL formula φ, and a transformation rule $R = (\mathcal{L}, \mathcal{R})$ applicable on M, the transformed LTS $T(M)$ obtained by applying R on M is guaranteed to satisfy the system of modal μ-calculus equations E_φ constructed with our technique.*

Proof sketch. We prove this by reasoning towards a contradiction. Consider an LTS M satisfying the ALTL formula φ, on which a rule $R = (\mathcal{L}, \mathcal{R})$ is applicable, but the transformed LTS $T(M)$ does not satisfy the system E_φ constructed with our technique. If this is the case, then there must be a counter-example for the new formula, i.e., a trace w in $L(T(M))$ which is not in $L(simpl(T(\mathcal{L}^\kappa_\varphi)_\sigma))$, and hence, since $L(simpl(T(\mathcal{L}^\kappa_\varphi)_\sigma)) = L(T(\mathcal{L}^\kappa_\varphi)_\sigma)$, not in $L(T(\mathcal{L}^\kappa_\varphi)_\sigma)$. This counter-example must correspond to some (invalid) trace w in $T(M) \times T(\mathcal{L}^\kappa_\varphi)_\sigma$ from an initial state to a cycle σ involving only non-accepting states. In $T(\mathcal{L}^\kappa_\varphi)_\sigma$, w must also be invalid, involving a cycle σ' with only non-accepting states.

The cycle σ' cannot have been introduced by transformation, since all non-accepting cycles in $T(\mathcal{L}^\kappa_\varphi)$ introduced by R have been removed from $T(\mathcal{L}^\kappa_\varphi)_\sigma$. Furthermore, since the transformation of a Büchi automaton does not transform accepting states into non-accepting states, σ' must correspond to some cycle σ'' in $\mathcal{L}^\kappa_\varphi$ that only involves non-accepting states. Since all states in an LTS are reachable from an initial state, σ'' is reachable from an initial state in $\mathcal{L}^\kappa_\varphi$. Since $\mathcal{L}^\kappa_\varphi = \mathcal{L}^\kappa \times \mathcal{B}^\varphi$, this means a counter-example w' exists for $\mathcal{L}^\kappa$ w.r.t. φ.

As w is invalid in $L(T(\mathcal{L}^\kappa \times \mathcal{B}^\varphi))$, we must have $w \in L(T(\mathcal{L}^\kappa))$, which corresponds to a trace $w' \in L(\mathcal{L}^\kappa)$. Since $w \in L(T(M))$, we must also have $w' \in L(M)$. We can identify the following relations between states:

- A simulation relation D between $\mathcal{L}^\kappa_Q$ and M_Q, as in Lemma 1;
- Matches $m : \mathcal{L} \to M$ and $m' : \mathcal{L} \to \mathcal{L}^\kappa$.

Note that the fact that a transition $p \xrightarrow{a}_{\mathcal{L}^\kappa} p'$ is transformed using m' implies that neither p nor p' is κ and $a \neq 1$, and also, that the transitions $q \xrightarrow{a}_M q'$, with $p \ D \ q$, $p' \ D \ q'$ are transformed using m. Stated differently, the fact that $\mathcal{L}$ can be matched on some states $Q \subseteq M_Q$ means that those states and their transitions adhere to the pattern $\mathcal{L}$, but that implies that they can be related to the states in $\mathcal{L}^\kappa_Q$ that adhere to that pattern as well.

Next, we prove by induction that, given that $w \in L(T(\mathcal{L}^\kappa))$, $w \in T(M)$, and $w' \in L(\mathcal{L}^\kappa)$, also $w' \in L(M)$. We do this by induction on the $w_i \in w$.

- $i = 0$: $p_0 \xrightarrow{w'_0}_{\mathcal{L}^\kappa} p_1$ leads to $p'_0 \xrightarrow{w_0}_{T(\mathcal{L}^\kappa)} p'_1$. We distinguish two cases:
  1. p_0 and p_1 are matched on by m', i.e., $m'(p_0) = p'_0$ and $m'(p_1) = p'_1$. In that case, by the above reasoning, for states $q_0, q_1 \in M_Q$, with $p_0 \ D \ q_0$ and $p_1 \ D \ q_1$, we have a transition that is transformed in the same way, and can be simulated by $p_0 \xrightarrow{w'_0}_{\mathcal{L}^\kappa} p_1$, hence $q_0 \xrightarrow{w'_0}_M q_1$.

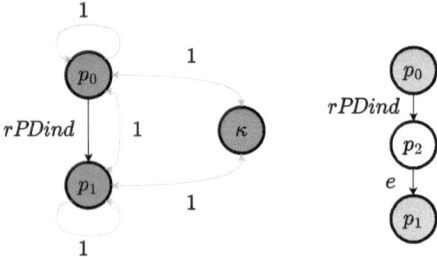

Fig. 4. κ-extended $\mathcal{L}$ (left) and $\mathcal{R}$ (right) of the transformation rule R_{rPDind}.

2. p_0 and p_1 are not matched on by m'. In that case, $w'_0 = w_0$. Since $w \in L(T(M))$, we must also have states $q_0, q_1 \in M_Q$ with $q_0 \xrightarrow{w_0}_M q_1$.

– $i+1$: By the induction hypothesis, transitions for the actions $w'_0, \ldots, w'_i$ exist in M. Existence of a transition with action w'_{i+1} that extends this path can be proven in the same way as for the base case.

The fact that $w' \in L(M)$ and that w' is a counter-example in $\mathcal{L}^\kappa \times \mathcal{B}^\varphi$ for φ implies that w' is also a counter-example in $M \times \mathcal{B}^\varphi$ for φ. But this contradicts with the assumption that M satisfies φ. □

6   A Progress Example

We have implemented our approach in a prototype tool, written in PYTHON 3.0. For the conversion of ALTL formulae into Büchi automata, we use the SPOT framework [12]. In order to further illustrate our approach and the practical use of it, we applied our tool on another example.

The model transformation specification we consider here has been used in the past to transform components in communication protocol specifications such as one for the IEEE-1394 protocol [24,28]. Figure 4 shows the transformation, formalized as a rule R_{rPDind}, in which the receiving of a message, represented by $rPDind$, is transformed to be followed by an (internal) step labelled e, representing the processing of the received message.

A typical property to be satisfied by a communication protocol is that actual progress is made, meaning that always eventually messages are received. In this example, this can be specified in ALTL as $\psi = \mathsf{G}\,\mathsf{F}\,rPDind$, where $rPDind$ is an action representing a message being received by the component on which the transformation is applied. The Büchi automaton for this is given in Fig. 5.

In step 1, we construct the product $\mathcal{L}^\kappa_\psi = \mathcal{L}^\kappa \times \mathcal{B}^\psi$. Figure 6 presents $\mathcal{L}^\kappa_\psi$ on the left, where, for readability, $rPDind$ has been abbreviated to r. When we check for matches of the left pattern of R_{rPDind} on $\mathcal{L}^\kappa_\psi$, we observe that $rPDind$-transitions between two different states can be matched on, but $rPDind$-selfloops of individual states *cannot* be matched on, due to the condition that a match must be an *injective* LTS morphism (see Definition 8), i,e, it is not allowed that multiple states in a pattern LTS are matched on the same state in the LTS or

Büchi automaton to be transformed. The effect of this is that, for instance, the $rPDind$-transition from state $\langle p_0, 0 \rangle$ to state $\langle p_1, 0 \rangle$ can be matched on, but the $rPDind$-transition from and to state $\langle p_0, 0 \rangle$ cannot.

We do not illustrate $T(\mathcal{L}_\psi^\kappa)$ here, but one can imagine what it looks like: every $rPDind$-transition between two different states is transformed to a sequence of two transitions, $rPDind$ followed by e, between the same two states, with a newly introduced state connecting the two transitions. The resulting Büchi automaton consists of 16 states.

R_{rPDind} introduces new states, and hence new transitions, but none of these transitions are in non-accepting cycles in $T(\mathcal{L}_\psi^\kappa)$, hence there is no need to introduce more accepting states. With the SPOT framework, we can minimize $T(\mathcal{L}_\psi^\kappa)$ to the Büchi automaton given on the right in Fig. 6.

Finally, when we apply our translation of a Büchi automaton to a system of modal μ-calculus equations, we obtain the following system E_ψ:

$$\mu X = X_0$$
$$\nu X_0 = (\langle rPDind \rangle \text{TRUE} \wedge [rPDind](X_0 \vee X_2) \vee$$
$$\quad (\langle \neg rPDind \rangle \text{TRUE} \wedge [\neg rPDind]X_1)$$
$$\mu X_1 = (\langle rPDind \rangle \text{TRUE} \wedge [rPDind]X_2) \vee (\langle \neg rPDind \rangle \text{TRUE} \wedge [\neg rPDind]X_1)$$
$$\mu X_2 = \langle e \rangle \text{TRUE} \wedge [e]X_0$$

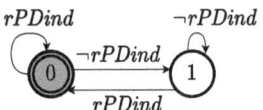

Fig. 5. $\mathcal{B}^\psi$: Büchi automaton corresponding to $\psi = \mathsf{G}\,\mathsf{F}\,rPDind$.

Our tool performed the whole computation in about five seconds.

Note that E_ψ and the simplified transformed Büchi automaton *almost* express that globally eventually $rPDind$ occurs with e following it. The self-loop of state 0 in the Büchi automaton on the right in Fig. 6 prevents this from being expressed, which is caused by the injectiveness condition of matches that we referred to. In practice, this means that if one knows that $rPDind$-selfloops do not occur in an LTS subjected to transformation, one may derive that the property being satisfied after transformation is essentially $\mathsf{G}\,\mathsf{F}(rPDind\,\mathsf{X}\,e)$.

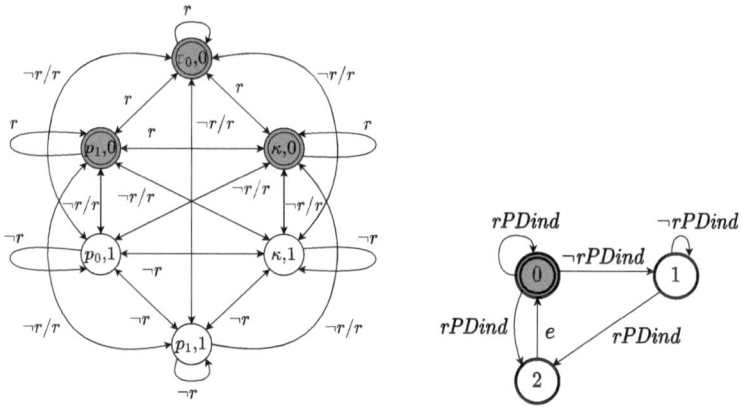

Fig. 6. $\mathcal{L}_{\psi}^{\kappa}$ (left) for the rule in Fig. 4, and $simpl(T(\mathcal{L}_{\psi}^{\kappa}))$ (right).

7   Conclusions

We have proposed a technique to construct temporal logic formulae that characterize the effect that formalized behavioural model transformations have on system components described by LTSs that satisfy a given ALTL property. It has been implemented, and can be applied efficiently in practice.

For future work, we can explore multiple directions. First of all, we aim to extend our approach to involve systems of model transformation rules, that are applied on multiple components at once. This setting has been investigated in [28, 33–35, 37] for property preservation, and is much more complex than reasoning about individual rules. Another direction is to consider input formulae expressed in other temporal logics, such as CTL or the modal μ-calculus itself.

References

1. Abadi, M., Lamport, L.: The existence of refinement mappings. Theor. Comput. Sci. **82**, 253–284 (1991)

2. Abrial, J.R., Butler, M., Hallerstede, S., Hoang, T., Mehta, F., Voisin, L.: Rodin: an open toolset for modelling and reasoning in Event-B. Softw. Tools Technol. Transf. **12**(6), 447–466 (2010)

3. Amrani, M., et al.: Formal verification techniques for model transformations: a tridimensional classification. J. Obj. Technol. **14**(3), 1–43 (2015). https://doi.org/10.5381/jot.2015.14.3.a1

4. Baier, C., Katoen, J.P.: Principles of Model Checking. The MIT Press (2008)

5. Baldan, P., Corradini, A., Ehrig, H., Heckel, R., König, B.: Bisimilarity and behaviour-preserving reconfigurations of open Petri nets. In: Mossakowski, T., Montanari, U., Haveraaen, M. (eds.) CALCO 2007. LNCS, vol. 4624, pp. 126–142. Springer, Heidelberg (2007). https://doi.org/10.1007/978-3-540-73859-6_9

6. Blech, J.O., Glesner, S., Leitner, J.: Formal verification of java code generation from UML models. In: 3rd International Fujaba Days, pp. 49–56. Fujaba Days (2005)

7. Braunstein, C., Encrenaz, E.: CTL-property transformation along an incremental design process. AVoCS. ENTCS **128**, 263–278 (2004)

8. Bresolin, D., Lanese, I.: Static and dynamic property-preserving updates. Inf. Comput.**279**, 104611 (2021)

9. Büchi, J.: On a decision method in restricted second order arithmetic. In: CLMPS, pp. 425–435. Stanford University Press (1962)

10. Combemale, B., Crégut, X., Garoche, P.L., Thirioux, X.: Essay on semantics definition in MDE - an instrumented approach for model verification. J. Softw. **4**(9), 943–958 (2009)

11. Cranen, S., Groote, J., Reniers, M.: A linear translation from CTL* to the first-order modal μ-calculus. Theor. Comput. Sci. **412**, 3129–3139 (2011)

12. Duret-Lutz, A., Lewkowicz, A., Fauchille, A., Michaud, T., Renault, É., Xu, L.: Spot 2.0 — a framework for LTL and ω-Automata manipulation. In: Artho, C., Legay, A., Peled, D. (eds.) ATVA 2016. LNCS, vol. 9938, pp. 122–129. Springer, Cham (2016). https://doi.org/10.1007/978-3-319-46520-3_8

13. Giannakopoulou, D.: Model Checking for Concurrent Software Architectures. Ph.D. thesis, University of London (1999)

14. Giese, H., Lambers, L.: Towards automatic verification of behavior preservation for model transformation via invariant checking. In: Ehrig, H., Engels, G., Kreowski, H.-J., Rozenberg, G. (eds.) ICGT 2012. LNCS, vol. 7562, pp. 249–263. Springer, Heidelberg (2012). https://doi.org/10.1007/978-3-642-33654-6_17

15. Hülsbusch, M., König, B., Rensink, A., Semenyak, M., Soltenborn, C., Wehrheim, H.: Showing full semantics preservation in model transformation - a comparison of techniques. In: Méry, D., Merz, S. (eds.) IFM 2010. LNCS, vol. 6396, pp. 183–198. Springer, Heidelberg (2010). https://doi.org/10.1007/978-3-642-16265-7_14

16. Kahsai, T., Roggenbach, M.: Property preserving refinement for CSP-CASL. In: Corradini, A., Montanari, U. (eds.) WADT 2008. LNCS, vol. 5486, pp. 206–220. Springer, Heidelberg (2009). https://doi.org/10.1007/978-3-642-03429-9_14

17. Karsai, G., Narayanan, A.: On the correctness of model transformations in the development of embedded systems. In: Kordon, F., Sokolsky, O. (eds.) Monterey Workshop 2006. LNCS, vol. 4888, pp. 1–18. Springer, Heidelberg (2008). https://doi.org/10.1007/978-3-540-77419-8_1

18. Kemp, T.: Translating LTL to the Equational μ-Calculus Using Büchi Automata Optimisations. University of Twente, Tech. rep. (2018)

19. Kleppe, A., Warmer, J., Bast, W.: MDA Explained: The Model Driven Architecture(TM): Practice and Promise. Addison-Wesley Professional (2003)

20. Kozen, D.: Results on the propositional μ-Calculus. Theor. Comput. Sc. **27**(3), 333–354 (1983)

21. Kundu, S., Lerner, S., Gupta, R.: Automated refinement checking of concurrent systems. In: ICCAD, pp. 318–325. IEEE (2007)

22. Kupferman, O., Vardi, M.: Freedom, weakness, and determinism: from linear-time to branching-time. In: Proceedings 13th IEEE Symposium on Logic in Computer Science, pp. 81–92 (1998)

23. Lano, K.: The B Language and Method. Springer, A Guide to Practical Formal Development (1996)

24. Luttik, S.: Description and Formal Specification of the Link Layer of P1394. Tech. Rep. SEN-R9706, CWI (1997)

25. Narayanan, A., Karsai, G.: Towards verifying model transformations. GT-VMT. ENTCS **211**, 191–200 (2008)

26. Pnueli, A.: The temporal logic of programs. In: 18th Annual Symposium on Foundations of Computer Science (FOCS), pp. 46–57. IEEE Computer Society (1977)

27. de Putter, S.M.J.: Verification of Concurrent Systems in a Model-Driven Engineering Workflow. Ph.D. thesis, Eindhoven University of Technology (2019)

28. de Putter, S.M.J., Wijs, A.J.: A formal verification technique for behavioural model-to-model transformations. Formal Aspects Comput. **30**(1), 3–43 (2018)

29. Ab. Rahim, L., Whittle, J.: A survey of approaches for verifying model transformations. Softw. Syst. Model. **14**(2), 1003–1028 (2015). https://doi.org/10.1007/s10270-013-0358-0

30. Sokolsky, O.V., Smolka, S.A.: Incremental model checking in the modal mu-calculus. In: Dill, D.L. (ed.) CAV 1994. LNCS, vol. 818, pp. 351–363. Springer, Heidelberg (1994). https://doi.org/10.1007/3-540-58179-0_67

31. Swamy, G.: Incremental Methods for Formal Verification and Logic Synthesis. Ph.D. thesis, University of California (1996)

32. Varró, D., Pataricza, A.: Automated formal verification of model transformations. In: CSDUML, pp. 63–78 (2003)

33. Wijs, A.: Define, verify, refine: correct composition and transformation of concurrent system semantics. In: Fiadeiro, J.L., Liu, Z., Xue, J. (eds.) FACS 2013. LNCS, vol. 8348, pp. 348–368. Springer, Cham (2014). https://doi.org/10.1007/978-3-319-07602-7_21

34. Wijs, A., Engelen, L.: Efficient property preservation checking of model refinements. In: Piterman, N., Smolka, S.A. (eds.) TACAS 2013. LNCS, vol. 7795, pp. 565–579. Springer, Heidelberg (2013). https://doi.org/10.1007/978-3-642-36742-7_41

35. Wijs, A., Engelen, L.: REFINER: towards formal verification of model transformations. In: Badger, J.M., Rozier, K.Y. (eds.) NFM 2014. LNCS, vol. 8430, pp. 258–263. Springer, Cham (2014). https://doi.org/10.1007/978-3-319-06200-6_21

36. Wijs, A.: Confluence detection for transformations of labelled transition systems. In: 1st Graphs as Models Workshop. EPTCS, vol. 181, pp. 1–15. Open Publishing Association (2015)

37. Wijs, A., Engelen, L.J.P.: Incremental formal verification for model refining. In: MoDeVVa, pp. 29–34. IEEE (2012)

Interpretation and Formalization of the Right-of-Way Rules

Victor A. Carreño[1]([envelope]) [iD], Mariano M. Moscato[2] [iD], Paolo M. Masci[2] [iD],
and Aaron M. Dutle[3] [iD]

[1] Compass Engineering, San Juan, PR, USA
`victor.carreno@ymail.com`
[2] National Institute of Aerospace, Hampton, VA, USA
`{mariano.moscato,paolo.masci}@nianet.org`
[3] NASA Langley Research Center, Hampton, VA, USA
`aaron.m.dutle@nasa.gov`

Abstract. This paper presents an interpretation and mathematical definition of the right-of-way rules as stated in USA, Title 14 of the Code of Federal Regulations, Part 91, Section 91.113 (14 CFR 91.113). In an encounter between two aircraft, the right-of-way rules defines which aircraft, if any, has the right-of-way and which aircraft must maneuver to stay well clear of the other aircraft.

The objective of the work presented in this paper is to give an unambiguous interpretation of the rules. From the interpretation, a precise mathematical formulation is created that can be used for analysis and proof of properties. The mathematical formulation has been defined in the Prototype Verification System (PVS) and properties of well formedness and core properties of the formalization have been mechanically proved. This mathematical formulation can be implemented digitally, so that right-of-way rules can be used in simulation or in future autonomous operations.

Keywords: Right-of-way · Safety · Regulations

1 Introduction

The right-of-way rules in 14 CFR 91.113 states that, weather conditions permitting, all operators of aircraft shall maintain vigilance so as to see and avoid other aircraft. This applies to both aircraft operating under instrument flight rules or visual flight rules. The rules contain operational considerations and physical considerations. Operational considerations include category of aircraft, phase of flight, and conditions such as aircraft in distress or refueling. Physical considerations embody the geometry of an encounter including position and velocity. The work presented in this paper focuses on the physical considerations of the rules.

There have been several previous efforts to characterize and mathematically formalize the right-of-way rules. In [1], the physical considerations of the rules are formalized, and safety and other properties are stated and proved. The rules are stated in [2] in a

S. L. Tapia Tarifa and J. Proença (Eds.): FACS 2022, LNCS 13712, pp. 59–73, 2022.
https://doi.org/10.1007/978-3-031-20872-0_4

quantified way with considerations for what it means to be "well clear" of other aircraft, definitions of "ambiguous" and "unambiguous" scenarios, angles and geometries. The International Civil Aviation Organization, Rules of the Air document [3], define international right-of-way rules and additionally provides geometric definitions to some terms that are not defined in 14 CFR 91.113 such as "overtaking." FAA, ATO (USA, Federal Aviation Administration, Air Traffic Organization) Order JO 7110.65 [4] provides definitions of terms regarding geometries and scenarios that might not be found in the regulations. In addition to the existing documents and technical papers, the authors have consulted with aviation experts in the interpretation of the rules [5, 6].

The work presented in this paper shares some of the mathematical definitions in [1]. However, crucial concepts are redefined to capture a different interpretation of the rules. For example, in this paper, "convergence" is not defined in terms of closure rate but rather takes into consideration the location, geometries, and whether one aircraft has crossed the track of the other. The definition of "head-on, or nearly so" is extended to include difference in tracks that are 180°, plus or minus an angle threshold. Also, the definitions consider the distance at which the aircraft will pass by projecting their trajectories. This is generally referred to as Horizontal Miss Distance (HMD) or projected distance at Closest Point of Approach (CPA). Even when the horizontal distance between the aircraft is decreasing, if the aircraft will pass at a sufficiently large distance, the scenario is not considered converging but rather "crossing". This interpretation is consistent with aviation expert's opinion [5, 6]. A detailed comparison of the formalization in [1] and the formalization in this paper can be found in Sect. 4.

More generally, the work presented here can be seen as the formalization of an airspace operational concept, including the verification of properties that are intended to hold. Similar work in this vein includes the analysis of the Small Aircraft Transportation System [7], and the specification and analysis of parallel landing scenarios [8]. This work is also related to the interpretation, specification, and analysis of ambiguous natural language requirements. Interpreting written requirements into a precise formulation is a well-known problem [9], and purpose-built methods both manual [10] and automatic [11] are currently being used and refined for doing so. This work uses the Prototype Verification System (PVS) [12] to specify the interpretation of the right-of-way rules, and verify several core properties. This formal specification and verification is also envisioned to be used as a component in a rule-compliant pilot model being developed for simulation purposes.

2 Formalization of Right-of-Way Rules

Section 14 CFR 91.113 defines the following right-of-way rules:

(a) Inapplicability. (Not used in the Right-of-Way Rules formalization.)
(b) General. When weather conditions permit, regardless of whether an operation is conducted under instrument flight rules or visual flight rules, vigilance shall be maintained by each person operating an aircraft so as to see and avoid other aircraft. When a rule of this section gives another aircraft the right-of-way, the pilot shall give way to that aircraft and may not pass over, under, or ahead of it unless well clear.

(c) In distress. (Not used in the Right-of-Way Rules formalization.).
(d) Converging. When aircraft of the same category are converging at approximately the same altitude (except head-on, or nearly so), the aircraft to the other's right has the right-of-way.
(e) Approaching head-on. When aircraft are approaching each other head-on, or nearly so, each pilot of each aircraft shall alter course to the right.
(f) Overtaking. Each aircraft that is being overtaken has the right-of-way and each pilot of an overtaking aircraft shall alter course to the right to pass well clear.

The right-of-way rules, as defined in the regulations, deal with pairs of aircraft and do not consider cases where more than two aircraft are involved in an encounter. Also, the required actions in paragraphs (e) and (f) are specified in the horizontal domain and no alternatives are given for the vertical domain. The formalization presented in this paper does not attempt to "extend," "change," or "improve" on the regulations. The formalization attempts to be a faithful interpretation of the regulations.

In the formalization presented in this paper, the airspace is modeled by a 2-dimensional Cartesian flat-earth projections. Aircraft are assumed to be "at approximately the same altitude" and represented by position and velocity vectors. It should be noted that when aircraft are vertically separated (not at approximately the same altitude), they are considered to be well clear and there is no need for the right-of-way rules to be applied. Furthermore, aircraft that need to maneuver to remain well clear are horizontally close (~5 nautical miles) and so the flat-earth assumption incurs negligible errors.

Consider two-dimensional position vectors $s_0, s_1, ..., s_n \in \mathbb{R}^2$ and two-dimensional non-zero (Sect. 6 discusses this constraint) velocity vectors $v_0, v_1, ..., v_n \in \mathbb{R}^2$. An aircraft A_0 is uniquely defined on the horizontal plane by its position and velocity vectors (s_0, v_0). For a vector $v_n \in \mathbb{R}^2$, define $v_n^{\perp}$ as the 90-degree clockwise rotation of $v_n = (v_{n,x}, v_{n,y})$ by $v_n^{\perp} = (v_{n,y}, -v_{n,x})$. The norm operator returning the magnitude of a vector s_n is represented by $\|s_n\|$. For any two vectors v_n, v_m the dot or scalar product is represented by $v_n \cdot v_m$.

2.1 Basic Definitions

The heading and position of an aircraft divide the horizontal plane into four quadrants as shown in Fig. 1. The following definitions formalize this notion.

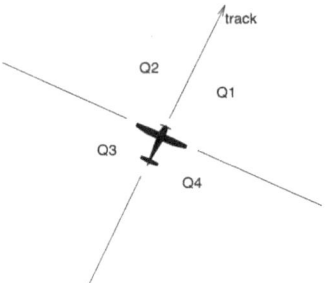

Fig. 1. Quadrants of the aircraft

Definition 1 (Quadrant). *Given an aircraft $A_0 = (s_0, v_0)$, a position s_1 is in aircraft's A_0 first, second, third, or fourth quadrant (Q1, Q2, Q3, or Q4 respectively) when the corresponding of the following predicates hold:*

$$Q1(A_0, s_1) := (s_1 - s_0) \cdot v_0^\perp > 0 \wedge (s_1 - s_0) \cdot v_0 \geq 0 \tag{1}$$

$$Q2(A_0, s_1) := (s_1 - s_0) \cdot v_0^\perp \leq 0 \wedge (s_1 - s_0) \cdot v_0 > 0 \tag{2}$$

$$Q3(A_0, s_1) := (s_1 - s_0) \cdot v_0^\perp < 0 \wedge (s_1 - s_0) \cdot v_0 \leq 0 \tag{3}$$

$$Q4(A_0, s_1) := (s_1 - s_0) \cdot v_0^\perp \geq 0 \wedge (s_1 - s_0) \cdot v_0 < 0 \tag{4}$$

Definition 2 (Track). The track of an aircraft $A_0 = (s_0, v_0)$ is defined as the angle between the north and the direction of an aircraft, measured in a clockwise direction. For example, an aircraft flying west in on a 270-degree track.

$$trk(A_0) := \tan^{-1}\left(\frac{v_{0,x}}{v_{0,y}}\right) \tag{5}$$

A trajectory is defined as the linear projection of the aircraft's velocity vector. The Closest Point of Approach (CPA) is the geometrical condition when the trajectories of two aircraft will be at the smallest distance or range. The time to CPA is the time from the current position to the moment when the distance is smallest when their velocity vectors are projected in a straight trajectory. The time is positive when the CPA will occur in the future and negative when the CPA has already occurred. If the velocity vectors are parallel in the same direction and their magnitudes are equal, the current distance between aircraft will not change and is the CPA. In this case, the time to CPA is defined to be zero.

Definition 3 (Time to Closest Point of Approach). *For aircraft $A_0 = (s_0, v_0)$ and $A_1 = (s_1, v_1)$ the time to closest point of approach is defined as:*

$$t_{CPA}(A_0, A_1) := \begin{cases} -\frac{(s_0 - s_1) \cdot (v_0 - v_1)}{(v_0 - v_1) \cdot (v_0 - v_1)} & \text{if } v_0 \neq v_1 \\ 0 & \text{if } v_0 = v_1 \end{cases} \tag{6}$$

Definition 4 (Horizontal Miss Distance). *The horizonal miss distance of two aircraft $A_0 = (s_0, v_0)$ and $A_1 = (s_1, v_1)$ is the distance at the Closest Point of Approach when their position and velocity vectors are projected in time:*

$$HMD(A_0, A_1) := \|(s_0 - s_1) + t_{CPA}(A_0, A_1)(v_0 - v_1)\| \tag{7}$$

The trajectories of two aircraft on a two-dimensional airspace with non-zero velocity vectors are either going to cross (in the future) or already crossed (in the past) unless their velocity vectors are parallel. Figure 2 illustrates the three cases of a generic crossing. In the first case, neither aircraft has passed the crossing point. In the second, only A_0 has crossed the trajectory of A_1. In the third, each aircraft has crossed the other's trajectory.

In order to define the crossing of aircraft, the relative position and the relative motion of an aircraft with respect to another are defined.

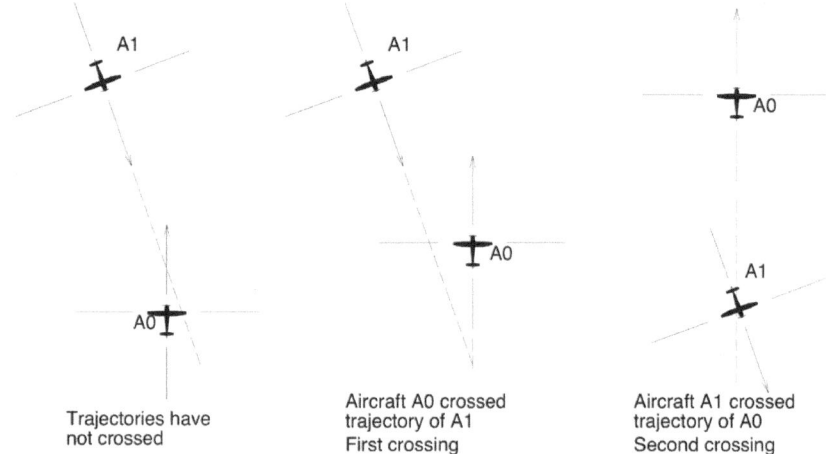

Fig. 2. Before and after trajectory crossing

Definition 5 (Relative Position). *A position* s_1 *is to the left of (respectively to the right of) an aircraft* $A_0 = (s_0, v_0)$ *when the following predicates hold (respectively):*

$$to\_the\_left\_of(A_0, s_1) := (s_1 - s_0) \cdot v_0^\perp < 0 \tag{8}$$

$$to\_the\_right\_of(A_0, s_1) := (s_1 - s_0) \cdot v_0^\perp > 0 \tag{9}$$

Definition 6 (Relative Motion). *An aircraft* $A_1 = (s_1, v_1)$ *is moving left to right (respectively right to left) with respect to aircraft* $A_0 = (s_0, v_0)$ *when the following predicates hold (respectively):*

$$left\_to\_right(A_0, A_1) := v_0 \cdot v_1^\perp < 0 \tag{10}$$

$$right\_to\_left(A_0, A_1) := v_0 \cdot v_1^\perp > 0 \tag{11}$$

Using the definitions above, it is possible to formalize situational notions describing scenarios in which aircraft have or have not crossed each other's trajectories.

Definition 7 (Going to Cross). *An aircraft* $A_1 = (s_1, v_1)$ *is going to cross the trajectory of* $A_0 = (s_0, v_0)$ *(in the future) if either it is to the left of* A_0 *and it is moving left to right of if it is to the right of* A_0 *and it is moving right to left:*

$$going\_to\_cross(A_0, A_1) := (to\_the\_left\_of(A_0, s_1) \wedge left\_to\_right(A_0, A_1)) \vee \atop (to\_the\_right\_of(A_0, s_1) \wedge right\_to\_left(A_0, A_1)) \tag{12}$$

Definition 8 (Already Crossed). *Aircraft* A_1 *already crossed the trajectory of* A_0 *(in the past) if it is to the left of* A_0 *and it is moving right to left or if it is to the right of* A_0 *and it is moving left to right:*

$$crossed(A_0, A_1) := (to\_the\_left\_of(A_0, s_1) \wedge right\_to\_left(A_0, A_1)) \vee \atop (to\_the\_right\_of(A_0, s_1) \wedge left\_to\_right(A_0, A_1)) \tag{13}$$

Definition 9 (Zero Crossed). *The trajectories of two aircraft are going to cross (in the future) when neither aircraft have crossed the trajectory of the other:*

$$zero\_crossed(A_0, A_1) := going\_to\_cross(A_0, A_1) \wedge going\_to\_cross(A_1, A_0) \quad (14)$$

Definition 10 (One crossed). *One aircraft has crossed the trajectory of the second but the second has not crossed the trajectory of the first:*

$$\begin{aligned} one\_crossed(A_0, A_1) := (going\_to\_cross(A_0, A_1) \wedge crossed(A_1, A_0)) \vee \\ (going\_to\_cross(A_1, A_0) \wedge crossed(A_0, A_1)) \end{aligned} \quad (15)$$

Definition 11 (Both crossed). *Both aircraft have crossed each other's trajectories:*

$$both\_crossed(A_0, A_1) := crossed(A_0, A_1) \wedge crossed(A_1, A_0) \quad (16)$$

Definition 12 (Parallel). *The trajectories of aircraft A_0 and A_1 are parallel when the following predicate holds:*

$$parallel(A_0, A_1) := v_0 \cdot v_1^{\perp} = 0 \quad (17)$$

Definition 13 (Same Orientation). *The following predicate holds when the angle between the trajectories of aircraft A_0 and A_1 is acute:*

$$same\_orientation(A_0, A_1) := v_0 \cdot v_1 > 0 \quad (18)$$

Definition 14 (Opposite Orientation). *The following predicate holds when the angle between the trajectories of aircraft A0 and A1 are obtuse:*

$$opposite\_orientation(A_0, A_1) := v_0 \cdot v_1 < 0 \quad (19)$$

Definition 15 (In Q1 and was in Q2). *An aircraft A_1 is in the first quadrant of aircraft A_0 and was in the second quadrant of aircraft A_0 when the following predicate holds:*

$$\begin{aligned} is\_in\_q1\_and\_was\_in\_q2(A_0, A_1) := (Q1(A_0, s_1) \wedge (Q4(A_1, s_0)) \vee \\ (opposite\_orientation(A_0, A_1) \wedge \\ left\_to\_right(A_0, A_1))) \end{aligned} \quad (20)$$

2.2 Convergence, Divergence, and Overtake

The concept of convergence, head-on or nearly so, and overtaking used in the right-of-way rules can be characterized by the relative quadrant location of each aircraft with respect to the other, the track angle between them, and their horizontal miss distance. Figure 3 shows examples of relative quadrant locations: aircraft A_1 in aircraft A_0's quadrant Q1 and aircraft A_0 in aircraft A_1's quadrants Q1 to Q4.

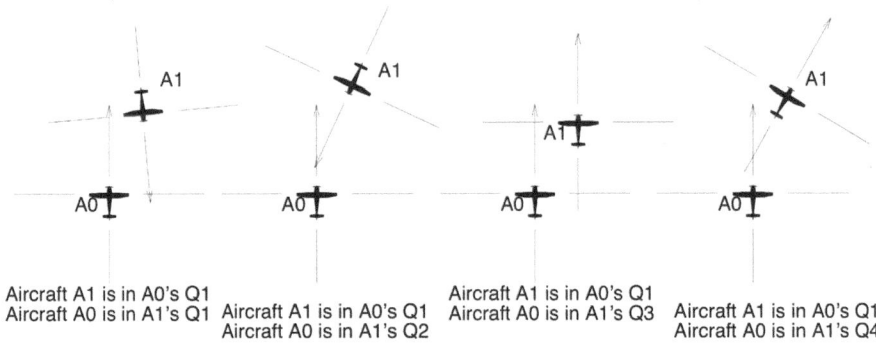

Aircraft A1 is in A0's Q1
Aircraft A0 is in A1's Q1

Aircraft A1 is in A0's Q1
Aircraft A0 is in A1's Q2

Aircraft A1 is in A0's Q1
Aircraft A0 is in A1's Q3

Aircraft A1 is in A0's Q1
Aircraft A0 is in A1's Q4

Fig. 3. Aircraft A_0 in aircraft A_1's quadrants Q1 to Q4

Table 1 shows the quadrant locations that are required for aircraft A_0 and A_1 to be converging, diverging, or overtaking. The combination of quadrants is necessary but not sufficient for convergence and overtake. For aircraft to be converging or overtaking, an HMD less than some threshold must also be satisfied. The HMD threshold is important because it makes an operational distinction between aircraft that are converging and aircraft whose trajectories are crossing but are not considered converging. For example, the trajectory of an aircraft over the Mediterranean Sea might be crossing the trajectory of an aircraft over Australia. However, these two aircraft are not operationally converging. The formalization leaves HMD as a parameter to be defined depending on the type of operation.

Table 1. Quadrant location for convergence and overtake requirement

A_0/A_1	Q1	Q2	Q3	Q4
Q1	Convergence	Convergence	Overtake	Overtake
Q2	Convergence	Convergence	Overtake	Overtake
Q3	Overtake	Overtake	Divergence	Divergence
Q4	Overtake	Overtake	Divergence	Divergence

Definition 16 (General convergence). *Let $A_0 = (\mathbf{s_0}, \mathbf{v_0})$ and $A_1 = (\mathbf{s_1}, \mathbf{v_1})$ be aircraft and δ_C a nonnegative real number. A_0 and A_1 are converging when the following predicate holds:*

$$converging(A_0, A_1)(\delta_C) := Q_C(A_0, A_1) \wedge (HMD(A_0, A_1) < \delta_C) \qquad (21)$$

where the predicate $Q_C(A_0, A_1)$ in the equation above are the quadrant locations of A_0 with respect to A_1 and the location of A_1 with respect to A_0:

$$Q_C(A_0, A_1) := (Q1(A_0, \mathbf{s_1}) \wedge Q1(A_1, \mathbf{s_0})) \vee (Q1(A_0, \mathbf{s_1}) \wedge Q2(A_1, \mathbf{s_0})) \vee$$
$$(Q2(A_0, \mathbf{s_1}) \wedge Q1(A_1, \mathbf{s_0})) \vee (Q2(A_0, \mathbf{s_1}) \wedge Q2(A_1, \mathbf{s_0})) \qquad (22)$$

The general convergence definition above includes cases where the aircraft could be head-on or nearly so. An additional constraint is put in place to exclude the head-on or nearly so cases.

Definition 17 (Convergence, not head-on). *Aircraft A_0 and A_1 are converging but not head-on when they are converging and the angle between the aircraft tracks is greater than180° plus θ_H or less than 180 ° minus θ_H, where θ_H is the angular threshold:*

$$conv\_not\_headon(A_0, A_1)(\delta_C, \theta_H) := Q_C(A_0, A_1) \wedge (HMD(A_0, A_1) < \delta_C) \wedge$$
$$(180 + \theta_H < |trk(A_0) - trk(A_1)| \vee \qquad (23)$$
$$|trk(A_0) - trk(A_1)| < 180 - \theta_H)$$

Definition 18 (Head-on, or nearly so). *Aircraft A_0 and A_1 are head-on, or nearly so, when they are converging and the difference in their tracks is 180° plus or minus θ_H:*

$$headon(A_0, A_1)(\delta_C, \theta_H) := \quad Q_C(A_0, A_1) \wedge (HMD(A_0, A_1) < \delta_C) \wedge$$
$$(180 - \theta_H \leq |trk(A_0) - trk(A_1)| \leq 180 + \theta_H) \qquad (24)$$

Definition 19 (Overtaking). *Aircraft A_0 is overtaking aircraft A_1 when aircraft A_1 is in quadrants Q1 or Q2 of aircraft A_0 and aircraft A_0 is in quadrants Q3 and Q4 of aircraft A_1 and the HMD is less than a threshold:*

$$overtaking(A_0, A_1)(\delta_O) := (Q1(A_0, s_1) \vee Q2(A_0, s_1)) \wedge$$
$$(Q3(A_1, s_0) \vee Q4(A_1, s_0)) \wedge \qquad (25)$$
$$(HMD(A_0, A_1) < \delta_O)$$

Definition 20 (Right-of-way). *Aircraft $A_1 = (s_1, v_1)$ has the right-of-way and aircraft $A_0 = (s_0, v_0)$ has to give way to A_1 when A_0 is overtaking A_1 or the aircraft are converging (except head-on or nearly so), A_1 is to the right of A_0, and their trajectories have not crossed:*

$$right\_of\_way(A_1, A_0)(\delta_O, \delta_C, \theta_H) := \quad overtaking(A_0, A_1)(\delta_O) \vee$$
$$(conv\_not\_headon(A_0, A_1)(\delta_C, \theta_H) \wedge$$
$$to\_the\_right\_of(A_0, s_1) \wedge \qquad (26)$$
$$zero\_crossed(A_0, A_1))$$

3 Properties of the Right-of-Way Rules

This section presents properties of the right-of-way rules. These properties ensure that the right-of-way rules, as interpreted above, do not conflict with the sense of the rules. They were each specified and proved in the Prototype Verification System (PVS).

Theorem 1 (Safety 1). *For all aircraft A_0 and A_1 and all $\delta_O, \delta_C, \theta_H \in \mathbb{R}_{\geq 0}$ it is never the case that both aircraft have the right-of-way at the same time.*

$$right\_of\_way(A_1, A_0)(\delta_O, \delta_C, \theta_H)$$

$$\Rightarrow \neg right\_of\_way(A_0, A_1)(\delta_O, \delta_C, \theta_H)$$

Theorem 2 (Safety 1a). *For all aircraft A_0 and A_1, if the aircraft have not crossed trajectories, then one and only one aircraft is to the right of the other.*

$$zero\_crossed(A_1, A_0)$$

$$\Rightarrow (to\_the\_right\_of(A_0, s_1) \wedge \neg to\_the\_right\_of(A_1, s_0))$$

$$\vee (to\_the\_right\_of(A_1, s_0) \wedge \neg to\_the\_right\_of(A_0, s_1))$$

Theorem 3 (Overtaking is Asymmetric). *For all aircraft A_0 and A_1 and all $\delta_O \in \mathbb{R}_{\geq 0}$, if A_0 is overtaking A_1 then A_1 is not overtaking A_0.*

$$overtaking(A_0, A_1)(\delta_O) \Rightarrow \neg overtaking(A_1, A_0)(\delta_O)$$

Theorem 4 (No right-of-way after crossing). *For all aircraft A_0 and A_1, after the first aircraft has crossed the trajectory of the second, but before the second has crossed the trajectory of the first, both of them will be to the right of each other or both will be to the left of each other.*

$$one\_crossed(A_0, A_1)$$

$$\Rightarrow (to\_the\_right\_of(A_0, s_1) \wedge to\_the\_right\_of(A_1, s_0))$$

$$\vee (to\_the\_left\_of(A_0, s_1) \wedge to\_the\_left\_of(A_1, s_0))$$

This is a geometrical property and not a right-of-way rules property. However, it has a right-of-way rules implication. If both aircraft are to the right of each other, then according to rule (d), they both have right-of-way. However, according to Theorem 1, both cannot have right-of-way simultaneously. If both are to the left of each other, then rule (d) does not apply. The implication is that when one aircraft has crossed the trajectory of the other, neither aircraft has right-of-way.

Theorem 5 (Awareness). *When any two aircraft A_0, A_1 are converging (not head-on or nearly so) it is sufficient for an aircraft to know if one of the aircraft has crossed the trajectory of the other and if the traffic is to its right to determine if it has right-of-way.*

$$conv\_not\_headon(A_0, A_1)(\delta_C, \theta_H)$$

$$\Rightarrow (zero\_crossed(A_1, A_0) \wedge to\_the\_right\_of(A_0, s_1)$$

$$\Rightarrow right\_of\_way(A_1, A_0)(\delta_O, \delta_C, \theta_H))$$

Theorem 6 (Had right-of-way). *For any two aircraft A_0, A_1, to be in each other's first quadrant, at least one had to cross from the other's second quadrant.*

$$Q1(A_0, s_1) \wedge Q1(A_1, s_0) \wedge \neg parallel(A_0, A_1)$$

$$\Rightarrow (is\_in\_q1\_and\_was\_in\_q2(A_0, A_1) \vee is\_in\_q1\_and\_was\_in\_q2(A_1, A_0))$$

The significance of this theorem is that if the aircraft are in each other's first quadrant, and neither have right of way because both are to the right of each other, then before the crossing, one of them was to the right and one was to the left and hence one had right of way and the other did not.

4   Comparison with Previous Formalization

This section presents a detailed account of the differences between the definitions presented in [1] and the interpretation and definitions in this work.

4.1   Converging

The definition of convergence in [1] is based on closure rate. If the horizontal distance between the aircraft is getting smaller, the aircraft are converging. The implication of this definition is that given the speeds of the aircraft are approximately equal, the following two cases in Fig. 4 are considered horizontally converging:

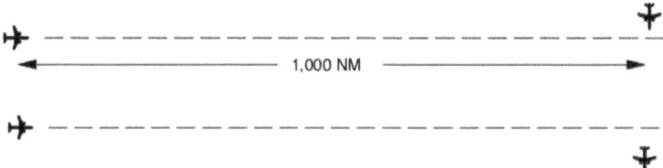

Fig. 4. Cases 1 and 2, cases considered converging as defined in [1]

To operational personnel, such as Air Traffic Controllers, these two cases are not converging scenarios, the first because the horizontal miss distance is large, and the second because the aircraft that has crossed the trajectory of the other is leaving the other aircraft behind.

In this paper, convergence is defined by the relative location of one aircraft with respect to the location and motion of the other and the Horizontal Miss Distance (HMD). Case 1 does not satisfy the definition of convergence because $HMD > \delta_C$ (assuming δ_C is significantly smaller that a distance in the order of 1,000 nautical miles) and Case 2 does not satisfy the definition of convergence because the aircraft to the left is in Quadrant 4 of the aircraft to the right and $Q_C(A_0, A_1)$ is false.

4.2 Head-On or Nearly So

Paragraph (e) of 14 CFR 91.113 alludes to the scenario as "head-on or nearly so." Narckawicz et al. [1] defines head-on as when the aircraft have trajectories strictly 180° opposite each other and when their trajectories are perfectly aligned with the line segment connecting their positions. This definition does not include the "or nearly so" part of the definition in the regulations.

In Fig. 5 below, Case 3 satisfy the definition of head-on in [1]. However, Case 4 and Case 5, where the aircraft are head-on or nearly so, do not satisfy the definition in [1].

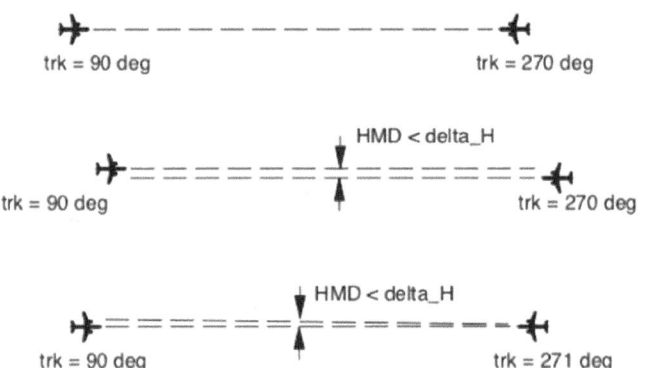

Fig. 5. Cases 3 to 5, cases illustrating strictly head-on and head-on or nearly so

In this paper, head-on or nearly so is defined by specifying the relative quadrant with respect to each other, a Horizontal Miss Distance and a range of angles. This definition is satisfied by cases 3, 4, and 5 above (assuming θ_H is greater or equal to one degree).

4.3 Overtaking

The definition of overtaking in [1] suffers from the same shortcoming as the definition of head-on. Only geometries where the tracks are perfectly aligned and the trajectory difference is zero are considered to be overtaking. Scenarios where the trajectories are displaced sideways by any distance greater than zero or the difference in trajectory angle is not zero do not satisfy the definition of overtaking.

This paper defines overtaking in terms of the relative quadrants and the Horizontal Miss Distance. Scenarios where the difference in track angles is greater than zero and the tracks are not perfectly aligned satisfy this definition, as long as the HMD is less than the given threshold.

4.4 Right-of-Way

The main objective of the right-of-way rules is to determine, based on geometry and state, which aircraft, in an encounter between a pair of aircraft, has the right-of-way, if any. There are major differences in the definitions of right-of-way in [1] and in this paper.

These differences lead to different aircraft having right-of-way in the same scenario, as it will be shown below.

There are four shortcomings with the right-of-way definition in [1]:

First, in scenarios where the aircraft are nearly head-on, this definition would give one of the aircraft the right-of-way when, in reality, neither aircraft should have right-of-way and both aircraft should turn right to stay well clear as stated in paragraph (e) of 14 CFR 91.113. Case 8 in Fig. 6 shows two aircraft, A1 and A2, in a nearly head-on encounter scenario. The right-of-way definition in [1] improperly gives aircraft A2 the right-of-way.

A1   A2
✈===== == — — — ===== ✈
trk = 90 deg   trk = 272 deg

Fig. 6. Case 8, Nearly head-on

Second, when two aircraft are converging and neither have crossed the other's trajectory, paragraph (d) of 14 CFR 91.113 gives one of them the right-of-way. After one crosses the trajectory of the other, they both will be to the right of each other or both will be to the left. However, the definition in [1] continues to give the right-of-way to the aircraft that was to the right before the crossing. In some scenarios, this could be problematic. Case 9 in Fig. 7 shows a converging crossing scenario where A2 is to the right of A1 and has the right-of-way. After A1 crosses the trajectory of A2, A2 will be mostly or completely out of the field of view of A1. At this point, A2 should not continue to have right-of-way. This state will be more applicable to an overtaking situation where the aircraft behind will have to give way to the aircraft in front.

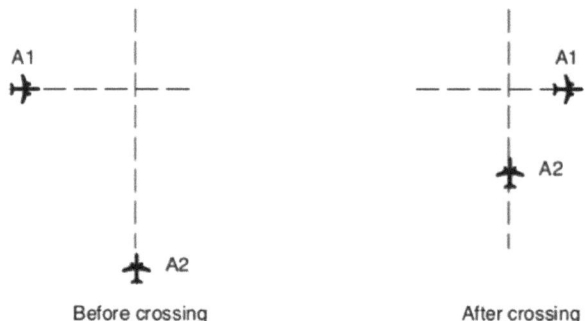

Before crossing   After crossing

Fig. 7. Case 9, before and after crossing and persistence of right-of-way

Third, the right-of-way definition in [1] gives right-of-way to an aircraft in scenarios where the aircraft are diverging. This is not covered in 14 CFR 91.113 regulations. That is, for diverging aircraft, the regulations do not give right-of-way to either aircraft.

Finally, in scenarios where an aircraft is overtaking another, but the difference in their trajectory angles is not zero and/or their trajectories are not perfectly aligned, the definition of right-of-way in [1] improperly gives the right-of-way to the aircraft that is to the right and not to the aircraft that is being overtaken. Case 10 in Fig. 8 shows a scenario where aircraft A1 is overtaking A2. However, according to the definition in [1], A1 has right of way instead of A2, which contradicts paragraph (f) of 14 CFR 91.113. The issue arises from the definition of overtaking in [1], which does not consider Case 10 of Fig. 8 as an overtaking scenario.

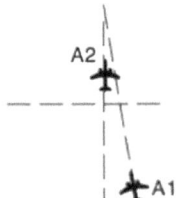

Fig. 8. Case 10, Overtaking

The right-of-way definition in this paper gives right-of-way to the aircraft that is being overtaken, but the definition of overtaking is not restricted to same trajectory, zero angle cases. It also gives right-of-way to the aircraft that is to the right in a converging scenario, but it excludes head-on or nearly so, and limits the definition of convergence to scenarios where aircraft are on quadrants Q1 or Q2 of each other.

For the right-of-way definition in this paper, aircraft A1 has right-of-way when A0 is overtaking A1 or when the aircraft are converging, are not head-on or nearly so, A1 is to the right of A0, and their trajectories have not crossed.

The authors believe that this is the correct definition of the right-of-way rules described in 14 CFR 91.113 and the interpretation of these regulations.

5   Aspects of the Right-of-Way Rules Formalization

This section discusses aspects of the right-of-way rules, the definitions, and formalization. In parts of the right-of-way rules, the following terms are used: "may not pass over, under, or ahead", "when aircraft of the same category are converging", "when aircraft are approaching each other", and "each aircraft that is being overtaken."

The rules are defined in terms of motion, either in relative terms or in absolute terms. In order for two aircraft to have relative motion, at least one of them has to be moving with respect to the other or with respect to a reference. Hence, when formalizing the right-of-way rules in mathematical terms, the constraint is put in place that there is relative motion between the aircraft. That is, that the velocity vector of at least one aircraft is non-zero. If there is no relative motion between the aircraft, it will be impossible to pass, to converge, to approach, or to overtake.

Other considerations in the formalization of the right-of-way rules is the notion of direction, location of the aircraft respect to the other (quadrant), head-on, and right or left. In general, an aircraft flies in the direction of its longitudinal axis. However, there are many instances when the motion of an aircraft is not aligned with its longitudinal axis. For example, a rotorcraft could be moving perpendicular to its longitudinal axis. A slow aircraft in the presence of strong winds could be moving, relative to ground, along its longitudinal axis but opposite of what is consider the front of the aircraft (moving backward with respect to the ground). An aircraft in the presence of a crossed wind will not have its longitudinal axis aligned with its direction of travel.

There are some instances that are not covered by the right-of-way rules. An encounter where an aircraft is stationary (zero velocity vector) is not covered. For example, when an aircraft A1 is approaching a stationary aircraft A2, it is not possible to determine whether the aircraft are converging head-on or being overtaken. It is assumed that the intent of the rules is that paragraph (b) applies and that "vigilance shall be maintained by each person operating an aircraft so as to see and avoid other aircraft" but that neither paragraph (e) nor (f) applies and that neither of the aircraft has right of way.

6 Summary and Conclusion

The paper presents an interpretation and a mathematical definition of the right-of-way rules as defined in the US Title 14 Code of Federal Regulations 91.113. The main contributions are: (i) a formalization of the regulations that align with the exact definition of the regulations and with the interpretation by operational experts; (ii) a mechanized analysis in the Prototype Verification System (PVS) of well formedness and core properties of the formalization; (iii) a detailed discussion of the differences between the formalization presented in this paper and the one found in [1]. Additional objectives are to use the mathematical formulation presented in this paper to develop a rule compliant virtual pilot that can be used in simulation experiments and possibly be used in future autonomous vehicles. Planned future work includes the formulation of stability properties of the right-of-way rules and the mechanized proof of these properties.

Acknowledgements. Funding for the first author has been provided by the National Aeronautics and Space Administration. Research by the second and third authors is supported by the National Aeronautics and Space Administration under NASA/National Institute of Aerospace Cooperative Agreement NNL09AA00A.

References

1. Narkawicz, A., Muñoz, C.A., Maddalon, J.: A mathematical analysis of air traffic priority rules. In: Proceedings of the 12th AIAA Aviation Technology, Integration, and Operations (ATIO) Conference, AIAA-2012–5544, Indianapolis, Indiana (2012)
2. RTCA DO-365A: Minimum Operational Performance Standards (MOPS) for Detect and Avoid (DAA) Systems, Appendix H, February 2020
3. International Civil Aviation Organization: Rules of the Air, Annex 2 to the Convention on International Civil Aviation, 10th edn. (2005)
4. U.S. Department of Transportation, Federal Aviation Administration, Air Traffic Organization, Order JO 7110.65X, October 2017
5. Arbuckle, D.: Personal communications regarding interpretation of convergence and divergence, March 2021

6. Minck, J.: Personal communications regarding interpretation of convergence and divergence, March 2021

7. Muñoz, C., Carreño, V., Dowek, G.: Formal analysis of the operational concept for the small aircraft transportation system. In: Butler, M., Jones, C.B., Romanovsky, A., Troubitsyna, E. (eds.) Rigorous Development of Complex Fault-Tolerant Systems. LNCS, vol. 4157, pp. 306–325. Springer, Heidelberg (2006). https://doi.org/10.1007/11916246_16

8. Carreño, V., Muñoz, C.: Formal analysis of parallel landing scenarios. In: 19th DASC. 19th Digital Avionics Systems Conference. Proceedings (Cat. No.00CH37126), vol. 1, pp. 1D6/1–1D6/6 (2000). https://doi.org/10.1109/DASC.2000.886893

9. Shah, U.S., Jinwala, D.C.: Resolving ambiguities in natural language software requirements: a comprehensive survey. ACM SIGSOFT Softw. Eng. Notes **40**(5), 1–7 (2015). https://doi.org/10.1145/2815021.2815032

10. Mavridou, A., Bourbouh, H., Giannakopoulou, D., Pressburger, T., Hejase, M., Garoche, P.L., Schumann J.: The ten lockheed martin cyber-physical challenges: formalized, analyzed, and explained. In: 2020 IEEE 28th International Requirements Engineering Conference (RE), pp. 300–310. IEEE (2020). https://doi.org/10.1109/RE48521.2020.00040

11. Ghosh, S., Elenius, D., Li, W., Lincoln, P., Shankar, N., Steiner, W.: ARSENAL: automatic requirements specification extraction from natural language. In: Rayadurgam, S., Tkachuk, O. (eds.) NFM 2016. LNCS, vol. 9690, pp. 41–46. Springer, Cham (2016). https://doi.org/10.1007/978-3-319-40648-0_4

12. Owre, S., Rushby, J.M., Shankar, N.: PVS: a prototype verification system. In: Kapur, D. (ed.) CADE 1992. LNCS, vol. 607, pp. 748–752. Springer, Heidelberg (1992). https://doi.org/10.1007/3-540-55602-8_217

Formal Model In-The-Loop for Secure Industrial Control Networks

Laurynas Ubys[1], Valeriu Nicolas Vancea[1], Tomas Kulik[1,2]([✉]),
Peter Gorm Larsen[1], Jalil Boudjadar[1], and Diego F. Aranha[1]

[1] DIGIT, Aarhus University, Nordre Ringgade 1, 8000 Aarhus, Denmark
tku@sweetgeeks.dk
[2] Sweet Geeks, Innovations Alle 3, 7100 Vejle, Denmark

Abstract. Current trends of digitalization are becoming significantly prevalent within the field of industrial control systems. While in recent history a typical industrial control system would have been isolated with rudimentary ways of extracting data, nowadays it is becoming expected that the control system could not only provide large amounts of data over the network but also receive firmware updates and patches. To this end it is important to secure the communication between the components of the system, as well as ensure that only approved components can communicate together. Secure communication and device authentication could be achieved by use of cryptographic keys and certificates. The system however must be able to securely manage the keys and certificates in order to ensure their authenticity and validity. In this paper we present a prototype of a pluggable key management device for industrial control systems with a key management protocol and integrated formal analysis of the running system – a *model in-the-loop*. This allows the system to continuously analyse the network traffic according to the protocol using VDM and hence assure compliance with several security properties. We use off-the-shelf hardware, custom key and device management protocol and VDM to ensure that the device satisfies requirements posed by our industrial partner.

Keywords: VDM · Aspect oriented programming · Formal model · Network security

1 Introduction

Modern industrial control systems are often based on several network-connected subsystems. Given that this network is often monitored remotely via Internet and through untrusted networks, such systems provide a valuable attack target to potential attackers. In order to provide protection from external attacks that could compromise the correct operation or even safety of the controlled system, the control network often utilizes security protocols to ensure the consistency and authenticity of the messages

This work is supported by the Manufacturing Academy of Denmark (MADE), for more information see https://www.made.dk.

exchanged within the network [13]. One of the important prerequisites for successful use of the majority of security protocols is utilization of robust secret key management.

In modern devices the secret keys are often generated by use of a specific hardware such as built-in, or add-on Trusted Platform Modules (TPMs) [12]. Furthermore, the communication can then be protected by use of asymmetric cryptography, utilizing public/private key pairs [25], where further structures such as certificates [23] can be used to convey authentication information between devices. It can also be beneficial for the system to possess an embedded certificate authority, or a centralized certificate management component that handles the necessary certificate issuance and renewal within the system in a secure manner [14]. An important aspect of a key management subsystem is trust and correctness. One way to help to ensure these aspects of the system is to use formal methods. This could be carried out in multiple ways: formal methods could be used to analyse the model of the system prior to its implementation to ensure that the design follows the specific requirements [16,33]. While this approach is often used, a downside is that it does not cover potential errors and deviations from the specification introduced during implementation [8], especially if robust code generation from a formal system is not available. To this end, different approaches could be utilized, where the formal method is integrated into the system itself, creating a model-in-the-loop analysis [32], runtime verification [1] or where the model is used to generate test cases to test the implementation [8].

In this paper we present a prototype key management system following a model-in-the-loop approach utilizing a formal model created using VDM [6]. We demonstrate two novel approaches of using the VDM model in-the-loop, where we develop several components necessary for connecting the live system with the VDM model. In the first case the model does not have visibility to the system implementation, what we have dubbed the *blackbox* approach, only utilizing the network communication traces captured between the devices. In the second case we employ Aspect Oriented Programming (AOP) [3] in Java, where we execute specific method calls directly through the VDM model utilizing the contracts built in the model to ensure the correct behaviour of the key management subsystem, what we have dubbed the *whitebox* approach. We present how the model reacts to the introduction of specific faults that could lead to potential vulnerabilities in the system. Furthermore, we compare the two approaches in terms of what they could capture and quantify during modelling time. In order to validate the approach we have built a prototype of the system using Raspberry PI boards with external TPMs to emulate specific industrial control system components. The system model design and the properties that the system shall satisfy have been determined together with our industrial partner[1] over the span of several years while designing the control system. The main focus has been put on the generation of private keys, issuance of certificates and following communication from a real world industrial control system.

The rest of the paper is organized as follows: Sect. 2 introduces the concepts of VDM, the model-in-the-loop approach, AOP and the basic elements of cryptography used within the system. Section 3 presents the architecture of the industrial control system and its prototype in both hardware and software. Section 4 then describes the components of the blackbox model as well as presents a specific network trace that has been

---

[1] Our industrial partner wishes to remain anonymous.

analysed. Section 5 introduces the concepts of AOP as applied in our work and then provides definitions and scenarios for our whitebox model. Section 6 discusses the results obtained from both modelling approaches and provides a comparison between them. Section 7 presents the related work. Finally Sect. 8 concludes the paper and introduces potential future work.

2   Background

The Vienna Development Method (VDM) [4] is one of the most mature formal methods originally developed at the IBM Laboratories in Vienna in the 1970's [21]. In this paper we make use of the VDM++ dialect [5]. This is an object-oriented notation including pre- and post-conditions, where definitions are organised in classes, and standard inheritance is possible. Standard libraries including Input-Output (IO) can be included, and essential functionality where a Java implementation is used can be declared using an "is not yet specified" principle. Inside each class it is possible to define instance variables, types, functions and operations. The difference between functions and operations is that functions have no visibility of instance variables at all, whereas operations can have an additional flag indicating that they are pure (i.e., cannot change any instance variables). Both of these can be declared either explicitly, or implicitly using pre- and post-conditions. The types include a number of basic types and type constructors for sets, sequences and mappings. Different tools have been developed supporting the different VDM dialects and the Overture tool on top of the Eclipse platform is probably the one most well-known (and the one used in this work) [18].

The technique discussed in this paper is known as "model in-the-loop". This technique utilizes a formal model connected to the live system to analyse data that the system generates during runtime. The system could invoke its calls through a formal model or interact with a formal model with the use of a specific interface. This is often called runtime verification [19], since the state of the system at the moment is understood by the model and could be checked against the desired properties. The benefits of runtime verification are the deep understanding of the actual behaviour of the system during execution and its subsequent run through the model. This could be used with a feedback loop to the system, either to adjust specific parameters as a result of the analysis, or, in our case to provide feedback about potential security issues. As the system gains new features, the model can be expanded to encompass them, providing scalability to an increasing feature set. The challenge within the runtime verification approaches is that the system is first analysed only past its deployment.

Another concept used within this work is Aspect Oriented Programming (AOP). This is a programming paradigm that is utilized to increase modularity by separating cross-cutting concerns such as security or logging, given that the functionality is reusable within different invocation points. The paradigm utilizes so called *advice*, that is a function that can modify other functions, i.e., changing a behaviour of existing implementation. In this work we use this approach to combine the execution of Java code with the operations of the VDM model. This is done by utilizing the AspectJ [10] AOP extension for Java and its join-point model allowing to specify *pointcuts*, specific places within the implementation where we shall call the VDM operations.

Cryptographic mechanisms can be generally classified in *symmetric* (or secret-key) and *asymmetric* (or public-key). Symmetric cryptography techniques are very efficient for encrypting and authenticating data in bulk, but have the main limitation of requiring parties to share a secret key securely before the communication. Asymmetric cryptography alleviates the problem by allowing secrets to be negotiated between parties which can exchange public keys and verify their authenticity. In such a public key infrastructure, the ownership of public keys can then be established/verified through certificates binding a public key to its owner through a digital signature computed by a certificate authority. For sensitive private keys involved in decryption or computing digital signatures, it is typical to involve TPM chips for their generation, storage and usage in a way that minimizes key *exposure* during system operation. TPM chips are dedicated cryptographic hardware packages with built-in physical security measures against attackers attempting to collect secret-correlated leakage through a side channel or more invasive alternatives, which give them a certain degree of tamper resistance.

3 System Architecture

The system under consideration consists of several components. In this section we present the wider architecture of the industrial control system and then focus on the key management subsystem. The different components of the system are (1) the operations terminal, providing an user interface for local access to the system; (2) embedded controllers executing the industrial control, where we consider a controller not yet connected to the control network as a spare part; (3) the keyvault, an embedded system providing the cryptographic key management functionality; and (4) an Ethernet switch providing connectivity between these different components. In order to be able to analyse the system we further employ (5) an embedded device that we use as a translator to translate a captured network trace to a trace that our VDM interpreter can understand. The key management subsystem that we focus on within this paper consists of several embedded controllers and a keyvault with the Ethernet switch and the translator as supporting components, as shown in Fig. 1.

In the prototype we have constructed to carry out the analysis, the controllers are two Raspberry Pi computers equipped with a TPM used for generating private keys, while the keyvault and the translator are Raspberry Pi computers without a TPM. The switch used is a managed switch that allows for port mirroring in order to capture traffic between the other components and provide it to the translator. The keyvault can be seen as an entry point to the system that the controller must initially contact to join the network. Furthermore the keyvault acts as a certificate authority for the system, with certificate structure as shown in Fig. 2. The system utilizes several certificates with **cert_m** being a (self-signed) certificate of the manufacturer, **cert_kv** being a certificate of the keyvault, **cert_ct** being a certificate of the controller, **cert_ca** being a (self-signed) certificate of the local certificate authority and finally **cert_eff** certificates being the effective certificates used for communication. Once the controller initiates joining into the network, it follows a specific joining protocol, where both the controller (spare part) and the keyvault mutually authenticate as shown in Fig. 3. In case that the controller has successfully joined the network, it can then communicate with other controllers within the network and is a subject to a rekeying process, where new encryption keys are generated and a new short term certificate is obtained by following a protocol illustrated in Fig. 4.

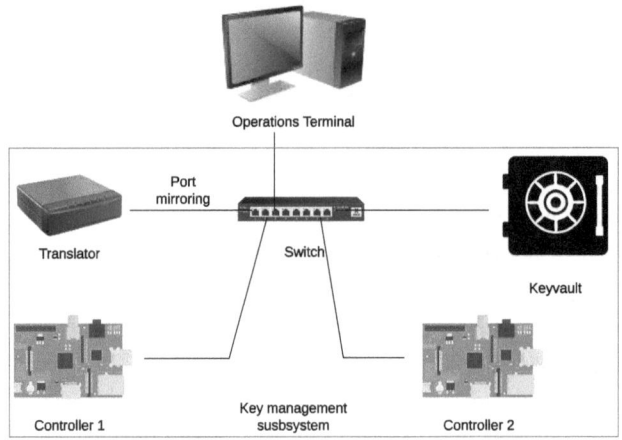

Fig. 1. Architecture of the key management subsystem of an industrial control system

It is important to note that the translator component is only utilized for the blackbox part of the study, where it translates the network traffic generated between the devices. For the white box portion of this study the translator is not needed since there is visibility directly to the method calls executed on the controllers that are then dispatched directly through the VDM model.

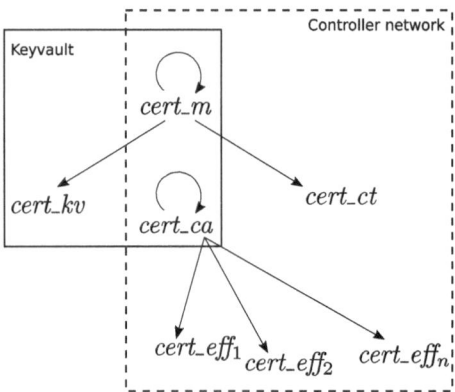

Fig. 2. Certificate tree used within the system, the downward arrows represent signatures of leaf certificates by the corresponding certificate authorities

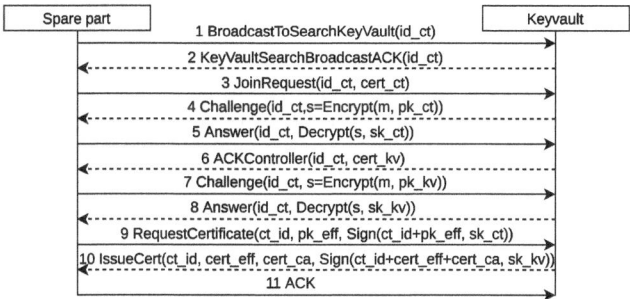

Fig. 3. The protocol followed by the spare part to become a controller

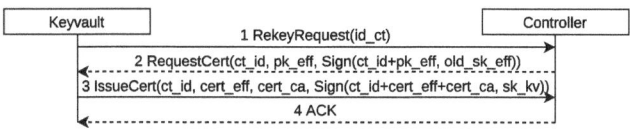

Fig. 4. The protocol followed by the controller to obtain a new effective certificate

4 Black Box Model

The blackbox model utilizes the translator component to translate a network trace captured from a communication between different components to a VDM trace. This trace is then run through a VDM model which captures the protocol properties. The traffic captured could be either the traffic of the joining or the rekeying process. The network trace is captured by use of the Python's Scapy module [9] used within the translator component and then translated to a trace understandable by the VDM interpreter using Python. An example of this is shown in Listing 1.1, showing two invocations of the getStatus call within the protocol with a unique id assigned to each request/response pair. Full translator implementation could be found at [31].

```
1  [{"id"|->1,"messageType"|->"getStatus"},
2      {"status"|->"Available","id"|->1,"messageType"|->"
          getStatusResponse"},
3      {"id"|->2,"messageType"|->"getStatus"},
4      {"status"|->"Busy","id"|->2,"messageType"|->"
          getStatusResponse"}]
```

Listing 1.1. Example output of the VDM trace translator

The captured messages are then run through the VDM model, where the VDM class defines a "MessageArgumentType", a type capturing a union of all the different types the captured trace element can be of. This together with the "MessageRead" type that maps a sequence of characters to the "MessageArgumentType" is used to match a single element of the captured trace file. We utilize the VDM IO package to read the network trace file. The VDM model contains a class for every message type that can be exchanged between the different components of the system, which is modelled as a subclass extending a base class with three basic parameters: *messageType* modelling the type of the message, *controllerId*, modelling the identity of the controller sending the message and a flag *isForKeyVault* stating if the message is aimed at the keyvault component. An example of a message exchanged between the controller and the keyvault during the joining process of the controller, the "JoinRequestMessage", is shown in Listing 1.2.

Once the captured traces are loaded it is possible to analyse them against several possible attack or malfunction scenarios. In this paper we present two scenarios, the *replay attack within a network* and a *faulty implementation*. In the *replay attack within a network* we expect an intruder to capture and replay (send again) one or several messages, a situation specifically requested to be mitigated by our industrial partner. To carry out this scenario we have instructed the controller to attempt to join the network twice, which was ignored by the keyvault, and the trace analysis has shown a precondition violation stating that the same *cert_ct* shall not be presented twice to the keyvault, where this precondition acts as a precondition on a constructor of a "JoinRequestMessage". This is shown in Listing 1.2.

```
1  class JoinRequestMessage is subclass of BaseMessage
2  instance variables
3      private static certificates: set of seq of char := {};
4  operations
5      public JoinRequestMessage: seq of char * seq of char *
6      bool * seq of char ==> JoinRequestMessage
7      JoinRequestMessage(messageType, controllerId,
8          isForKeyVault, certificate) == (
9          AddField("controllerCertificate", certificate);
10         certificates := certificates union {certificate};
11         BaseMessage(messageType, controllerId, isForKeyVault
              );
12     )
13     pre certificate not in set certificates
14     post GetMessageType() = "JoinRequest";
15
16     public pure GetControllerCertificate: () ==> seq of char
17     GetControllerCertificate() ==
18         return GetField("controllerCertificate");
19 end JoinRequestMessage
```

Listing 1.2. Join request message send by the controller.

As for the *faulty implementation* scenario, we have added a fault to the controller software where it will always provide the same value for the *pk_eff*, hence not complying with a protocol specification in regards to the rekeying. To carry out this scenario we have set the keyvault to request rekeying every 15 s and observe several successful rekey attempts. This is because the keyvault did not implement a check for uniqueness of the *pk_eff*. Running the trace through the model however captures this issue as a precondition validation due to key rotation not happening already at the first instance of rekeying. The model capturing this issue is shown in Listing 1.3.

```
1  class SendNewEffectivePublicKeyMessage is subclass of
       BaseMessage
2  instance variables
3      private static publicKeys: set of seq of char := {};
4      private static hashes: set of seq of char := {};
5  operations
6      public SendNewEffectivePublicKeyMessage: seq of char *
7      seq of char * bool * seq of char * seq of char
8          ==> SendNewEffectivePublicKeyMessage
9      SendNewEffectivePublicKeyMessage(messageType,
           controllerId,
10     isForKeyVault, effectivePublicKey, hash) == (
11         AddField("effectivePublicKey", effectivePublicKey);
12         AddField("hash", hash);
13         publicKeys := publicKeys union {effectivePublicKey};
14         hashes := hashes union {hash};
15         BaseMessage(messageType, controllerId, isForKeyVault
               );
16     )
17     pre effectivePublicKey not in set publicKeys and
18         hash not in set hashes and len hash = 512
19     post GetMessageType() = "SendNewEffectivePublicKey";
20
21     public pure GetEffectivePublicKey: () ==> seq of char
22     GetEffectivePublicKey() ==
23         return GetField("effectivePublicKey");
24
25     public pure GetHash: () ==> seq of char
26     GetHash() == return GetField("hash");
27  end SendNewEffectivePublicKeyMessage
```

Listing 1.3. Key rotation under the new effective public key.

The blackbox model was hence used primarily to analyse the network communication between the devices. While during the case study the trace was first captured and stored within a file, it is possible to feed the trace directly from the capture to the VDM model through a simple file interface. The blackbox approach poses several challenges discussed in Sect. 6.

5 White Box Model

The whitebox model utilizes AspectJ and Java to directly interact with the VDM model. This allows the model to get insight into the different methods that are called directly on the different system components at runtime. In order to achieve this, AspectJ and Java annotations are used to instruct the Java methods that shall be analysed during runtime via dispatch of parameters through a VDM model. The full Java implementation can be found at [30]. In order to implement the aspect weaving (integration of aspect code to the classes using it) we have created a custom Java annotation, @VDMOperation. Any Java method containing this annotation causes three files to be generated during compilation, a Java file and two VDM files, where the Java file provides the Java code necessary to link to the VDM file containing the annotated methods as skeletons within the VDM model, where these are simply set as is not yet specified acting as an interface between the model and the implementation. The final generated VDM file contains a VDM class responsible for invocation of the Java implementations of the methods present in the VDM skeleton file. In this way, it is possible to utilize the VDM pre- and post-conditions as well as invariants for runtime verification of a Java program. The AspectJ extension was utilized to change the behaviour of Java methods containing @VDMOperation at runtime. Specifically we use it to run the program in two different ways, (1) where we launch the Java program using a classic Java main method in which case it is running as a standard Java program or (2) using a main method present in the Overture Tool Library, where the aspect creates a VDM interpreter object, which is then used to run the annotated methods. To finalize the link between Java and VDM, we have created type conversions between Java and VDM for several supported types.

To demonstrate our approach we have annotated parts of the Java implementation of the controller and keyvault protocol, specifically the joining and the rekeying protocol. An important part is sending a message between the devices. These exchanged messages need to follow the protocol rules, hence we have created a mapping enforcing the message order. An excerpt of this is shown in Listing 1.4, where it could be seen that in several instances it is possible to repeat a message, for example the <CHALLENGE_ANSWER> message.

```
1  allowedMessageOrder : map Message'MessageType to set of
       Message'
2  MessageType = {
3  <UNKNOWN> |-> {<JOIN_REQUEST>},
4  <JOIN_REQUEST> |-> {
5    <JOIN_REQUEST>, <CHALLENGE_ANSWER>},
6  <CHALLENGE_ANSWER> |-> {
7    <CHALLENGE_ANSWER> , <CHALLENGE_SUBMISSION>, <JOIN_REQUEST
       >},
8  <SIGNING_ACK> |-> {
9    <SIGNING_REQUEST>, <CONTROLLER_CERTIFICATE_UPDATE>,
10  <DUMMY_MESSAGE>}};
```

Listing 1.4. Extract from message ordering rules in VDM

We have then annotated the SendMessage Java method by utilizing the custom @VDMOperation annotation in order to generate the linking files. The implementation of the send message method is shown in Listing 1.5. This then corresponds to the skeleton VDM implementation, shown in Listing 1.6 and the twin VDM file containing the logic for calling the Java implementation with associated VDM pre and post conditions is shown in Listing 1.7. The generated Java class responsible for type conversion between VDM and Java is for space reasons omitted.

Listing 1.5. Java implementation of the protocol send message

```
1    @VDMOperation
2    public void SendMessage(String address, String type,
         String contents) {
3        JSONObject message = new JSONObject(contents);
4        message.put(MessageField.TYPE.Value(), type);
5        this.sender.SendMessage(address, message.toString())
           ;}
```

```
1  class generated_vdm_VDMControllerProtocolContext
2  operations
3  public setJavaObject: (seq of char) ==> ()
4  setJavaObject(vdmObjectName) == is not yet specified;
5  ...
6  public SendMessage:(seq of char) * (seq of char) * (seq of
       char) ==> ()
7  SendMessage(address,type,contents) == is not yet specified;
8  ...
9  end generated_vdm_VDMControllerProtocolContext
```

Listing 1.6. Skeleton (interface) for the VDM model generated from the method in Listing 1.5

```
1  public GENERATED_SendMessage:(seq of char) * (seq of char) *
       (seq of char) ==> ()
2  GENERATED_SendMessage(address,type,contents) == (
3      previousMessage := Message`StringToMessageType(type);
4      return javaObject.SendMessage(address,type,contents);
5  ) pre ( let messageType : Message`MessageType = Message`
       StringToMessageType(type) in (
6          messageType in set dom allowedMessageOrder and
7          messageType in set allowedMessageOrder(
               previousMessage)));
```

Listing 1.7. Full VDM representation of the Java method from Listing 1.5

Using the whitebox approach we have annotated 20 Java methods that are present within our case study with the @VDMOperation annotation, where 10 of them were part of the controller and 10 were part of the keyvault component. This was to cover the method calls responsible for the protocol exchange. We have revisited the two attack or malfunction scenarios from the blackbox approach, the *replay attack within a network* and the *faulty implementation*. In the first case the model did not report any violation, since the keyvault would simply ignore any replayed join request as the controller with the same identity has already joined the network. Due to this no annotated operation is dispatched. As for the second case we have detected a violation of an invariant stating that all *cert_eff* certificates shall be unique, ensuring that the secrets are not being reused. The implementation and its VDM specification for this case is partially shown in Listing 1.8, with deviation occurring due to the saving of old certificate and/or key (lines 8–10) and Listing 1.9 with invariants capturing this deviation (lines 3–7).

Listing 1.8. Annotated method responsible for saving of effective keys

```
1   @VDMOperation
2   public void SaveEffectiveKeys(String
        effectiveCertificateString, String
        effectivePublicKeyString,
3          String effectivePrivateKeyString) {
4          this.effectiveCertificateString.set(
               effectiveCertificateString);
5          byte[] effectiveCertificate = Common.
               StringToByteArray(effectiveCertificateString
               );
6          byte[] effectivePublicKey = Common.
               StringToByteArray(effectivePublicKeyString);
7          ...
8          Common.WriteToFile(effectiveCertificate,
               ControllerProtocolContext.CERT_EFF_FILE_PATH
               );
9          Common.WriteToFile(effectivePublicKey,
               ControllerProtocolContext.PK_EFF_FILE_PATH);
10         this.hasJoined.set(true);}
```

```
1   ...
2   effectiveCertificates : seq of seq of char := [];
3   inv forall i, j in set inds effectiveCertificates & i <> j
       => effectiveCertificates(i) <> effectiveCertificates(j);
4
5   effectivePublicKeys : seq of seq of char := [];
6   inv forall i, j in set inds effectivePublicKeys & i <> j =>
       effectivePublicKeys(i) <> effectivePublicKeys(j);
7   ...
8   public GENERATED_SaveEffectiveKeys:(seq of char) * (seq of
       char) * (seq of char) ==> ()
9   GENERATED_SaveEffectiveKeys(effectiveCertificateString,
       effectivePublicKeyString,effectivePrivateKeyString) == (
10    effectiveCertificates := effectiveCertificates ^ [
         effectiveCertificateString];
11    effectivePublicKeys := effectivePublicKeys ^ [
         effectivePublicKeyString];
12    ...
```

```
13    javaObject.SaveEffectiveKeys(effectiveCertificateString,
         effectivePublicKeyString,effectivePrivateKeyString);
14  ) pre effectivePublicKeyString = pendingEffectivePublicKey
       and effectivePrivateKeyString =
       pendingEffectivePrivateKey;
15  ...
```

Listing 1.9. VDM operation for Saving of the effective keys and associated invariants

The whitebox approach to runtime validation omits any file interfaces between the implementation and the model. It however poses several challenges in terms of coverage and applicability discussed in Sect. 6.

6 Results and Comparisons

Both of the presented approaches have provided either live system analysis results with a certain delay (the blackbox approach) or a direct runtime validation (the whitebox approach) and have provided several answers for our industrial partner.

In case of the blackbox approach we have focused on at first collecting a network trace capturing the communication between components and then carrying out the analysis. This has given us several full runs of the protocol where we could observe the behaviour of our formal VDM model. The translator was able to obtain all of the protocol messages and we could read them from a file for the analysis. This has shown to be a viable approach to analysis of a communication protocol, while there are however several downsides. The first challenge is the need to utilize a switch capable of port mirroring in order to capture the traffic between the devices. Due to this, in very large systems it might be possible to overwhelm the mirroring capacity of the switch, dropping mirrored messages and potentially leading to a false issue detection. Another challenge is ensuring that the model can read the messages from the file fast enough in case that this approach should be utilized at runtime of the system. Finally, while this approach can analyse the network exchange, it has no vision to computation carried out on the device itself, what could lead to omission of implementation details during the analysis. It is also important to note that, in its current iteration, the VDM trace translator only supports basic types and does not allow for hierarchical data representations. The benefit of this approach is that it does not require being embedded into the implementation itself and any changes to the protocol could be handled by modifications to the trace translator in order to capture the updated protocol flow.

The whitebox approach on the other hand has shown to be able to carry out the analysis at runtime with ease while providing vision into the computation carried out on the device itself and hence is suitable for use with internal components for example cryptographic key generation. The approach of providing a Java annotation for developers increases the ease with which the analysis could be connected with the implementation. In its current iteration the annotations are only developed for the *VDM Operations*, hence it is not possible to generate other VDM constructs such as *VDM Functions* for example. The implemented annotations allow for specifying of pre- and post-conditions by allowing for encoding them as string attributes, however other analysis constructs, for example invariants, need to be manually included in the VDM model. Another limitation is that the VDM interpreter is only capable of executing one method at a time, meaning that in case that an annotated method in Java is executed from another annotated method, only the first method will be executed through the VDM, while the other method will be executed normally in Java despite the annotation. Finally a minor limitation is that currently only the VDM++ dialect is supported, this could however easily be extended to other VDM dialects by use of Overture tools. This approach also focuses primarily on a single component, hence a full system picture could only be obtained by combination of results of analysis of the different components. The benefit of the approach is simple inclusion of the formal analysis due to the annotations for systems developed in Java with the potential to be extended to other object oriented languages such as C#.

Overall both approaches provide a way of utilizing VDM against an actual live system and selecting the approach depends on the analysis requirements at hand. In order to obtain higher coverage both approaches could be used in conjunction, what however comes at a price of increased complexity in system setup as both network mirroring and annotated Java code needs to be utilized. We also note that the approaches are not strictly limited to use within industrial control systems and could be applied to different types of systems within wide range of domains and as such are generally applicable.

7 Related Work

Due to its criticality for maintaining secure communication and functionality of modern industrial control systems, security infrastructures and key management protocols have been thoroughly analysed in the state of the art [2, 11, 14, 15, 17, 20, 24, 28, 29]. Likewise, different design alternatives have been considered to build secure-by-design systems such as AOP [7, 22, 26].

The authors of [20] proposed an extended Authentication and Key Agreement protocol to enable dynamic user management and revocation using Schnorr signatures. They also conducted a formal security analysis following the user dynamics and connectivity for predefined attack scenarios, conducted using a simulation-based proof in the random oracle model. Compared to that, we consider the actual system runtime and analyse it against security properties rather than specific attack types so that we can identify unexpected attacks as behaviours not satisfying the security properties.

In [14, 15], the authors proposed a formal verification setup for security analysis of industrial control systems with a keyvault management protocol. The system models are expressed as formal specifications in VDM and UPPAAL respectively. However, the models rely on a high abstraction level to prevent state space explosion. Using AOP, our paper tackles the abstraction concerns through different granularities of the system behaviour (black box, grey box and actual implementation) to deliver an evidence-based security verification.

The authors of [27] introduced a formal security analysis framework for cyber-physical systems. The systems behaviour is expressed in ASLan++ where a tool-assisted verification is conducted using CL-AtSe. The system model is formed by different communicating agents and attack scenarios to be verified against defined security attributes.

To enable a thorough and formal description of security properties, the authors of [28] specified a set of security properties in VDM and used VDM-SL toolbox to perform formal analysis of the resulting models. However, the analysis is static and only considers syntax checking and type checking. Compared to that, our formal security analysis enables a semantic analysis of the different model behaviours driven by the actual system implementation.

The authors of [22] proposed a formal approach to develop secure access to sensitive resources using AOP. They start with an UML model, integrating the functionality and security requirements, which is translated to a formal B specification, from which a relational-like B implementation is derived using a refinement process. Similarly, our paper considers different abstraction levels for the security system verification and implementation however to increase the confidence and reliability of our security analysis our specification model is interactive and mostly driven by the network data from the implementation.

This paper proposes a model-in-the-loop framework for formal security analysis of a key management protocol for an industrial control system. The model is built in VDM and calibration of the model to the actual system implementation is conducted to secure high fidelity and actual data inputs. AOP is then specifically used in terms of white box implementation of an actual security device to collect actual behaviour traces and states and demonstrate the security analysis accuracy.

8   Conclusion

This paper presented an approach for model in-the-loop and runtime verification for the analysis of security features of an industrial control system with embedded certificate authority. The considered system architecture was based on a system under development by our industrial partner. The system under consideration is dynamic, meaning that different components could leave and join the system at any time. To this end, we have found that the use of formal models with security analysis based on data generated at system runtime is a viable approach that could discover potential design and implementation issues. We have specifically shown that it is possible to analyse the system at different levels of granularity. The blackbox approach presented has shown that simply by attaching an extra monitoring component to the system we can gain insights into potential issues via use of a formal model without any modification to the implementation of the system. We have also demonstrated use of AOP in order to provide a view directly into the implementation of the system via the whitebox approach. VDM has shown to provide a good interoperability with the different approaches due to its nature of guided analysis. To demonstrate our approach we have analysed the system via the blackbox and whitebox methodologies against faulty implementation and a replay attack, discovering that the approach would either detect these issues or in a case of the issue being removed due to implementation it would be ignored, as was the case within the whitebox approach.

As future work, we plan to subject our system to more attacks as well as improve the interoperability of the Java implementation with the VDM model, especially in the cases of nested method calls. We also plan to provide more coverage of the system by checking more security and privacy properties and utilizing security properties based on industrial cyber security standards. Finally we plan to utilize our method against systems from different domains with different security threats to demonstrate the general applicability of our approaches.

References

1. Bartocci, E., Falcone, Y., Francalanza, A., Reger, G.: Introduction to runtime verification. In: Bartocci, E., Falcone, Y. (eds.) Lectures on Runtime Verification. LNCS, vol. 10457, pp. 1–33. Springer, Cham (2018). https://doi.org/10.1007/978-3-319-75632-5_1
2. Dojen, R., Zhang, F., Coffey, T.: On the formal verification of a cluster based key management protocol for wireless sensor networks. In: 2008 IEEE International Performance, Computing and Communications Conference, pp. 499–506
3. Elrad, T., Filman, R.E., Bader, A.: Aspect-oriented programming: introduction. Commun. ACM **44**(10), 29–32 (2001)
4. Fitzgerald, J.S., Larsen, P.G., Verhoef, M.: Vienna development method. In: Wah, B. (ed.) Wiley Encyclopedia of Computer Science and Engineering. Wiley, Hoboken (2008)
5. Fitzgerald, J., Larsen, P.G., Mukherjee, P., Plat, N., Verhoef, M.: Validated Designs for Object-Oriented Systems. Springer, New York (2005). https://doi.org/10.1007/b138800,http://overturetool.org/publications/books/vdoos/
6. Fitzgerald, J.S., Larsen, P.G., Verhoef, M.: Vienna Development Method, pp. 1–11. Wiley, Hoboken (2008). https://doi.org/10.1002/9780470050118.ecse447
7. Gao, S., Deng, Y., Yu, H., He, X., Beznosov, K., Cooper, K.: Applying aspect-orientation in designing security systems: a case study, August 2004
8. Gargantini, A., Heitmeyer, C.: Using model checking to generate tests from requirements specifications, pp. 146–162, January 1999
9. Hansen, Y.: Python Scapy Dot11: Python Programming for Wi-Fi Pentesters, 2nd edn. CreateSpace Independent Publishing Platform, North Charleston, SC, USA (2018)
10. Hilsdale, E., Hugunin, J.: Advice weaving in AspectJ. In: Proceedings of the 3rd International Conference on Aspect-Oriented Software Development, AOSD 2004, pp. 26–35. ACM, New York (2004). https://doi.org/10.1145/976270.976276
11. Kahya, N., Ghoualmi, N., Lafourcade, P.: Key management protocol in WIMAX revisited. In: Wyld, D., Zizka, J., Nagamalai, D. (eds.) Advances in Computer Science, Engineering & Applications. AISC, vol. 167, pp. 853–862. Springer, Heidelberg (2012). https://doi.org/10.1007/978-3-642-30111-7_82
12. Kinney, S.L.: Trusted Platform Module Basics: Using TPM in Embedded Systems. Elsevier, Amsterdam (2006)
13. Knowles, W., Prince, D., Hutchison, D., Disso, J.F.P., Jones, K.: A survey of cyber security management in industrial control systems. Int. J. Crit. Infrastruct. Prot. **9**, 52–80 (2015). https://doi.org/10.1016/j.ijcip.2015.02.002
14. Kulik, T., Boudjadar, J., Aranha, D.: Towards formally verified key management for industrial control systems. In: 8th International Conference on Formal Methods in Software Engineering, pp. 119–129, October 2020. https://doi.org/10.1145/3372020.3391555
15. Kulik, T., Boudjadar, J., Tran-Jørgensen, P.W.V.: Security verification of industrial control systems using partial model checking. In: FormaliSE 2020, pp. 98–108. ACM, New York (2020). https://doi.org/10.1145/3372020.3391558
16. Kulik, T., et al.: A survey of practical formal methods for security. Form. Asp. Comput. (2022). https://doi.org/10.1145/3522582

17. Kulik, T., Tran-Jørgensen, P.W.V., Boudjadar, J.: Formal security analysis of cloud-connected industrial control systems. In: Lanet, J.-L., Toma, C. (eds.) SECITC 2018. LNCS, vol. 11359, pp. 71–84. Springer, Cham (2019). https://doi.org/10.1007/978-3-030-12942-2_7

18. Larsen, P.G., Battle, N., Ferreira, M., Fitzgerald, J., Lausdahl, K., Verhoef, M.: The overture initiative - integrating tools for VDM. SIGSOFT Softw. Eng. Notes **35**(1), 1–6 (2010). https://doi.org/10.1145/1668862.1668864

19. Leucker, M., Schallhart, C.: A brief account of runtime verification. J. Log. Algebraic Program. **78**(5), 293–303 (2009). The 1st Workshop on Formal Languages and Analysis of Contract-Oriented Software (FLACOS'07)

20. Li, W., Li, X., Gao, J., Wang, H.: Design of secure authenticated key management protocol for cloud computing environments. IEEE Trans. Dependable Secur. Comput. **18**(3), 1276–1290 (2021). https://doi.org/10.1109/TDSC.2019.2909890

21. Lucas, P.: On the formalization of programming languages: early history and main approaches. In: Bjørner, D., Jones, C.B. (eds.) The Vienna Development Method: The Meta-Language. LNCS, vol. 61, pp. 1–23. Springer, Heidelberg (1978). https://doi.org/10.1007/3-540-08766-4_8

22. Mammar, A., Nguyen, T.M., Laleau, R.: A formal approach to derive an aspect oriented programming-based implementation of a secure access control filter. Inf. Softw. Technol. **92**, 158–178 (2017). https://doi.org/10.1016/j.infsof.2017.08.001

23. Myers, M., Adams, C., Solo, D., Kemp, D.: Internet x. 509 certificate request message format. Request for Comments **2511** (1999)

24. Naoui, S., Elhdhili, M.E., Saidane, L.A.: Security analysis of existing IoT key management protocols. In: 2016 IEEE/ACS 13th International Conference of Computer Systems and Applications (AICCSA), pp. 1–7 (2016)

25. Pointcheval, D.: Asymmetric cryptography and practical security. J. Telecommun. Inf. Technol., 41–56 (2002)

26. Rahli, V., Guaspari, D., Bickford, M., Constable, R.L.: Formal specification, verification, and implementation of fault-tolerant systems using EventML. Electron. Commun. Eur. Assoc. Softw. Sci. Technol. **72** (2015)

27. Rocchetto, M., Tippenhauer, N.O.: Towards formal security analysis of industrial control systems. In: Proceedings of the 2017 ACM on Asia Conference on Computer and Communications Security, ASIA CCS 2017, pp. 114–126. ACM, New York (2017)

28. Tahir, H.M., Shouket, A., Hussain, S., Nadeem, M., Raza, Z., Zafar, N.A.: Formalization of security properties using VDM-SL. In: 2015 International Conference on Information and Communication Technologies (ICICT), pp. 1–6 (2015)

29. Tomas Kulik, Hugo Daniel Macedo, P.T., Larsen, P.G.: Modelling the HUBCAP sandbox architecture in VDM: a study in security. In: The 18th OVERTURE Workshop, pp. 1–15 (2020)

30. Vancea, V., Ubys, L.: Java implementation (2021). https://github.com/valeriuvancea/Key-Managemnt-Protocol-Java-VDM

31. Vancea, V., Ubys, L.: VDM Trace translator (2021). https://github.com/valeriuvancea/Network-Trace-to-VDM-Trace-Translator

32. Weyns, D., Iftikhar, M.U., de la Iglesia, D.G., Ahmad, T.: A survey of formal methods in self-adaptive systems. In: Proceedings of the Fifth International C* Conference on Computer Science and Software Engineering, C3S2E 2012, pp. 67–79. ACM, New York (2012). https://doi.org/10.1145/2347583.2347592

33. Wing, J.M.: A specifier's introduction to formal methods. Computer **23**(9), 8–22 (1990)

Debugging of BPMN Processes Using Coloring Techniques

Quentin Nivon[(✉)] and Gwen Salaün

Univ. Grenoble Alpes, CNRS, Grenoble INP, Inria, LIG, 38000 Grenoble, France
`quentin.nivon@inria.fr`

Abstract. A business process is a collection of related tasks organized in a specific order whose overall execution solves a specific service or product. BPMN has become the standard workflow-based notation for developing business processes. Designing business processes using BPMN is however error-prone. Recent works have proposed verification techniques for analyzing processes and for detecting possible issues. In particular, model checking is an established technique for automatically verifying that a model (e.g., a BPMN process) satisfies a given temporal property. When the model violates the property, the model checker returns a counterexample, which is a sequence of actions leading to a state where the property is not satisfied. Understanding this counterexample for debugging the process is not an easy task, especially if the counterexample is not expressed using the original notation (BPMN here). In this paper, we focus on the model checking of BPMN processes. When properties are violated, we propose to transform counterexamples back on to the original BPMN process in order to simplify the debugging steps. To do so, we rely on coloration techniques. The approach proposed in this paper is fully automated using several tools and was validated on many examples.

1 Introduction

For decades, business processes used by industries and companies have not stopped evolving and becoming more complicated. This had the consequence of enlarging the already existing gap between business analysts, developers and end users. Reducing this gap has been one of the main motivations for the BPMN notation to be invented. Business Process Model and Notation is a notation allowing to graphically describe a business process, with the goal of making it easily understandable for any type of users. In some companies, especially the ones dealing with safety-critical or safety-related systems, it is very important to perform verifications of the behaviour of the BPMN model(s) describing their process(es). Model checking is a well-known technique to perform such tasks, widely used since its beginnings in the 1980's. Indeed, if the property is verified in each point of the model, the model checker returns *True*. Otherwise it returns *False* and gives a counterexample corresponding to a path in the specification that does not satisfy the property. Nonetheless, understanding this counterexample can be complex, even for expert users. It is even more complex

S. L. Tapia Tarifa and J. Proença (Eds.): FACS 2022, LNCS 13712, pp. 90–109, 2022.
https://doi.org/10.1007/978-3-031-20872-0_6

for beginners and BPMN users, as the counterexample returned by the model checker is either textual (sequence of actions violating the property), or visual (graphical representation of the sequence of actions violating the property), but it is not expressed in the BPMN notation.

In this paper, we focus on the model checking of BPMN processes. The main goal of this work is to transform counterexamples back on to the original BPMN process in order to simplify the debugging steps. To do so, we rely on some already existing techniques. The first one, called VBPMN [19], is used for converting BPMN into LNT. The second one is model checking [25], that we use for analyzing and debugging LNT specifications with respect to some temporal logic properties. The last one, called CLEAR [3] is used for generating a colored LTS representing the full set of traces violating the property, called a counterexample LTS (CLTS) [3]. This CLTS can however be hard to understand, for several reasons: (i) the CLTS notation strongly differs from the BPMN notation, (ii) the CLTS can be quite large if there are lots of bugs or if the BPMN contains certain types of constructs, such as parallel gateways, and (iii) the CLTS does not show exactly the source of the bugs.

In practice, the CLTS may give enough information regarding the bug(s) for expert users, but is not sufficient for beginner users, for whom the size of the CLTS and its syntactic differences with the BPMN notation may be a problem. To handle this problem, the main idea of our approach is to transform the information available in the CLTS model on to the BPMN model for simplifying the debugging phase. To do so, the approach consists of three cases or steps, which are handled one after the other. The first step, called *matching analysis*, aims at verifying whether the satisfaction/violation of the property can be directly represented on the initial BPMN process. If this is the case, we color this BPMN process and give it back to the user. This is the best case, because no modification of the initial BPMN process is required. If this can not be done, we generate a new BPMN process. This second step is called *unfolding*. This generated process is semantically equivalent to the initial one. Moreover, it has, by construction, the advantage of allowing us to color it to represent the satisfaction and the violation of the property. However, this generated BPMN process can possibly be quite large, in terms of number of nodes. In this case, understanding this process can be difficult for the user. To limit this, we propose a solution in which we try to reduce the size of this BPMN process by restructuring parts of the process using, for instance, parallel gateways. This third step preserves the semantics of the input process and is called *folding*.

Note that the class of temporal properties handled in this work is the class of safety properties, which is widely used in the verification of real-time and critical systems. Safety properties state that *"something bad must not happen"*.

Figure 1 gives an overview of the contributions presented in this paper. The approach takes as input the initial BPMN process and the temporal logic property to verify. The first step is performed by the VBPMN tool [19], which takes the BPMN process given as input and transforms it into LNT. Then, the LNT specification and the temporal logic property are given to CLEAR [5], which

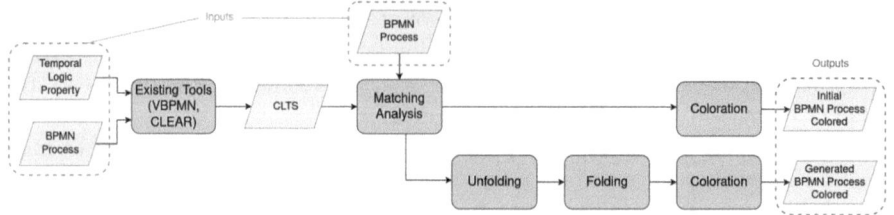

Fig. 1. Overview of the approach

generates a CLTS. Starting from this CLTS and the BPMN process, we perform the matching analysis. If a matching is found between the CLTS and the BPMN process, the BPMN process is colored according to the CLTS. Otherwise, we generate a new BPMN process with the help of the CLTS, we try to fold it, and finally we color the resulting process.

All the steps of the approach are fully automated by a prototype tool we implemented. As for evaluation, we carried out two experimental studies in order to assess the usability and performance of the approach on real-world and large examples.

The rest of this paper is organized as follows. Section 2 introduces several useful notions regarding BPMN and model checking. Section 3 presents the main steps of our approach for visualizing counterexamples at the BPMN level. Section 4 describes the tool support and experiments. Section 5 surveys related work and Sect. 6 concludes the paper.

2   Background

In this section, we introduce some notions needed to understand this work. First, we describe BPMN, which serves as input model to our approach. Then, we give an overview of Labelled Transition Systems and model checking. Finally, we explain what is a Counterexample LTS (CLTS).

Business Process Model and Notation (BPMN) is a business process modeling method aiming at describing business processes used by industries and companies. The main goal of this notation is to provide a way of representing precisely the process used by the company, and to make it understandable to every one (business analysts, developers, end users). This notation has been first introduced in 2004 by the Business Process Management Initiative and has become an ISO/IEC standard in 2013 [1]. It is now widely used across the world. The current version of BPMN is BPMN 2.0 and more details about it can be found in [1].

In this work, we focus on a subset of the BPMN syntax. This subset, described in Fig. 2, takes into account initial and end events, tasks and gateways. A gateway is either a split or a merge. A split gateway consists of a single incoming flow and multiple outgoing flows. A merge gateway consists of multiple incoming flows and a single outgoing flow. Several types of gateways are available, such as exclusive, parallel, and inclusive gateways. An exclusive gateway corresponds to a choice among several flows. A parallel gateway executes all possible flows at the same time. An inclusive gateway executes one or several flows. A study made in [18] on more that 800 real-world BPMN processes shows that the subset used in this paper suffices to build more than 90% of these real-world BPMN processes.

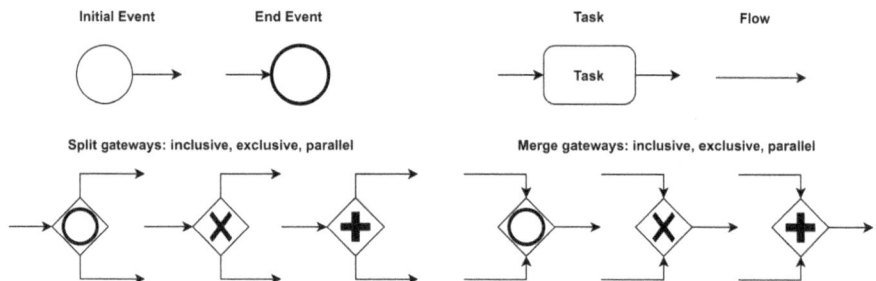

Fig. 2. Supported BPMN syntax

Definition 1. *(BPMN Graph) A BPMN process can be represented as a graph* $B = (S_N, S_E)$ *where* S_N *is a set of nodes (tasks, event, gateways, ...) and* S_E *is a set of edges (sequence flows).*

In this work, a Labelled Transition System (LTS) is used as input for the model checker, and, in a slightly different version called CLTS, used as one of the inputs of the proposed approach. An LTS is represented by a set of states, and a set of transitions connecting these states.

Definition 2. *(LTS) An LTS is a tuple* $M = (S, s^0, \Sigma, T)$ *where* S *is a finite set of states,* $s^0 \in S$ *is the initial state,* Σ *is a finite set of labels,* $T \subseteq S \times \Sigma \times S$ *is a finite set of transitions. A transition is represented as* $s \xrightarrow{l} s' \in T$*, where* $l \in \Sigma$*.*

An LTS can be seen as the semantical model of a BPMN process. Thus, both models have the same semantics. Indeed, all the elements of the BPMN process have a corresponding element in the LTS, and vice-versa. The correspondence is the following: each state of the LTS is either a sequence flow or an exclusive gateway in the BPMN process, and each transition is a task in the BPMN process. An LTS exhibits all possible executions of a system. One specific execution is called a *trace*.

Definition 3. *(Trace) Given an LTS* $M = (S, s^0, \Sigma, T)$*, a trace of size* $n \in \mathbb{N}$ *is a sequence of labels* $l_1, l_2, \ldots, l_n \in \Sigma$ *such that* $s^0 \xrightarrow{l_1} s_1 \in T, s_1 \xrightarrow{l_2} s_2 \in T, \ldots, s_{n-1} \xrightarrow{l_n} s_n \in T$*. A trace is either infinite because of loops or finite when the last state* s_n *has no outgoing transitions. The set of all traces of* M *is written as* $t(M)$*.*

Model checking consists in verifying that a behavioural model or specification (LOTOS New Technology (LNT) [9] in this work) satisfies a given temporal property P, which specifies some expected requirement of the system. Temporal properties are usually divided into two main families: safety and liveness properties [25]. In this work, we focus on safety properties, which are widely used in the verification of real-world systems. Safety properties state that *"something bad must not happen"*. A safety property is usually formalized using a temporal logic. We use MCL (Model Checking Language) [22] in this work, which is an action-based, branching-time temporal logic suitable for expressing properties of concurrent systems. MCL is an extension of alternation-free μ-calculus with regular expressions, data-based constructs, and fairness operators. A safety property can be semantically characterized by an infinite set of traces t_P, corresponding to the traces that violate the property P in an LTS. If the LTS model does not satisfy the property, the model checker returns a *counterexample*, which is one of the traces belonging to t_P.

Definition 4. *(Counterexample) Given an LTS $M = (S, s^0, \Sigma, T)$ and a property P, a counterexample is any trace which belongs to $t(M) \cap t_P$.*

The approach presented in [3,6] takes as input an LNT specification, which compiles into an LTS model, and a temporal property. The original idea of this work is to identify decision points where the specification (and the corresponding LTS model) goes from a (potentially) correct behaviour to an incorrect one. These choices turn out to be very useful to understand the source of the bug. These decision points are called *faulty states* in the LTS model.

In order to detect these faulty states, we first need to categorize the transitions in the model into different types. The transition type allows to highlight the compliance with the property of the paths in the model that traverse that given transition. Transitions in the counterexample LTS can be categorized into three types:

- *correct transitions*, which only belong to paths in the model that represent behaviours which always satisfy the property.
- *incorrect transitions*, which only belong to paths in the model that represent behaviours which always violate the property.
- *neutral transitions*, which belong to portions of paths in the model which are common to correct and incorrect behaviours.

The information concerning the detected transitions type (correct, incorrect and neutral transitions) is added to the initial LTS in the form of tags. The set of transition tags is defined as $\Gamma = \{correct, incorrect, neutral\}$. Given an LTS $M = (S, s^0, \Sigma, T)$, a tagged transition is represented as $s \xrightarrow{(l, \gamma)} s'$, where s, $s' \in S$, $l \in \Sigma$ and $\gamma \in \Gamma$. Thus, an LTS in which each transition has been tagged with a type is called a *tagged LTS*.

Definition 5. *(Tagged LTS) Given an LTS $M = (S, s^0, \Sigma, T)$, and the set of transition tags Γ, the tagged LTS is a tuple $M_T = (S_T, s_T^0, \Sigma_T, T_T)$ where $S_T = S$, $s_T^0 = s^0$, $\Sigma_T = \Sigma$, and $T_T \subseteq S_T \times \Sigma_T \times \Gamma \times S_T$.*

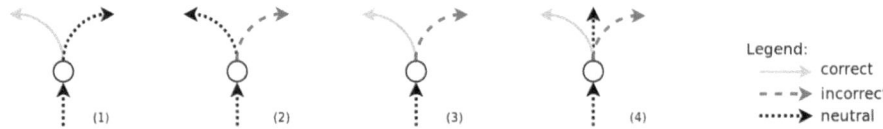

Fig. 3. The four types of faulty states

The tagged LTS where transitions have been typed allows us to identify faulty states, which are those in which an incoming neutral transition is followed by a choice between at least two transitions with different types (correct, incorrect, neutral). Such a faulty state consists of all the neutral incoming transitions and all the outgoing transitions.

Definition 6. *(Faulty State) Given the tagged LTS $M_T = (S_T, s_T^0, \Sigma_T, T_T)$, a state $s \in S_T$, such that $\exists t = s' \xrightarrow{(l,\gamma)} s \in T_T$, t is a neutral transition, and $\exists t' = s \xrightarrow{(l,\gamma)} s'' \in T_T$, t' is a correct or an incorrect transition, the faulty state s consists of the set of transitions $T_{nb} \subseteq T_T$ such that for each $t'' \in T_{nb}$, either $t'' = s' \xrightarrow{(l,\gamma)} s \in T_T$ or $t'' = s \xrightarrow{(l,\gamma)} s''' \in T_T$.*

By looking at outgoing transitions of a faulty state, we can identify four categories of faulty states (Fig. 3):

1. with at least one correct transition and one neutral transition (no incorrect transition),
2. with at least one incorrect transition and one neutral transition (no correct transition),
3. with at least one correct and one incorrect transition (no neutral transition), and
4. with at least one correct, one incorrect, and one neutral transition.

Finally, a CLTS can be defined as a tagged LTS with faulty states.

3   BPMN Coloration

In this section, we first give an overview of the whole approach for simplifying the debugging of BPMN processes. Second, we present in details the *folding* step, because it is the most complex step of the approach. More details about the whole approach can be found in [26].

3.1   Overview

In this subsection, we give an overview of the three main steps of the approach, which are summarized in Fig. 1.

Matching Analysis. The goal of this first step is to determine whether the counterexamples of the temporal logic property appearing in the CLTS can be represented directly on the initial BPMN process. This verification proceeds in two
steps. In the first step, we verify whether all the transitions of the CLTS having
the same label have the same color. Such a situation can happen whenever we
reach a BPMN task that can either violate and satisfy the property, regarding
its position in the execution flow. If it is not the case, we know that the initial
BPMN process can not be colored directly. Indeed, if two transitions with the
same label have different colors, the corresponding (unique) task of the BPMN
process should have two different colors at the same time, which is not possible. To solve this issue, we duplicate this task, making by the same time the
initial BPMN process not colorable directly. Otherwise, we perform the second
step. In this step, we try to associate each faulty state of the CLTS to a unique
node of the BPMN process. Such an association is found whenever a BPMN
node has ancestor tasks (resp. descendant tasks) for which labels correspond to
the labels of the incoming transitions (resp. outgoing transitions) of the current
faulty state, and the current BPMN node has not already been associated to
a faulty state. If this association is found for each faulty state of the CLTS,
then a matching exists between the BPMN process B and the CLTS M, noted
$Match(B, M)$. Then, we color the initial BPMN process, and return it to the
user.

Figure 4 illustrates this second step. Note that the four faulty states, namely
states 3, 4, 5 and 6, have been mapped to four unique nodes of the initial BPMN
process. Thus, a matching was found between the initial BPMN process and the
CLTS, so the initial BPMN process is colored and returned to the user. If no
such matching can be found, we perform a step called *unfolding*.

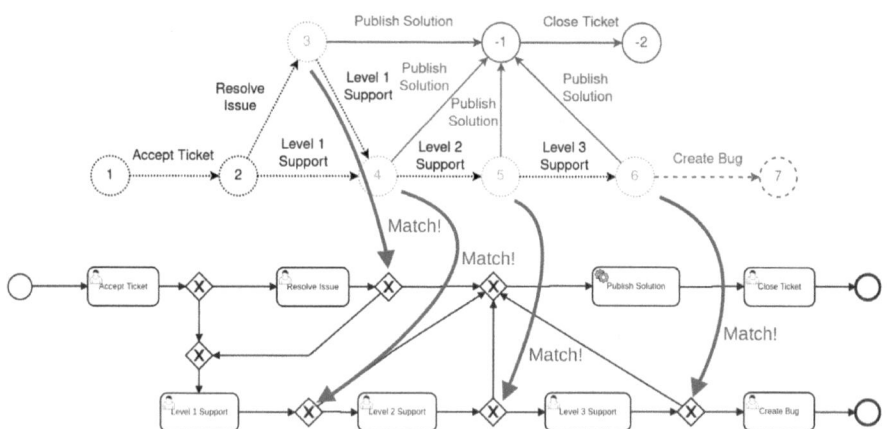

Fig. 4. Example of matching between the CLTS and the BPMN process

Unfolding. In this second step, we generate a new BPMN process that is semantically equivalent to the initial one. This equivalence is preserved because each element of the CLTS has a corresponding element in the BPMN process. More precisely, a transition in the CLTS corresponds to a task in the BPMN process, and a state in the CLTS corresponds either to a sequence flow or to an exclusive split gateway in the BPMN process. As this generated BPMN process is semantically equivalent to the CLTS, and the CLTS is colored, this BPMN process has the advantage of being colorable. Nonetheless, some constructions in the initial BPMN process may generate large BPMN processes during this phase, such as parallel gateways that will be rewritten in their exclusive versions according to Milner's theorem [2][1]. This phenomenon is accentuated by an implemental step called *flattening*, which occurs during the unfolding. In this step, each state of the CLTS having more than one incoming transition will be duplicated. More precisely, if the CLTS contains states s such that $\exists s' \xrightarrow{l_1} s$ and $s'' \xrightarrow{l_2} s$, then the flattening step will generate a new state s''' equivalent to s such that $s'' \xrightarrow{l_2} s'''$, and discard the transition $s'' \xrightarrow{l_2} s$. The flattening is necessary to be able to detect unfolded parallel gateways and replace them by their folded version, while remaining semantically equivalent to the initial BPMN process. Figure 5 illustrates the unfolding phase, along with the flattening step. As the reader can see, each transition of the CLTS has become a BPMN task, while each state has become a sequence flow or an exclusive gateway. Moreover, one can see that the number of nodes in the generated BPMN process is greater than the number of states and transitions in the CLTS. To mitigate this state explosion, we perform a third and last step called *folding*.

Folding. This third and last step aims at reducing the size (in terms of number of nodes) of the generated BPMN process, while increasing the understandability of this process. To do so, we focus on the detection of specific patterns, that we call *interleavings* or *diamonds*. Figure 6 illustrates with an example of such diamond.

As the reader can see, the first task of the upper path corresponds to the second task of the bottom path, and vice versa. In fact, such diamonds result of the rewriting of parallel gateways into their exclusive versions, according to Milner's theorem. Milner's theorem also states that the reverse transformation is possible, allowing us to rewrite this diamond into a parallel gateway containing tasks T_1 and T_2. This transformation may seem of minor interest in this example, as it only reduces the number of tasks from four to two. However, detecting unfolded parallel gateways containing three or more elements may lead to significant reductions of the total number of tasks of the process. Once such gateways have been detected, we generate their semantically equivalent folded version, and we replace the unfolded version by the folded one in the generated BPMN process. This transformation is of prime interest, because it reduces the number

---

[1] A parallel execution of two actions 'a' and 'b' for instance is equivalent to a choice between executing 'a' followed by 'b' or 'b' followed by 'a'.

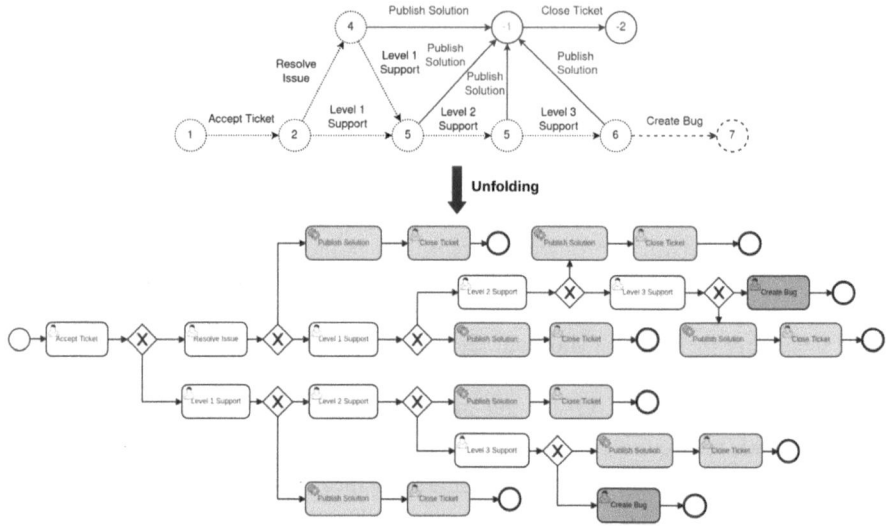

Fig. 5. Example of generation of a BPMN process from a CLTS (Unfolding)

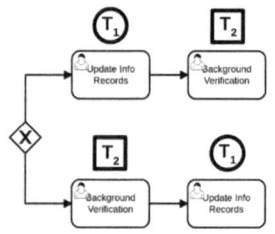

Fig. 6. Example of diamond pattern

of nodes of the BPMN process, while making it syntactically closer to the initial one, and consequently, more understandable for the user.

3.2 Folding

In this section, we focus on the folding step, which is the most crucial and complex one in our approach. In the following, only exclusive gateways are considered, as inclusive and parallel gateways have been rewritten in their exclusive versions according to Milner's theorem. In this context, the word *gateway* will abusively refer to exclusive gateways all along this subsection. Second, the term *folding* will be used to refer to the transformation of a set of tasks belonging to an exclusive gateway into a set of tasks belonging to a parallel gateway. Finally, the word *BPMN process* will refer to the unfolded version of the initial BPMN process.

Before detailing the folding approach, we need to differentiate two types of gateways: the *nested gateways* and the *outer gateways*. The difference between them is that a nested gateway is included in at least another gateway, while an outer gateway is not included in any other gateway. Note that this a strict categorization, because any gateway is either a nested gateway, or an outer gateway.

Definition 7. *(Nested Gateway)* *Let* $B = (S_N, S_E)$ *be a BPMN process, and* $S_{G_N} \subset S_N$ *the set of gateways composing B.* $\forall g \in S_{G_N}$, *g is a nested gateway if and only if* $\exists$ *path* $p = (e_1, ..., e_n)$ *s.t.* $e_n = g$, $\exists i \in [1...(n-1)]$ *where* e_i *is a split gateway and* $\exists p' = (e'_1, ..., e'_m)$ *starting from* e_i *for which* $g \in p'$.

Definition 8. *(Outer Gateway)* *Let* $B = (S_N, S_E)$ *be a BPMN process, and* $S_{G_N} \subset S_N$ *the set of gateways composing B.* $\forall g \in S_{G_N}$, *g is an outer gateway if and only if g is not a nested gateway.*

Figure 7 illustrates nested gateways. As the reader can see, gateways SG2 and MG2, circled in red, are nested gateways, because they are part of the gateways SG1 and MG1. Conversely, these last are outer gateways, are they are not included in any other gateways.

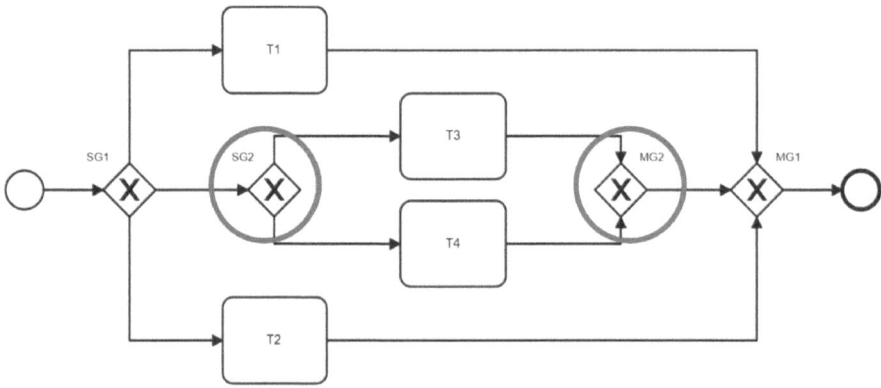

Fig. 7. Example of nested gateways

The folding approach proposed is composed of four main steps:

- **Step 1** retrieves all the outer gateways of the BPMN process that contain only nodes with identical colors, and separate each outer gateway and its subnodes into groups.
- **Step 2** computes, for each gateway of each group, a positive integer value representing the foldability that it can not exceed, *i.e.* the maximum size of the equivalent parallel gateway. We call this information *maximum foldability*. This information is then put in a data structure called *metadata*, that will contain information about the foldability of each gateway of the BPMN process.

- **Step 3** fills the metadata of each gateway of each group, in order to know if it is reducible, and, if yes, what is its composition.
- **Step 4** builds the most reduced version of each gateway group with the help of the metadata, and replace it by the generated one in the BPMN process.

In the rest of this section, we focus on **Step 3**, which is the core step of this approach. Indeed, this step determines if and how each gateway of the generated BPMN process can be folded. Before going further, it is worth noticing that the notion of *maximum foldability* represents the maximum size of the equivalent folded gateway. This notion is a maximum because, in practice, the effective foldability of a gateway can be lower than its maximum foldability, or even null if the gateway is not foldable.

Now, let us detail **Step 3**. **Step 3** is decomposed into two phases: one for gateways of maximum foldability 2, and one for gateways of maximum foldability strictly higher than 2. In this paper, we detail the first phase, managing gateways of maximum foldability 2. In this phase, we analyse all the paths starting from the current gateway in order to extract three pieces of information about it. The first two concern the foldability of the gateway. To decide whether a gateway is foldable or not, we analyze its outgoing paths. The outgoing paths is a set containing all possible paths that can be taken from a node, and which finish either by an element that is already in the path (loop), or by an end event. During this step, we check whether there exists a couple of paths for which the first node of the first path corresponds to the second node of the second path, and vice versa. Such paths are called *size-2 diamond-shaped paths*.

Proposition 1. *(Size-2 Diamond-Shaped Paths)* *Let G_E be an exclusive split gateway and $Out(G_E)$ the set of outgoing paths of G_E.*
$\forall (i,j) \in [1; |Out(G_E)|], i \neq j$, if $\exists p_i = (e_{i,1}, ..., e_{i,n})$, $p_j = (e_{j,1}, ..., e_{j,m}) \in Out(G_E), n, m \geq 2$ s.t. $e_{i,1} = e_{j,2}$ and $e_{i,2} = e_{j,1}$, then p_i and p_j are size-2 diamond-shaped paths.

Figure 6 illustrates a gateway having size-2 diamond-shaped paths. Note that the first node of the first path corresponds to the second node of the second path, and vice versa. If such paths are found, we compute the second information. To compute it, we differenciate the outgoing paths of the current gateway in three sets: those starting with the first size-2 diamond-shaped path, those starting with the second size-2 diamond-shaped path, and the others that are ignored. We keep the first two sets, and remove the first two tasks from each path of each set. The remaining paths in each set are called *out-of-scope paths*.

Proposition 2. *(Out-of-scope Paths)* *Let G_E be an exclusive split gateway of maximum foldability 2, (p_1, p_2) the size-2 diamond-shaped paths of G_E, and DSP the set of outgoing paths of G_E starting with p_1 or p_2. $\forall p = (e_1, ..., e_n) \in DSP$, if $size(p) \geq 3$, then the path $p' = (e_3, ..., e_n)$ is an out-of-scope path.*

Figure 8 shows an example of four out-of-scope paths belonging to two distincts sets of out-of-scope paths. One can see that round tasks and square tasks

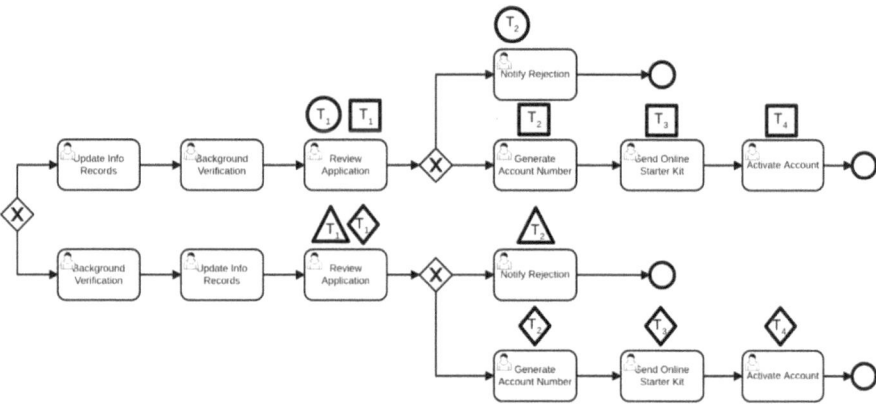

Fig. 8. Example of gateway having 2 sets of 2 out-of-scope paths

form two distinct out-of-scope paths, belonging to the same set. The same remark applies to triangle and diamond tasks.

Once we have computed those two sets of out-of-scope paths, we compare them. If they are identical, we conclude that the current gateway is foldable, with an effective foldability value of 2. We then mark this gateway as foldable in a parallel gateway of size 2. The tasks belonging to this parallel gateway correspond to the first two tasks of the size-2 diamond-shaped paths found. We also store all the out-of-scope paths computed.

If the gateway has been marked as foldable, we compute the third information. This information is used to know whether all the children of the current gateway are parallelizable or not. In practice, if the number of children of a gateway exceeds its effective foldability value, then we know that some of these children are not parallelizable. We called this information *gateway purity*.

Definition 9. *(Gateway Purity) Let G_{FE} be a foldable exclusive split gateway of effective foldability $n \in \mathbb{N}_{\geq 2}$. G_{FE} is pure if and only if $size(G_{FE}) = n$.*

Once the purity characteristic has been evaluated, we fill the metadata with the paths making the gateway not pure, if they exist. We call them *impure paths*. Figure 9 shows an example of impure path. In this example, the path containing the tasks *Inspect Documents* and *Perform Data Analysis* is an impure path. Indeed, this gateway can be reduced to a parallel gateway of size 2 containing the tasks *Update Info Records* and *Background Verification*, so the path circled in red on the figure is an impure path.

When all this information has been computed, we know precisely if and how each gateway of maximum foldability 2 is foldable. Then, we make use of this information to compute the foldability of gateways of maximum foldability 3, then 4, and so on. For these gateways, the approach is slightly different than the one we have presented above. Indeed, in such cases, we make use of the knowledge we have about the foldability of the inner gateways of maximum foldability lower

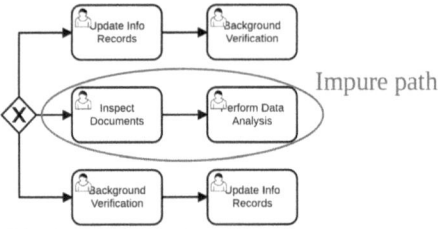

Fig. 9. Example of gateway having 1 impure path

than the maximum foldability of the current gateway by 1 to know whether it is foldable. A gateway of maximum foldability $f \geq 3$ is foldable if and only if it has f children and for which each child is parent of a pure gateway of maximum foldability $f-1$ containing all the children of the initial gateway except the current child. Figure 10 shows equivalence relationships between gateways illustrating how this step works in practice. As the reader can see, the first picture contains only exclusive gateways. After computing information about gateways of maximum foldability 2, we are able to generate the BPMN process shown in the second picture. With this knowledge, we understand that the whole gateway is a parallel gateway containing three tasks that have been unfolded. Thus, we are able to fold it back, as shown in picture 3.

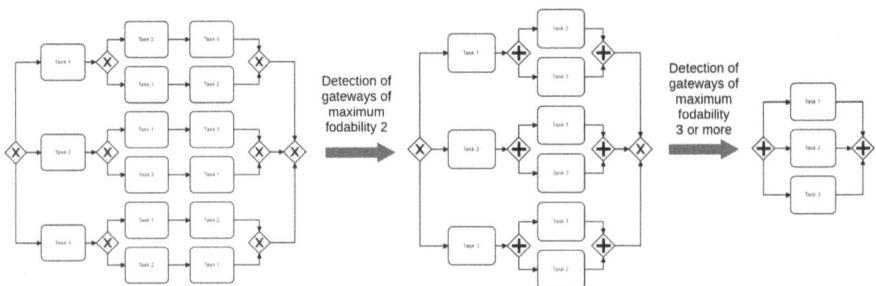

Fig. 10. Equivalence relationships during the folding process

Now, we have a full representation of the foldabilities that can be processed, thanks to the metadata. Figure 11 illustrates this knowledge by giving a textual representation of the metadata of the 3 gateways G_1, G_2 and G_3 visible on the figure. Following this information, we fold the foldable gateways and replace them in the unfolded BPMN process. Then, we return this BPMN process to the user. Figure 12 gives an overview of the whole folding process.

It is worth noticing that, starting from a BPMN process containing 24 tasks, the folding step returned a semantically equivalent process consisting of 7 tasks. In the end, we give back to the user a colored BPMN process, containing either

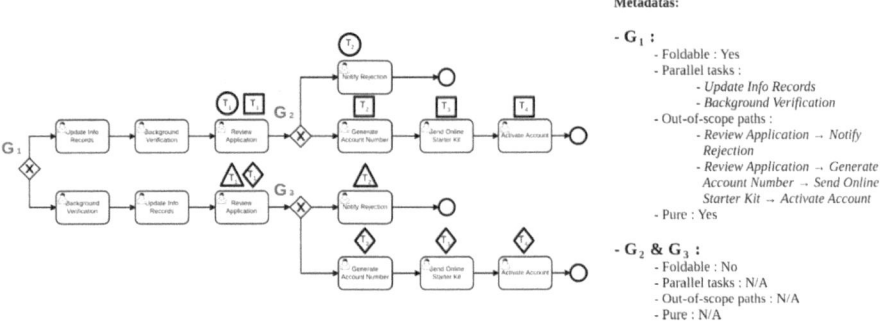

Fig. 11. Textual representation of the gateways' metadata

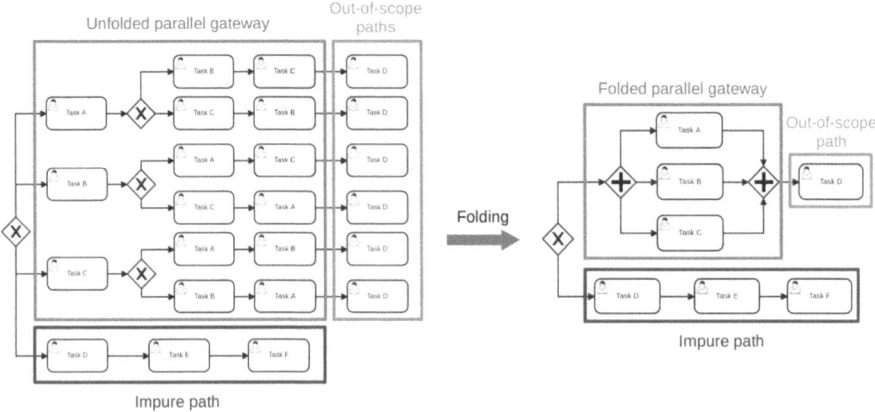

Fig. 12. Example of the whole folding process

black, green, or red tasks. From this process, the user knows that all the paths containing red tasks violate the given property, while paths containing green tasks satisfy it.

4   Tool and Experiments

4.1   Tool

The prototype tool implementing the approach was written in Java and consists of about 10,000 lines of code. It is worth noting that the prototype tool contains a lot of processing steps which have not been described previously in this paper, and which mostly consists in rewriting elements in other formats, such as BPMN to graph, BPMN to CLTS, CLTS to graph, etc. The tool allows the user to perform all the steps detailed in the previous section. It is worth noting that the tool has been tested and validated on about 150 BPMN processes.

4.2   Empirical Study

The goal of this study was to evaluate the usability of our approach in practice by quantifying the number of BPMN processes that are directly colorable, and if they are not, the number of processes that can be folded after the unfolding step. We used for this study 13 real-world processes taken from the literature, and corresponding to various application domains (such as transport, finance, or metallurgy). These processes have been given to the tool along with one well-suited safety property of the kind "T_1 *must not be executed before* T_2". The results of this study are the following: 21% of the BPMN processes were directly colorable, and 29% were successfully folded after unfolding. Overall, in 50% of the cases, we were either able to color the initial BPMN process or to improve the unfolded one.

4.3   Performance Study

This study aimed at evaluating the performance of our approach following two different points of view: (i) assessing the scalability of the prototype tool over real-world examples, and (ii) evaluating the scalability of the prototype tool when increasing the size of the BPMN processes (in terms of nodes). Therefore, the BPMN processes used in this study are of three different kinds:

1. real-world examples of the empirical study presented in the previous section.
2. handcrafted examples that contain a high number of nodes.
3. handcrafted examples that are highly parallel.

   The results of this performance study are presented in Table 1 and 2. Table 1 consists of three columns containing a short description of the BPMN process, the number of nodes in this process, and the execution time of the prototype tool.
   As far as real-world BPMN processes are concerned, the empirical study shows that the execution time is rather low. The highest time in the table is reached for the first process (vacations booking), for which the tool takes three seconds to execute all the steps of our approach. Performance of the tool is thus good for real-world BPMN processes.
   Table 2 presents the results of the study for larger examples. This table contains a first column containing the identifier of the BPMN process, followed by three columns presenting the sizes of the BPMN process and the LTS, and ends with four columns containing the execution times of each element of the toolchain and the global execution time.
   As for the processes with many nodes (rows 1 to 4), one can see that the main source of computation time is due to the use of VBPMN and CLEAR, which take more than 99% of the execution time. As an example, this the case for process 4 in the table, which consists of hundreds of nodes, and for which our tool takes about 3 s to complete. Let us now focus on the examples with a high level of parallelism (rows 5 to 9). The unfolding phase of our approach generates

Table 1. Results of the performance study on real-world BPMN processes

BPMN process	Number of nodes	Execution time of the prototype tool (s)
1. Vacations booking	13	3.436
2. Account opening	22	1.209
3. Publication process	20	1.351
4. Steel transformation	42	1.478
5. Plane entry	27	2.235
6. Mortgage application	15	1.974
7. Denoising	16	1.756
8. Credit offer	15	1.004
9. Buying process	29	1.540
10. Business process	15	1.325
11. Job hiring	29	1.798
12. Login process	17	1.204
13. Support ticket	22	1.431

Table 2. Results of the performance study on two types of handcrafted BPMN processes

BPMN process	Number of BPMN nodes	Number of states in the LTS	Number of nodes in the unfolded BPMN process	VBPMN execution time (s)	CLEAR execution time (s)	Prototype tool execution time (s)	Global execution time (s)
1. Generated large 1	62	140	62	15.98	0.164	1.881	18.03
2. Generated large 2	126	284	126	32.04	2.080	1.376	33.50
3. Generated large 3	254	572	254	<u>165.4</u>	6.461	2.440	<u>174.2</u>
4. Generated large 4	510	1,148	510	**902.3**	<u>124.3</u>	3.643	**1030**
5. Generated parallel 1	19	529	11,582	13.07	0.086	19.76	32.92
6. Generated parallel 2	20	606	23,165	13.15	0.089	82.68	95.92
7. Generated parallel 3	21	683	43,614	13.23	0.088	<u>466.8</u>	<u>480.1</u>
8. Generated parallel 4	22	770	87,229	13.61	0.082	**3948**	**3961**
9. Generated parallel 5	23	857	165,307	17.81	0.122	**10100**	**10117**

processes with a high number of nodes (up to almost two hundreds of thousands nodes in the table for row 9). In such extreme situations, our prototype tool takes minutes or even hours to execute and complete its tasks. However, it is worth reminding that we have not found any real-world process with such a high level of parallelism. One may also mention that performing only partial unfolding would lead to a significant decrease of the execution time of the prototype on highly parallel processes. Nonetheless, the proposed approach is based on model checking techniques that perform a full unfolding of the BPMN process in order to verify the given property. As our starting point is the output of the model checker, we can not perform this partial unfolding.

5   Related Work

Several previous works have focused on providing formal semantics and verification techniques for BPMN processes using a rewriting of BPMN into Petri nets, such as [11–13,21,27]. In these works, the focus is mostly on the verification of behavioural or syntactic problems of the BPMN, such as *deadlock* or *livelock*. As far as rewriting logic is concerned, in [15], the authors propose a translation of BPMN into rewriting logic with a special focus on data objects and data-based decision gateways. They provide new mechanisms to avoid structural issues in workflows such as flow divergence by introducing the notion of well-formed BPMN process. Rewriting logic is also used in [14] for analyzing BPMN processes with time using simulation, reachability analysis, and model checking to evaluate timing properties such as degree of parallelism and minimum/maximum processing times.

Let us now concentrate on those using process algebras for formalizing and verifying BPMN processes, which are closer to the approach proposed in this paper. The authors of [29] present a formal semantics for BPMN by encoding it into the CSP process algebra. They show in [30] how this semantic model can be used to verify compatibility between business participants in a collaboration. This work was extended in [31] to propose a timed semantics of BPMN with delays. [8,23] focus on the semantics proposed in [29,31] and propose an automated transformation from BPMN to timed CSP. In [16] the authors have proposed a first transformation from BPMN to LNT, targeted at checking the realizability of a BPMN choreography. In [11], the authors propose a new operational semantics of a subset of BPMN focusing on collaboration diagrams and message exchange. The BPMN subset is quite restricted (no support of the inclusive merge gateway for instance) and no tool support is provided yet.

The approach presented in [19] proposes verification and comparison techniques based on model checking. The BPMN process is translated into LNT, which is then mapped to LTS before being given as input to a model checker. This approach is close to ours, in the sense that it allows the user to verify a safety temporal logic property over a BPMN process, while providing him a counterexample in the form of an LTS if the property is violated. Other works, such as [17] and [24] are also making use of model checking to perform verifications of temporal logic properties over BPMN processes.

All the approaches presented beforehand in this section are close to what we propose since they focus on the verification of BPMN processes. However, none of these works provide any solution to support the debugging of BPMN processes or to visualize counterexamples at the BPMN level.

Finally, we present the approach proposed in [28], which is the closest to ours. This approach aims at representing visually violations of temporal logic properties regarding a given BPMN process. The authors focus on what they called *containment checking*. Containment checking consists of verifying properties built from high-level BPMN processes (*e.g.*, representations of BPMN processes without many details) over low-level models (*e.g.*, complete representation of BPMN processes). In this work, the high-level BPMN process is translated into Linear Temporal Logic properties [20] while the low-level model is translated into SMV [7]. Both are given as input to the nuSMV [10] model checker which generates counterexamples if the property is violated. Then, the results are given to the visualization engine that represents, in BPMN notation, the counterexamples given by nuSMV. Both works differ in terms of verification: model checking of safety properties for us whereas [28] performs conformance verification of a model regarding a higher-level model. Moreover, they visualize the counterexample only if it can be represented on the initial BPMN process (*i.e.*, if a matching exists), while we go beyond this case by providing additional coloration solutions if a matching does not exist.

6   Concluding Remarks

In this paper, we have proposed a way of improving and simplifying the comprehension and the visualization of safety property violations. The main objectives of the approach were (i) to give a visual feedback of the violation expressed in BPMN notation, (ii) to stay as syntactically close as possible to the initial BPMN process while remaining semantically equivalent, and (iii) to be as minimal as possible in terms of BPMN nodes in the final process. To reach these goals, we chose the coloration of BPMN processes as visualization technique. More precisely, we have first presented a solution in which the original BPMN process is directly colored according to the satisfaction/violation of the property given as input. However, even if this is the best solution because no modification of the original BPMN process is performed, there are cases in which it is not applicable. Therefore, we have proposed a complementary approach which, in a first step, generates a new BPMN process from the CLTS (*unfolding*), and in a second step, performs some minimization over it to lower the number of nodes (*folding*). The different steps of the approach presented in this paper were implemented in a tool written in Java consisting of about 10,000 lines of code. In order to evaluate the usability and the scalability of this prototype tool, we performed two studies: an empirical one to get insights on the behaviour of this approach on real-world examples, and a performance one aiming at ensuring that the prototype runs in reasonable time on real-world examples, and at verifying its scalability on large BPMN processes.

The main perspective of this work is to support liveness properties. In this case, counterexamples and counterexample LTSs have different shapes (lassos) [4], and the approach thus deserves to be revisited to take this specificity into account.

References

1. Information technology - Object Management Group Business Process Model and Notation (2013)
2. Baier, C., Katoen, J.-P.: Principles of Model Checking. MIT Press (2008)
3. Barbon, G., Leroy, V., Salaün, G.: Debugging of concurrent systems using counterexample analysis. In: Dastani, M., Sirjani, M. (eds.) FSEN 2017. LNCS, vol. 10522, pp. 20–34. Springer, Cham (2017). https://doi.org/10.1007/978-3-319-68972-2_2
4. Barbon, G., Leroy, V., Salaün, G.: Counterexample simplification for liveness property violation. In: Johnsen, E.B., Schaefer, I. (eds.) SEFM 2018. LNCS, vol. 10886, pp. 173–188. Springer, Cham (2018). https://doi.org/10.1007/978-3-319-92970-5_11
5. Barbon, G., Leroy, V., Salaün, G.: Debugging of behavioural models with CLEAR. In: Vojnar, T., Zhang, L. (eds.) TACAS 2019. LNCS, vol. 11427, pp. 386–392. Springer, Cham (2019). https://doi.org/10.1007/978-3-030-17462-0_26
6. Barbon, G., Leroy, V., Salaün, G.: Debugging of behavioural models using counterexample analysis. IEEE Trans. Softw. Eng. 47(6), 1184–1197 (2021)
7. Bérard, B., et al.: SMV — symbolic model checking. In: Bérard, B., et al. (eds.) Systems and Software Verification, pp. 131–138. Springer, Heidelberg (2001). https://doi.org/10.1007/978-3-662-04558-9_12
8. Capel, M.I., Morales, L.E.M.: Automating the transformation from BPMN models to CSP+T specifications. In: Proceedings of SEW, pp. 100–109. IEEE (2012)
9. Champelovier, D., et al.: Reference manual of the LNT to LOTOS translator (2005)
10. Cimatti, A., Clarke, E.M., Giunchiglia, F., Roveri, M.: NUSMV: a new symbolic model checker. Int. J. Softw. Tools Technol. Transfer, 410–425 (2000)
11. Corradini, F., Polini, A., Re, B., Tiezzi, F.: An operational semantics of BPMN collaboration. In: Braga, C., Ölveczky, P.C. (eds.) FACS 2015. LNCS, vol. 9539, pp. 161–180. Springer, Cham (2016). https://doi.org/10.1007/978-3-319-28934-2_9
12. Decker, G., Weske, M.: Interaction-centric modeling of process choreographies. Inf. Syst. 36, 292–312 (2011)
13. Dijkman, R., Dumas, M., Ouyang, C.: Semantics and analysis of business process models in BPMN. Inf. Softw. Technol. 50, 1281–1294 (2008). Butterworth-Heinemann
14. Durán, F., Salaün, G.: Verifying timed BPMN processes using Maude. In: Jacquet, J.-M., Massink, M. (eds.) COORDINATION 2017. LNCS, vol. 10319, pp. 219–236. Springer, Cham (2017). https://doi.org/10.1007/978-3-319-59746-1_12
15. El-Saber, N., Boronat, A.: BPMN formalization and verification using Maude. In: Proceedings of the BM-FA, pp. 1–8. ACM (2014)
16. Güdemann, M., Poizat, P., Salaün, G., Ye, L.: VerChor: a framework for the design and verification of choreographies. IEEE Trans. Serv. Comput. 9, 647–660 (2016)
17. Kherbouche, O., Ahmad, A., Basson, H.: Using model checking to control the structural errors in BPMN models. In: IEEE International Conference on Research Challenges in Information Science (RCIS), vol. 7 (2013)

18. Krishna, A., Poizat, P., Gwen, S.: Checking business process evolution. Sci. Comput. Program. **170**, 1–26 (2019)
19. Krishna, A., Poizat, P., Salaün, G.: VBPMN: automated verification of BPMN processes (tool paper). In: Polikarpova, N., Schneider, S. (eds.) IFM 2017. LNCS, vol. 10510, pp. 323–331. Springer, Cham (2017). https://doi.org/10.1007/978-3-319-66845-1_21
20. Kröger, F., Merz, S.: Temporal Logic and State Systems. Springer, Heidelberg (2008). https://doi.org/10.1007/978-3-540-68635-4
21. Martens, A.: Analyzing web service based business processes. In: Cerioli, M. (ed.) FASE 2005. LNCS, vol. 3442, pp. 19–33. Springer, Heidelberg (2005). https://doi.org/10.1007/978-3-540-31984-9_3
22. Mateescu, R., Thivolle, D.: A model checking language for concurrent value-passing systems. In: Cuellar, J., Maibaum, T., Sere, K. (eds.) FM 2008. LNCS, vol. 5014, pp. 148–164. Springer, Heidelberg (2008). https://doi.org/10.1007/978-3-540-68237-0_12
23. Mendoza-Morales, L., Capel, M., Pérez, M.: Conceptual framework for business processes compositional verification. Inf. Softw. Technol. **54**, 149–161 (2012)
24. Messaoud Maarouk, T., El Habib Souidi, M., Hoggas, N.: Formalization and model checking of BPMN collaboration diagrams with DD-LOTOS. Comput. Inform. **40**, 1080–1107 (2021)
25. Milner, R.: Communication and Concurrency. Prentice Hall International, Hoboken (1989)
26. Nivon, Q.: Model checking and debugging of BPMN processes using coloration techniques. Master thesis (2022)
27. Raedts, I., Petkovic, M., Usenko, Y.S., van der Werf, J.M., Groote, J.F., Somers, L.: Transformation of BPMN models for behaviour analysis. In: Proceedings of the MSVVEIS 2007, pp. 126–137 (2007)
28. UL Muram, F., Tran, H., Uwe, Z.: Counterexample analysis for supporting containment checking of business process model. In: International Workshop on Process Engineering (IWPE), vol. 1 (2015)
29. Wong, P., Gibbons, J.: A process semantics for BPMN. In: Proceedings of the ICFEM 2008, pp. 355–374 (2008)
30. Wong, P., Gibbons, J.: Verifying business process compatibility. In: Proceedings of the QSIC 2008, pp. 126–131 (2008)
31. Wong, P.Y.H., Gibbons, J.: A relative timed semantics for BPMN. Electron. Notes Theor. Comput. Sci. **229**, 59–75 (2009)

WEASY: A Tool for Modelling Optimised BPMN Processes

Angel Contreras, Yliès Falcone, Gwen Salaün[✉], and Ahang Zuo

Univ. Grenoble Alpes, CNRS, Grenoble INP, Inria, LIG, 38000 Grenoble, France
gwen.salaun@inria.fr

Abstract. Business Process Model and Notation (BPMN) is a standard modelling language for workflow-based processes. Building an optimised process with this language is not easy for non-expert users due to the lack of support at design time. This paper presents a lightweight modelling tool to support such users in building optimised processes. First, the user defines the tasks involved in the process and possibly gives a partial order between tasks. The tool then generates an abstract graph, which serves as a simplified version of the process being specified. Next, the user can refine this graph using the minimum and maximum execution time of the whole graph computed by the tool. Once the user is satisfied with a specific abstract graph, the tool synthesises a BPMN process corresponding to that graph. Our tool is called WEASY and is available as an open-source web application.

1 Introduction

The Business Process Model and Notation [5] (BPMN) is a workflow-based notation published as an ISO standard. BPMN is currently the popular language for business process modelling. However, specifying processes with BPMN is difficult for non-experts and remains a barrier to the wide adoption of BPMN in the industry. While process mining techniques [7] help to infer processes from execution logs automatically, they do not provide a solution to make users more comfortable with BPMN. Another solution presented in [6] aims at converting text to BPMN, but the text considered as input is not user-friendly and must respect a precise grammar. Moreover, optimization is a tricky phase, which should be considered as soon as possible during the development process. Early integration of optimization allows for building efficient processes, hence reducing the process execution time and the associated costs.

In this paper, we propose a semi-automated approach for supporting users in the modelling of business processes to build optimised BPMN processes at design time. The main idea is to start with a rough model of the process-to-be and refine it by introducing further details in the process step by step. More precisely, as a first step, we expect from the user that (s)he defines the set of tasks involved in the process. Each task comes with a range of minimum and maximum durations.

© The Author(s), under exclusive license to Springer Nature Switzerland AG 2022
S. L. Tapia Tarifa and J. Proença (Eds.): FACS 2022, LNCS 13712, pp. 110–118, 2022.
https://doi.org/10.1007/978-3-031-20872-0_7

Given a set of tasks and a partial order between some of these tasks, the tool automatically generates an abstract graph, which serves as a first version of the process. We use an abstract graph for modelling purposes because it avoids introducing gateways and possible complex combinations of gateways necessary for expressing looping behaviour, for example. Given such an abstract graph, the tool can compute the minimum and maximum times for executing the whole graph. This information regarding time analysis is particularly interesting for optimization purposes and the user can rely on this information for refining the abstract graph, particularly by reducing this execution time and thus the associated costs. When the user is satisfied with a specific abstract graph and its corresponding execution times, (s)he can decide to generate the corresponding BPMN process automatically.

All these features are implemented in Python and integrated into a web application, which serves as a front-end UI where the user can call these functionalities and visualise the results. The tool, called WEASY[1], was applied to several case studies for evaluation purposes and results were satisfactory in terms of usability and performance.

The rest of the paper is organised as follows. Section 2 introduces BPMN and other models used in this work. Section 3 surveys WEASY's functionalities. Section 4 presents some experimental results. Section 5 illustrates how our tool works on a case study. Section 6 concludes.

2 Models

In this work, we consider a subset of BPMN [5] focusing on behavioural aspects (including start/end events, tasks, flows, exclusive/parallel gateways) and time (task duration). As far as time is concerned, each task defines a range of durations indicating the minimum and maximum duration it takes to execute that task. Once the task completes, its outgoing flow is triggered.

WEASY relies on a notion of partial order between tasks. A partial order consists of a set of tasks (all tasks involved in a process) and a relation between tasks. When two tasks are related, it means that the first one must execute before the second one in the final process. This notion of partial order is used to simplify the modelling of BPMN processes.

Abstract graphs are primary models of WEASY, and serve as abstract representations of BPMN processes. An abstract graph consists of a set of nodes and a set of directed edges. A node is defined as a set of tasks and a set of graphs. A graph is thus a hierarchical structure. Abstract graphs are simpler than BPMN processes because they do not rely on nodes and gateways, which can yield intricate structures when they are nested or express specific behaviours such as loops or unbalanced structures. Note that nodes are defined using parallel execution semantics. This means that all the tasks and graphs in a node execute in parallel. A choice in the graph can be expressed by several edges outgoing from the

---

[1] The source code and the instructions for installing and using WEASY are available at: https://github.com/ahzm/weasy.

same node. Looping behaviour can be expressed using an edge going back to a predecessor node.

3   Tool

The WEASY tool consists of two parts. The first part of the tool implements in Python the main features of the approach: (i) transformation from a partial order of tasks to an abstract graph, (ii) computation of the minimum/maximum execution time of an abstract graph, and (iii) transformation of an abstract graph to BPMN. More precisely, the user first gives as input a set of tasks and a partial order of tasks, then Algorithm (i) processes this set of inputs and returns a hierarchical abstract graph. Algorithm (ii) goes through this abstract graph and computes its minimum and maximum execution times. Next, the user decides whether to refine this abstract graph by adjusting the position of nodes (tasks) based on this returned result. Finally, Algorithm (iii) transforms the abstract graph into a standard BPMN model. The reader interested in more details regarding the algorithms behind these features should refer to [3]. Figure 1 shows the classes implemented in Python. These classes encode the different models used in our approach (partial order, abstract graph, BPMN graph) as well as the algorithms used for automating the main steps of the proposed modelling technique.

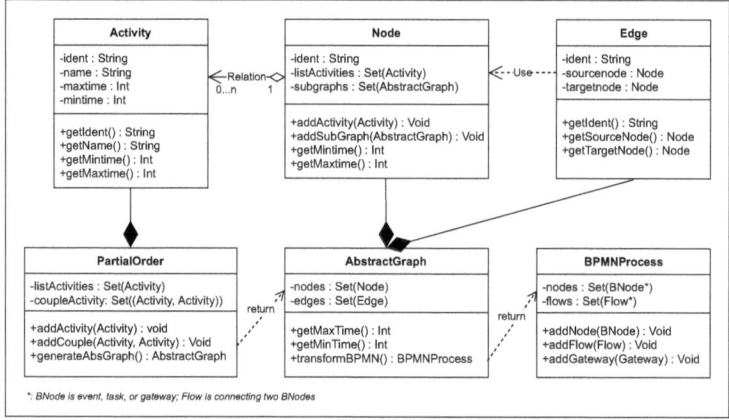

Fig. 1. Class diagram of the tool features implemented in Python.

The second part of the tool corresponds to a web application or User Interface, which can be used to design a process by defining a partial order, refining abstract graphs and finally generating the BPMN process. All the algorithms implemented in Python can be called using buttons, and all the results (abstract graphs, BPMN) can be visualised in the web application. This web application was implemented by using the following JavaScript libraries: *React, MxGraph* and *bpmn-js*.

4   Experiments

In this section, we report on experiments to evaluate performance of the tool and usability of the approach with respect to manual modelling.

Performance. We evaluated the performance of the three main features of our implementation, in practice, by applying the tool to input models of increasing size. These experiments were run on a Windows 10 machine with a Core i5@1.70 GHz processor and 16 GB of RAM. Table 1 presents the execution time of the algorithm generating an abstract graph from a partial order. We vary the number of tasks (first column) and the size of the partial order given as input. The second column gives the percentage of tasks related to another one according to the partial order. For instance, for 1000 tasks, 90% means that 900 tasks (out of 1000) are used in the partial order relation set. The last column shows the time required to execute the algorithm. When increasing the number of tasks (up to 1000), the time remains reasonable (less than 0.5 s). For very large applications involving 10 000 tasks, we can note the different execution times for generating the initial abstract graph. When the total number of tasks remains the same and the size of the partial orders increases, then its execution time decreases. This stems from the fact that the constraints on the graph to be generated augment with the number of tasks in the partial order. The more constraints on the graph, the less computation is required. For example, when the proportion of tasks appearing in the partial order is 30%, the execution time takes about 28 s. Conversely, when the percentage of tasks is 90%, the execution takes about 4 s.

Table 1. Execution time of the algorithm converting a partial order to abstract graphs.

# Tasks	Partial order	Time (s)
10	90%	≈0
100	90%	≈0
500	90%	0.02
1000	90%	0.05
10 000	30%	27.76
	60%	15.36
	90%	4.34

Table 2 presents the time for computing the minimum/maximum time according to different abstract graphs. The table contains five columns exhibiting the graph identifier, the number of nodes, the number of edges, the number of tasks, and the algorithm execution time. The execution time is less than a second for graphs containing up to 300 nodes and edges. However, for larger graphs with 500 nodes or more, it takes more than a second (about 16 s for the largest example with 1000 nodes in the table). This increase mainly comes from the graph size, that is, the number of nodes and edges that all need to be traversed by this algorithm. Execution time also depends on the structure of the graph. For instance, note that for graphs $G5$ and $G6$, it takes more time to execute the algorithm for 100 nodes ($G5$) than for 300 nodes ($G6$), because $G5$ exhibits more paths to be explored (due to interleaving and loops) than $G6$.

Table 2. Execution time for the minimum/maximum time computation.

Identifier	Nodes	Flows	Tasks	Time (s)
$G1$	5	4	10	≈0
$G2$	30	37	50	≈0
$G3$	30	38	100	≈0
$G4$	50	63		0.01
$G5$	100	120	500	0.62
$G6$	300	313		0.11
$G7$	500	524	1000	13.81
$G8$	1000	1020	10 000	16.08

The algorithm transforming abstract graphs to BPMN processes is very efficient (less than a second for all the entries in Table 2).

To summarise, it is worth noting that, even if we have shown examples with hundreds or thousands of tasks, nodes, and flows/edges in this section, in practice, BPMN processes/graphs are usually rather small (less than 100 nodes). For this size, all our algorithms are very efficient (less than a second).

Table 3. Empirical study results.

Group	Correctness	Max time	Modelling time
Manual	50%	32 days	33 min
Semi-auto	100%	30 days	10 min

Usability. We carried out an empirical study to assess the usefulness of our approach compared to classic manual modelling. We have provided two groups of three people (all being non-expert users) with an informal description of a business process (an employee hiring process in a company, see Sect. 5). This description makes explicit the list of activities (with minimum/maximum duration) and the set of ordering constraints to be respected by some of these activities. The goal of the exercise was to provide a BPMN model corresponding to this problem. The BPMN process should (i) use all tasks, (ii) satisfy all ordering constraints, and (iii) be as optimised as possible, that is, the maximum execution time of the process should be as short as possible. The first group of people did not use any tool to support the modelling phase, whereas the second group used the approach and tool proposed in this paper. It was also clearly stated in the exercise that several successive versions of the process could be built, but only the final version was sent as the result.

We have evaluated the results provided by each person using three criteria: correctness of the result, maximum time of the final process, and modelling time. As far as the correctness is concerned, this criterion checks whether the model satisfies the ordering constraints. It is worth noting that, for the given problem, a single process was correct with respect to the given ordering constraints. The second criterion is the maximum execution of the process, which must be as short as possible (the employee hiring process lasts a maximum of 30 days in its most optimised version). Modelling time is the time the person takes to build the final BPMN process. Table 3 details the results for each group of people. Each line in the table presents the average of the results for each group.

Interestingly, the quality of the model (correctness with respect to initial requirements) is much lower using manual modelling, which is indeed more prone to mistakes. The maximum execution time of the resulting process is slightly higher using manual modelling, showing that optimisation is more difficult without using a tool systematically computing this time. Finally, it is much longer (more than three times) to obtain the proposed final process compared to our semi-automated approach.

5   Case Study

We illustrate how our tool works with an employee recruitment process. This process focuses on the different tasks to be carried out once the employee has successfully passed the interview. The employee has to complete some paperwork. (S)He has to see the doctor for a medical check-up. If the employee needs a visa, (s)he should also apply for a work visa. At some point, (s)he should submit all documents. If these documents are unsatisfactory, the company may ask for them again. If everything is fine, all documents are accepted as is. All provided documents are archived properly once validated. The employee is also added to the personnel database and Human Resources (HR) anticipate wage payment while an assistant prepares the welcome kit (office, badge, keys, etc.).

According to this short description of the expected process, the user (someone from the HR staff for example) first needs to define the corresponding tasks and gives an approximate duration for each task as follows:

- T_1: "Fill-in form", [1 day–2 days]
- T_2: "Medical check-up", [1 day–5 days]
- T_3: "Visa application", [7 days–14 days]
- T_4: "Submit documents", [1 day–2 days]
- T_5: "Documents accepted", [1 day–2 days]
- T_6: "Documents rejected", [1 day–2 days]

- T_7: "Archive all documents", [1 day–3 days]
- T_8: "Update personnel database", [1 day–2 days]
- T_9: "Anticipate wages", [3 days–10 days]
- T_{10}: "Prepare welcome kit", [3 days–5 days]

The user can then define an order between some of these tasks. In the case of this example, the following ordering constraints are defined by the user:

- submitting documents can only appear after filling forms, medical check-up, and visa application ⤳ (T_1, T_4), (T_2, T_4), and (T_3, T_4);
- documents are accepted or rejected once they have been submitted ⤳ (T_4, T_5) and (T_4, T_6);
- archiving documents, updating database, anticipating wages, and preparing welcome kit can appear only after validation of the documents ⤳ (T_5, T_7), (T_5, T_8), (T_5, T_9), and (T_5, T_{10});
- updating personnel database should be executed before anticipating wages and preparing welcome kit ⤳ (T_8, T_9) and (T_8, T_{10}).

The definition of tasks and ordering constraints is achieved using WEASY on dedicated interfaces, as illustrated in Fig. 2 (left).

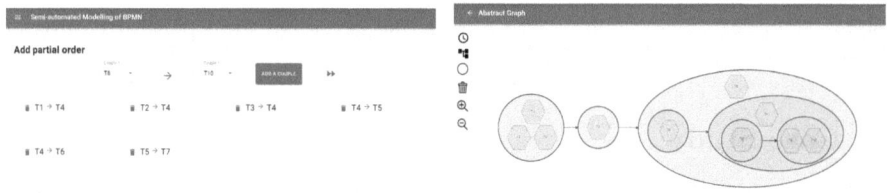

Fig. 2. Definition of partial orders using WEASY (left), and generation and visualization of the first graph (right).

The tool then takes this data (set of tasks and set of pairs) as input, and returns the abstract graph given in Fig. 2 (right). The generation algorithm works as follows. First, it detects that tasks T_1, T_2, and T_3, have as common successor task T_4. Then T_4 is the shared task, splitting the directed graph into two parts. Therefore, the algorithm creates a node to store T_1, T_2, and T_3 and a second node to store T_4. Since T_4 has two successor tasks T_5 and T_6, the algorithm then creates a new node to store them. Since T_5 has some successor tasks, and T_6 does not have any successor task, the algorithm creates a subgraph in this new node and moves T_5 to this subgraph. The rest of the algorithm execution extends this subgraph to integrate the remaining tasks and finally returns the graph given in Fig. 2 (right).

We can then click in the menu on the clock icon (left top corner) in order to compute execution times for this graph. As shown in Fig. 3 (left), the tool returns 13 as minimum execution time and 30 as maximum execution time.

This abstract graph contains three nodes. The algorithm first computes the minimum/maximum execution time of the nodes. These times for the first node are 7 and 14, 1 and 2 for the second node, and 5 and 14 for the last node. Since the nodes in the graph are executed sequentially, the execution times of these three nodes are summed to compute the final result.

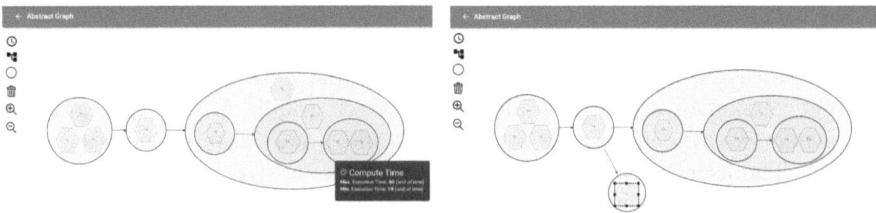

Fig. 3. Computation of times for the first graph (left), and refinement of the graph by moving a task (right).

In most cases, the initial graph can be improved, and this is the goal of the refinement steps. Consider the abstract graph shown in Fig. 3 (left). For this abstract graph, the refinement process consists of two steps. Tasks T_5 and T_6 (documents accepted or rejected) should appear in two different nodes. To do so, in the first step, we add a new node to the graph. In a second step, we move the task T_6 to this new node as shown in Fig. 3 (right). We then recompute the execution times of this new abstract graph. Its minimum execution time is 9 and its maximum execution time is 30. After the refinement step, the maximum time is the same, but the minimum time has improved (going from 13 to 9).

Finally, from the abstract graph given in Fig. 3 (right), we call the BPMN generation algorithm for obtaining the resulting BPMN process shown in Fig. 4. We can see that the first node with three tasks (T_1, T_2, T_3) transforms into a

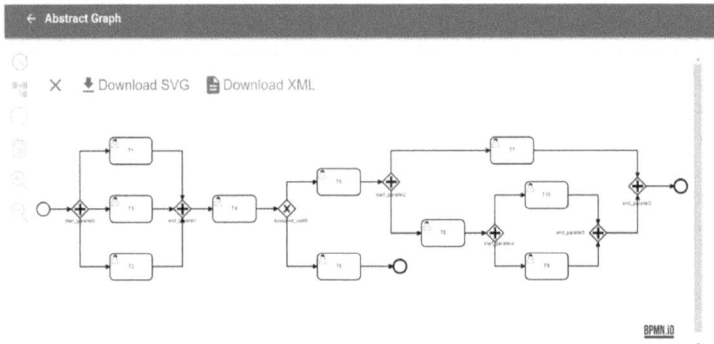

Fig. 4. Transformation of the final graph to BPMN.

split and a join parallel gateway. The algorithm also generates a split exclusive gateway right after T_4 because the corresponding node has two outgoing edges. After T_5, there are again several parallel gateways because there are multiple tasks within the same node.

6 Concluding Remarks

In this paper, we have presented a tool that facilitates the modelling of optimised BPMN processes. Our solution relies on a simple notation for describing an abstract version of a process called an abstract graph. An abstract graph is first generated by defining the set of tasks and a partial order between some of them. The user can then successively refine this abstract graph. Refinement is guided by the minimum and maximum execution times needed for executing the whole process. Once the user is satisfied, the last step transforms this graph into a BPMN process. All the steps of our approach have been implemented in the WEASY tool, which was successfully applied to a series of examples for validation purposes. The main perspective of this work aims at enlarging the considered BPMN subset to take additional constructs into account, such as data, conditions or probabilities in exclusive gateways, see [1, 2, 4] for instance.

Acknowledgements. This work was supported by the Région Auvergne-Rhône-Alpes within the "*Pack Ambition Recherche*" programme.

References

1. Durán, F., Rocha, C., Salaün, G.: Symbolic specification and verification of data-aware BPMN processes using rewriting modulo SMT. In: Rusu, V. (ed.) WRLA 2018. LNCS, vol. 11152, pp. 76–97. Springer, Cham (2018). https://doi.org/10.1007/978-3-319-99840-4_5
2. Durán, F., Rocha, C., Salaün, G.: A rewriting logic approach to resource allocation analysis in business process models. Sci. Comput. Program. 183, **102303** (2019)
3. Falcone, Y., Salaün, G., Zuo, A.: Semi-automated modelling of optimized BPMN processes. In: Proceedings of SCC 2021. IEEE (2021)
4. Falcone, Y., Salaün, G., Zuo, A.: Probabilistic model checking of BPMN processes at runtime. In: ter Beek, M.H., Monahan, R. (eds.) IFM 2022. LNCS, vol. 13274, pp. 191–208. Springer, Cham (2022). https://doi.org/10.1007/978-3-031-07727-2_11
5. ISO/IEC: International Standard 19510, Information technology - Business Process Model and Notation (2013)
6. Ivanchikj, A., Serbout, S., Pautasso, C.: From text to visual BPMN process models: design and evaluation. In Proceedings of MoDELS 2020, pp. 229–239. ACM (2020)
7. van der Aalst, W.M.P.: Process mining. Commun. ACM **55**(8), 76–83 (2012)

Logics and Semantics

Embeddings Between State and Action Based Probabilistic Logics

Susmoy Das$^{(\boxtimes)}$ and Arpit Sharma

Department of Electrical Engineering and Computer Science, Indian Institute
of Science Education and Research Bhopal, Bhopal, India
{susmoy18,arpit}@iiserb.ac.in

Abstract. This paper defines embeddings between state-based and action-based probabilistic logics which can be used to support probabilistic model checking. First, we propose the syntax and semantics of an action-based Probabilistic Computation Tree Logic (APCTL) and an action-based PCTL* (APCTL*) interpreted over action-labeled discrete-time Markov chains (ADTMCs). We show that both these logics are strictly more expressive than the probabilistic variant of Hennessy-Milner logic (prHML). Next, we define an embedding *aldl* which can be used to construct an APCTL* formula from a PCTL* formula and an embedding *sldl* from APCTL* formula to PCTL* formula. Similarly, we define the embeddings *aldl'* and *sldl'* from PCTL to APCTL and APCTL to PCTL, respectively. Finally, we prove that our logical embeddings combined with the model embeddings enable one to minimize, analyze and verify probabilistic models in one domain using state-of-the-art tools and techniques developed for the other domain.

Keywords: Probabilistic · Markov chain · Logic · Embeddings · Model checking

1 Introduction

Discrete time Markov chains (DTMCs) [6,28,29] are widely used to capture the behavior of systems that are subject to uncertainties, e.g. performance and dependability evaluation of information processing systems. DTMCs can be broadly classified into two categories, namely, state-labeled discrete-time Markov chains (SDTMCs) and action-labeled discrete-time Markov chains (ADTMCs). SDTMCs are typically used by the model checking community to establish the correctness of probabilistic systems. Several expressive logics like Probabilistic Computation Tree Logic (PCTL) [6,21], PCTL* [4,6] and PCTL with rewards [2] have been defined to specify interesting properties which can be verified using tools such as Probabilistic Symbolic Model Checker (PRISM) [30], Markov Reward Model Checker (MRMC) [27] and Storm [16]. In contrast, ADTMCs are more commonly used as semantic model for amongst others ACP [5], CCS [20], CSP [34], LOTOS [36] and Petri nets [35]. To compare the behavior of ADTMCs,

S. L. Tapia Tarifa and J. Proença (Eds.): FACS 2022, LNCS 13712, pp. 121–140, 2022.
https://doi.org/10.1007/978-3-031-20872-0_8

a wide range of linear-time and branching-time relations have been defined, e.g. weak and strong variants of bisimulation equivalence [7,8,11,19,23,26,32] and simulation pre-orders [8,17,25].

Recently, a formal framework has been proposed which bridges the gap between these two modeling communities by relating the ADTMC and SDTMC models [14]. More specifically, this paper defines embeddings ald and sld which can be used to construct an ADTMC from an SDTMC and an SDTMC from an ADTMC, respectively. Additionally, inverse of these embeddings, i.e. ald^{-1} and sld^{-1} have been defined which can reverse the effects of applying ald and sld, respectively. These embeddings enable one to use the state-of-the-art tools developed in one setting for model minimization in the other setting as they preserve bisimulation, backward bisimulation and trace equivalence relations. Although these embeddings are very useful from a practical point of view, no effort has been made to investigate if they can be used to support probabilistic model checking of randomized systems. In other words, it is not known if transformation functions can be defined between probabilistic logics which can be used to construct equivalent properties on the embedded model and thus enabling one to leverage the model checking related advancements made by the other community.

This paper focuses on bridging this gap by proposing transformation functions between state-based and action-based probabilistic logics. We start by slightly modifying the embeddings proposed in [14] which allows us to take into account the invisible computation steps, i.e. τ action transitions. Next, we propose the syntax and semantics of an action-based PCTL* (APCTL*) and an action-based PCTL (APCTL) both of which are interpreted over ADTMCs. We use an auxiliary logic of actions for defining APCTL* and APCTL as this logic allows one to express interesting modalities in the action-based probabilistic setting. We show that both APCTL* and APCTL are strictly more expressive than the probabilistic variant of Hennessy-Milner logic (prHML) defined in [32]. We define an embedding $aldl$ which can be used to move from PCTL* to APCTL* and an embedding $sldl$ from APCTL* to PCTL*. In a similar vein, we define the embeddings $aldl'$ and $sldl'$ from PCTL to APCTL and APCTL to PCTL, respectively. Next, we prove that for an ADTMC $\mathcal{D}$ and APCTL* formula Φ: $\mathcal{D} \models_A \Phi \implies sld(\mathcal{D}) \models_S sld(\Phi)$ and vice versa, i.e. for an SDTMC $\mathcal{D}$ and PCTL* formula Φ: $\mathcal{D} \models_S \Phi \implies ald(\mathcal{D}) \models_A aldl(\Phi)$. Here, $\models_A$ and $\models_S$ have been used to denote the satisfaction relations in the action and state-labeled domains, respectively. Similarly, we prove that if an ADTMC $\mathcal{D}$ satisfies an APCTL formula Φ, then the corresponding SDTMC $sld(\mathcal{D})$ will satisfy the PCTL formula $sldl'(\Phi)$ and vice versa, i.e. for SDTMC $\mathcal{D}$ and PCTL formula Φ: $\mathcal{D} \models_S \Phi \implies ald(\mathcal{D}) \models_A aldl'(\Phi)$. Finally, we briefly discuss the relationship between our logics and several variants of probabilistic modal μ-calculus [13,31,33,37]. Note that specifying properties using our logics is easier as compared to variants of probabilistic modal μ-calculus which involve fix-point operators and have limited tool support.

This work is motivated by the fundamental results of De Nicola and Vaandrager [39,40] where model and logical embeddings have been defined for nondeterministic systems. Our results on logical embeddings combined with the modified model embeddings and results proved in [14] provide a complete theoretical framework for minimizing, analyzing and verifying models of one domain using the tools and techniques developed for the other domain. For example, the behavior of a randomized system can be captured using mCRL2 [12] or CADP [18] and interesting properties can be specified using APCTL. The model and property can then be transformed into SDTMC and PCTL using our model and logical embeddings, respectively. In the last step, the SDTMC can be minimized (e.g. bisimulation minimization) and verified using a state-based probabilistic model checker, e.g. PRISM [30] or Storm [16]. Moreover, unlike probabilistic modal μ-calculus [12] where model checking requires exponential time, APCTL model checking can be solved in polynomial time.

Organisation of the Paper. Section 2 briefly recalls the main concepts of SDTMCs and ADTMCs. Section 3 defines the modified embeddings *sld* and *ald*. Section 4 recalls the syntax and semantics of PCTL and PCTL*. Section 5 provides the syntax and semantics of APCTL* and APCTL. Section 6 defines the logical embeddings and proves the preservation results. Section 7 presents the related work. Finally, Sect. 8 concludes the paper and provides pointers for future research.

2   Discrete-Time Markov Chains

Definition 1. *A state-labeled discrete-time Markov chain (SDTMC) is a tuple* $\mathcal{D} = (S, AP, P, s_0, L)$ *where:*

- S *is a countable, nonempty set of states,*
- AP *is the set of atomic propositions,*
- P *is the transition probability function satisfying* $P : S \times S \rightarrow [0, 1]$ *s.t.* $\forall s \in S: \sum_{s' \in S} P(s, s') = 1$,
- s_0 *is the initial state, and*
- $L : S \rightarrow 2^{AP}$ *is a labeling function.*

$\mathcal{D}$ is called finite if S and AP are finite. Let $\rightarrow = \{(s, p, s') \mid P(s, s') = p{>}0\}$ denote the set of all transitions for an SDTMC $\mathcal{D}$. We denote $s \xrightarrow{p} s'$ if $(s, p, s') \in \rightarrow$. Let $s \xrightarrow{p}$ denote that $\exists s' \in S$ s.t. $P(s, s') = p > 0$. Similarly, let $\xrightarrow{p} s$ denote that $\exists s' \in S$ s.t. $P(s', s) = p > 0$.

A sequence of states $s_0, s_1, s_2, \ldots$ is an infinite path in SDTMC. We denote a path by π. A finite path is of the form: $s_0, s_1, s_2, \ldots, s_n$. The length of a finite path, denoted by $len(\pi)$ is given by the number of transitions along that path. Length of the finite path given above is $len(\pi) = n$. For an infinite path π, we have, $len(\pi) = \infty$. We denote the n-th state along a path π by $\pi[n-1]$ ($\pi[0]$ denotes the first state from which the path starts). We denote the concatenation of two paths π_1 and π_2 as juxtaposition $\pi = \pi_1 \pi_2$ which is only defined if π_1 is finite and the last visited state of π_1 is same as $\pi_2[0]$. The suffix (proper suffix resp.) of a path π, say θ is denoted by $\pi \leq \theta$ ($\pi < \theta$). Let $Paths(s)$ denote the set of all infinite paths starting in s. Let $Paths_{fin}(s)$ denote the set of all finite paths starting in s.

Definition 2 (Cylinder set [43]). *Let* $s_0, \ldots, s_k \in S$ *with* $P(s_i, s_{i+1}) > 0$ *for* $0 \leq i < k$. $Cyl(s_0, \ldots, s_k)$ *denotes the cylinder set consisting of all paths* $\pi \in Paths(s_0)$ *s.t.* $\pi[i] = s_i$ *for* $0 \leq i \leq k$.

Intuitively the cylinder set spanned by the finite path π consists of all infinite paths that start with π. Let $\mathcal{F}(Paths(s_0))$ be the smallest σ-algebra on $Paths(s_0)$ which contains all sets $Cyl(s_0, \ldots, s_k)$ s.t. $s_0, \ldots, s_k$ is a state sequence with $P(s_i, s_{i+1}) > 0$, $(0 \leq i < k)$.

Definition 3. *The probability measure* Pr *on* $\mathcal{F}(Path(s_0))$ *is the unique measure defined by induction on* k *in the following way. Let* $\Pr(Cyl(s_0)) = 1$ *and for* $k > 0$:

$$\Pr(Cyl(s_0, \ldots, s_k, s')) = \Pr(Cyl(s_0, \ldots, s_k)) \cdot P(s_k, s')$$

Definition 4. *An action-labeled discrete-time Markov chain (ADTMC) is a tuple* $\mathcal{D} = (S, Act, P, s_0)$ *where:*

- *S is a countable, nonempty set of states,*
- *Act is a set of actions which contains the special action τ,*
- *$P : S \times Act \times S \to [0,1]$ is the transition probability function satisfying:* $\sum_{s' \in S, a \in Act} P(s, a, s') = 1 \ \forall s \in S$, *and*
- *s_0 is the initial state.*

$\mathcal{D}$ is called finite if S and Act are finite. τ is a special action used to denote an invisible computation. Let $\to = \{(s, a; p, s') \mid P(s, a, s') = p > 0\}$ denote the set of all transitions for an ADTMC $\mathcal{D}$. We denote $s \xrightarrow{a,p} s'$ if $(s, a; p, s') \in \to$. Let $s \xrightarrow{a,p}$ denote that $\exists s' \in S$ and $\exists a \in Act$ s.t. $P(s, a, s') = p > 0$. If a state s cannot perform a particular action, say a, we denote it by $s \not\xrightarrow{a}$. Similarly, let $\xrightarrow{a,p} s$ denote that $\exists s' \in S$ and $\exists a \in Act$ s.t. $P(s', a, s) = p > 0$. If a state s cannot be reached in one step by performing a particular action, say a, then we denote it by $\not\xrightarrow{a} s$.

An alternating sequence of states and actions $s_0 \alpha_1, s_1 \alpha_2$, $s_2 \alpha_3, \ldots$ is an infinite run in an ADTMC. We denote a run by ρ. A finite run is of the form: $s_0 \alpha_1, s_1 \alpha_2, s_2 \alpha_3, \ldots, s_{n-1} \alpha_n, s_n$. The length of a finite run, denoted by $len(\rho)$ is given by the number of transitions along that run. Length of the finite run given above is $len(\rho) = n$. For an infinite run ρ, we have, $len(\rho) = \infty$.

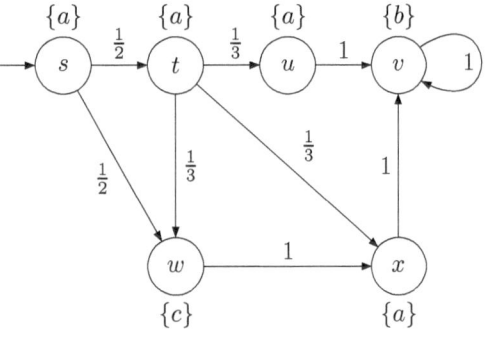

Fig. 1. SDTMC $\mathcal{D}$

We denote the n-th state along a run ρ by $\rho[n-1]$ ($\rho[0]$ denotes the first state from which the run starts) and the n-th action along the run by $\rho_\alpha[n]$ ($\rho_\alpha[1]$ denotes the first transition label of the run). We denote the concatenation of two runs ρ_1 and ρ_2 as juxtaposition $\rho = \rho_1\rho_2$ which is only defined if ρ_1 is finite and the last visited state of ρ_1 is same as $\rho_2[0]$. The suffix (proper suffix resp.) of a run ρ, say θ is denoted by $\rho \leq \theta$ ($\rho < \theta$). Let $Runs(s)$ denote the set of all infinite runs starting in s. Let $Runs_{fin}(s)$ denote the set of all finite runs starting in s.

Definition 5 (Cylinder set [14]). Let $s_0, \ldots, s_k \in S$ with $P(s_i, a_{i+1}, s_{i+1}) > 0$ for $0 \leq i < k$. $Cyl(s_0a_1, \ldots, s_{k-1}a_k, s_k)$ denotes the cylinder set consisting of all runs $\rho \in Runs(s_0)$ s.t. $\rho[i] = s_i$ for $0 \leq i \leq k$ and $\rho_\alpha[i] = a_i$ for $1 \leq i \leq k$.

Intuitively the cylinder set spanned by the finite run ρ consists of all infinite runs that start with ρ. Let $\mathcal{F}(Runs(s_0))$ denote the smallest σ-algebra on $Runs(s_0)$ which contains all sets of the form $Cyl(s_0a_1, \ldots, s_{k-1}a_k, s_k)$ s.t. $s_0a_1, \ldots, s_{k-1}a_k, s_k$ is an alternating sequence of states and actions with $P(s_i, a_{i+1}, s_{i+1}) > 0$, $(0 \leq i < k)$.

Definition 6. The probability measure $\Pr$ on $\mathcal{F}(Runs(s_0))$ is the unique measure defined by induction on k in the following way. Let $\Pr(Cyl(s_0)) = 1$ and for $k > 0$:

$$\Pr(Cyl(s_0a_1, \ldots, s_{k-1}a_k, s')) = \Pr(Cyl(s_0a_1, \ldots, s_{k-2}a_{k-1}, s_{k-1})) \cdot P(s_{k-1}, a_k, s')$$

3 Modified Embeddings *sld* and *ald*

This section presents the model embeddings which can be used to construct an ADTMC from an SDTMC and vice versa. We slightly modify the embeddings proposed in [14]. This modification allows us to take into account the invisible computation steps, i.e. τ action transitions. *sld* allows one to construct an SDTMC from an ADTMC and *ald* is used for constructing an ADTMC from an SDTMC. Note that all the preservation results proved in [14] still hold for our modified embeddings. In other words, these embeddings preserve strong forward, strong backward bisimulation and trace equivalence relations. Additionally, when combined with inverse embeddings a model can be minimized in one setting by minimizing its embedded model in the other setting and taking the inverse of the embedding. Note that these embeddings create additional states and transitions for storing important information about the original model which would be exploited by the inverse embeddings to revert back to the original model. We omit the definitions of reversibility criteria and inverse embeddings as they are not required for developing the formal framework of logical embeddings which is the main focus of our work. Interested readers are referred to [14] for further details.

Definition 7 (sld). Let $\mathcal{D} = (S, Act, P, s_0)$ be an ADTMC. The embedding $sld :$ $ADTMC \rightarrow SDTMC$ is formally defined as $sld(\mathcal{D}) = (S', AP', P', s_0', L')$ s.t.

- $S' = S \cup \{(a, t) \mid P(s, a, t) > 0 \text{ for some } s, t \in S \text{ and } a \neq \tau\}$,

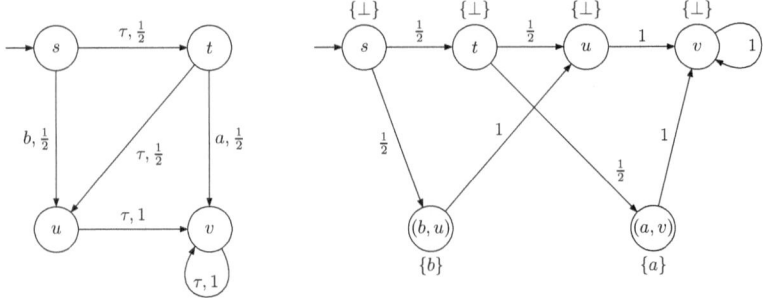

Fig. 2. ADTMC $\mathcal{D}$ and the corresponding $sld(\mathcal{D})$

- $AP' = (Act \setminus \{\tau\}) \cup \{\bot\}$ where $\bot \notin Act$,
- The rate function P' is defined by:

$$
\begin{aligned}
P'(s,(a,t)) &= P(s,a,t) && \text{for all } s,t \in S \text{ s.t. } P(s,a,t) > 0 \text{ and } a \neq \tau, \\
P'(s,t) &= P(s,\tau,t) && \text{for all } s,t \in S \text{ s.t. } P(s,\tau,t) > 0, \text{ and} \\
P'((a,t),t) &= 1 && \text{for all } (a,t) \in S' \setminus S,
\end{aligned}
$$

- $s'_0 = s_0$, and
- $L'(s) = \{\bot\}\ \forall s \in S$ and $L'((a,t)) = \{a\}$.

Definition 8 (ald). Let $\mathcal{D} = (S, AP, P, s_0, L)$ be an SDTMC. The embedding $ald : SDTMC \rightarrow ADTMC$ is formally defined as $ald(\mathcal{D}) = (S', Act', P', s'_0)$ s.t.

- $S' = S \cup \{\bar{s} \mid s \in S\}$,
- $Act' = 2^{AP} \cup \{\tau, \bot\}$,
- P' is defined by:

$$
\begin{aligned}
P'(\bar{s}, \bot, t) &= P(s,t) && \text{for all } s,t \in S \text{ s.t. } P(s,t) > 0 \text{ and } L(s) \neq L(t), \\
P'(\bar{s}, \tau, t) &= P(s,t) && \text{for all } s,t \in S \text{ s.t. } P(s,t) > 0 \text{ and } L(s) = L(t), \\
P'(s, L(s), \bar{s}) &= 1 && \text{for all } s \in S, \text{ and}
\end{aligned}
$$

- $s'_0 = s_0$.

Example 1 (Modified embeddings). Consider the SDTMC $\mathcal{D}$ shown in Fig. 1. The ADTMC obtained by applying ald, i.e. $ald(\mathcal{D})$ is shown in Fig. 3. Next, consider the ADTMC $\mathcal{D}$ shown in Fig. 2 (left). The SDTMC obtained by applying sld, i.e. $sld(\mathcal{D})$ is shown in Fig. 2 (right).

4   State Based Logics

We briefly recall the syntax and semantics of PCTL [21] and PCTL* [4] whose interpretation domains are SDTMCs.

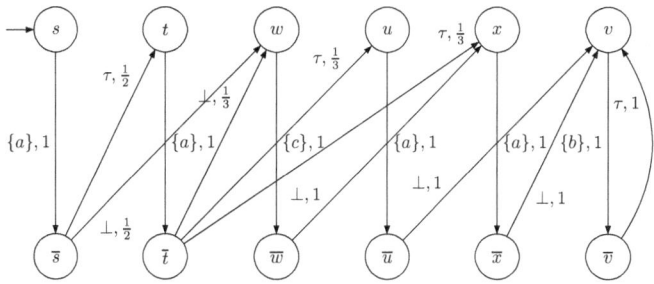

Fig. 3. $ald(\mathcal{D})$ corresponding to SDTMC $\mathcal{D}$ of Fig. 1

4.1   PCTL

Probabilistic Computation Tree Logic (PCTL) is a probabilistic branching-time temporal logic that allows one to express the probability measures of satisfaction for a temporal property by a state in an SDTMC. The syntax is given by the following grammar where $\Phi, \Phi', \ldots$ range over PCTL state formulae and $\Psi, \Psi', \ldots$ range over path formulae:

- State Formulae:
  $\Phi ::== \mathbf{true} \mid a \mid \neg \Phi \mid \Phi \wedge \Phi' \mid P_J(\Psi)$, for some $a \in$ AP
- Path Formulae:
  $\Psi ::== \mathbf{X}\Phi \mid \Phi \mathbf{U}\Phi'$

where $J \subseteq [0,1] \subset \mathbb{R}$ is an interval. The operators $\mathbf{X}$ and $\mathbf{U}$ are called the ne**X**t and the **U**ntil operators, respectively. The satisfaction relation for PCTL can be given as follows. Let $\mathcal{D}$ be an SDTMC. Satisfaction of a PCTL state formula Φ by a state s or a path formula Ψ by a path π, notation, $s \models_S \Phi$ or $\pi \models_S \Phi$ is defined inductively by:

$$s \models_S \mathbf{true} \qquad \text{always;}$$
$$s \models_S a \qquad \text{iff } a \in L(s);$$
$$s \models_S \neg\Phi \qquad \text{iff } s \not\models_S \Phi;$$
$$s \models_S \Phi \wedge \Phi' \qquad \text{iff } s \models_S \Phi \text{ and } s \models_S \Phi';$$
$$s \models_S P_J(\Psi) \qquad \text{iff} \sum_{\pi \in Paths(s), \pi \models_S \Psi} Pr(\pi) \in J;$$
$$\pi \models_S \mathbf{X}\Phi \qquad \text{iff } \pi[1] \models_S \Phi; \text{ and}$$
$$\pi \models_S \Phi \mathbf{U}\Phi' \qquad \text{iff } \exists k \geq 0, \text{ s.t. } \pi[k] \models_S \Phi' \text{ and } \forall 0 \leq i < k, \pi[i] \models_S \Phi.$$

4.2   PCTL*

PCTL* extends PCTL by relaxing the requirement of the rule that any temporal operator must be preceded by a state formula. It also allows nesting and Boolean combination of path formulae. The syntax of PCTL* is defined by the following grammar where $\Phi, \Phi', \ldots$ range over PCTL* state formulae and $\Psi, \Psi', \ldots$ range over path formulae:

- State Formulae:
  $\Phi ::== \mathbf{true} \mid a \mid \neg \Phi \mid \Phi \wedge \Phi' \mid P_J(\Psi)$, for some $a \in$ AP
- Path Formulae:
  $\Psi ::== \Phi \mid \Psi \wedge \Psi' \mid \neg \Psi \mid \mathbf{X}\Psi \mid \Psi \mathbf{U}\Psi'$

where $J \subseteq [0,1] \subset \mathbb{R}$ is an interval. The satisfaction relation for PCTL* can be given as follows. Let $\mathcal{D}$ be an SDTMC. Satisfaction of a PCTL* formula Φ by a state s or a path formula Ψ by a path π, notation, $s \models_S \Phi$ or $\pi \models_S \Phi$ is defined inductively by:

$$s \models_S \mathbf{true} \qquad \text{always;}$$
$$s \models_S a \qquad \text{iff } a \in L(s);$$
$$s \models_S \neg\Phi \qquad \text{iff } s \not\models_S \Phi;$$
$$s \models_S \Phi \wedge \Phi' \qquad \text{iff } s \models_S \Phi \text{ and } s \models_S \Phi';$$
$$s \models_S P_J(\Psi) \qquad \text{iff } \sum_{\pi \in Paths(s), \pi \models_S \Psi} Pr(\pi) \in J;$$
$$\pi \models_S \Phi \qquad \text{iff } \pi[0] \models \Phi;$$
$$\pi \models_S \Psi \wedge \Psi' \qquad \text{iff } \pi \models_S \Psi \text{ and } \pi \models_S \Psi';$$
$$\pi \models_S \neg\Psi \qquad \text{iff } \pi \not\models_S \Psi;$$
$$\pi \models_S \mathbf{X}\Psi \qquad \text{iff } \exists \pi', \theta \text{ s.t. } \pi = \pi'\theta, \text{ where } len(\pi') = 1, \text{ and } \theta \models_S \Psi; \text{ and}$$
$$\pi \models_S \Psi \mathbf{U}\Psi' \qquad \text{iff } \exists \theta \text{ s.t. } \pi \leq \theta \text{ and } \theta \models_S \Psi' \text{ and } \forall \pi \leq \eta < \theta : \eta \models_S \Psi.$$

5 Action Based Logics

This section introduces the probabilistic action-based computation tree logics for ADTMCs and also discusses their expressiveness. First, we briefly recall the syntax and semantics of a probabilistic variant of Hennessy-Milner logic (prHML) [32]. Next, we propose the syntax and semantics of an action-based PCTL* (APCTL*). We also introduce an auxiliary logic of actions which can be used to express more powerful and interesting modalities in the action-based probabilistic setting. Next, we propose the syntax and semantics of an action-based PCTL (APCTL) and compare it with APCTL*. Finally, we prove that both APCTL* and APCTL are strictly more expressive than prHML.

5.1 prHML

prHML is a probabilistic extension of the Hennessy-Milner logic [22] originally defined for the case of probabilistic transition systems. The syntax of prHML is defined as follows:

$$\Phi ::== \mathbf{true} \mid \neg\Phi \mid \Phi \wedge \Phi' \mid <a>_p \Phi$$

where Φ and Φ' are prHML formulas, $a \in Act$ and $p \in [0,1]$ is a real number. Next, we define the satisfaction relation of prHML. Let $\mathcal{D}$ be an ADTMC. Satisfaction of a prHML formula Φ by a state s, notation, $\mathcal{D}, s \models_A \Phi$ or just $s \models_A \Phi$ is defined inductively by:

$$s \models_A \mathbf{true} \qquad\qquad\qquad \text{always;}$$
$$s \models_A \neg\Phi \qquad\qquad\quad \text{iff } s \not\models_A \Phi;$$
$$s \models_A \Phi \wedge \Phi' \qquad\qquad \text{iff } s \models_A \Phi \text{ and } s \models_A \Phi'; \text{ and}$$
$$s \models_A <a>_p \Phi \qquad\qquad \text{iff} \sum_{\rho \in Runs(s), \rho[1] \models_A \Phi, \rho_\alpha[1]=a} P(s, a, \rho[1]) \le p.$$

Next, we establish the connection between bisimilarity and satisfaction of prHML formulas.

Proposition 1. *For each pair of states s and s' of an ADTMC $\mathcal{D}$; s and s' satisfy the same prHML formulas iff s and s' are forward bisimilar.*

5.2   APCTL*

The syntax of the logic APCTL* is defined by the following grammar where $\Phi, \Phi', \ldots$ range over APCTL* state formulae and $\Psi, \Psi', \ldots$ range over run formulae:

– State Formulae:
   $\Phi ::== \mathbf{true} \mid \neg \Phi \mid \Phi \wedge \Phi' \mid P_J(\Psi)$
– Run Formulae:
   $\Psi ::== \Phi \mid \Psi \wedge \Psi' \mid \neg\Psi \mid \mathbf{X}\Psi \mid \mathbf{X}_a\Psi \mid \Psi\mathbf{U}\Psi'$, for every $a \in Act \setminus \{\tau\}$

where $J \subseteq [0,1] \subset \mathbb{R}$ is an interval. Next, we define the satisfaction relation of APCTL*. Let $\mathcal{D}$ be an ADTMC. Satisfaction of an APCTL* state formula Φ by a state s or a run formula Ψ by a run ρ, notation, $\mathcal{D}, s \models_A \Phi$ ($\mathcal{D}, \rho \models_A \Phi$ resp.) or just $s \models_A \Phi$ ($\rho \models_A \Phi$ resp.) is defined inductively by:

$$s \models_A \mathbf{true} \qquad\qquad \text{always;}$$
$$s \models_A \neg\Phi \qquad\qquad \text{iff } s \not\models_A \Phi;$$
$$s \models_A \Phi \wedge \Phi' \qquad\quad \text{iff } s \models_A \Phi \text{ and } s \models_A \Phi';$$
$$s \models_A P_J(\Psi) \qquad\qquad \text{iff} \sum_{\rho \in Runs(s), \rho \models_A \Psi} Pr(\rho) \in J;$$
$$\rho \models_A \Phi \qquad\qquad\quad \text{iff } \rho[0] \models \Phi;$$
$$\rho \models_A \Psi \wedge \Psi' \qquad\quad \text{iff } \rho \models_A \Psi \text{ and } \rho \models_A \Psi';$$
$$\rho \models_A \neg\Psi \qquad\qquad \text{iff } \rho \not\models_A \Psi;$$
$$\rho \models_A \mathbf{X}\Psi \qquad\qquad \text{iff } \exists \rho', \theta \text{ s.t. } \rho = \rho'\theta, \text{ where } len(\rho') = 1 \text{ and } \theta \models_A \Psi;$$
$$\rho \models_A \mathbf{X}_a\Psi \qquad\qquad \text{iff } \exists \rho', \theta \text{ s.t. } \rho = \rho'\theta, \text{ where } len(\rho') = 1, \rho_\alpha[1] = a \text{ and } \theta \models_A \Psi; \text{ and}$$
$$\rho \models_A \Psi\mathbf{U}\Psi' \qquad\quad \text{iff } \exists \theta \text{ s.t. } \rho \le \theta \text{ and } \theta \models_A \Psi' \text{ and } \forall \rho \le \eta < \theta : \eta \models_A \Psi.$$

Some auxiliary notations for APCTL* are as follows:

$s \models_A \mathbf{false}$ for $s \models_A \neg\mathbf{true}$;

$s \models_A \Phi \vee \Phi'$ for $s \models_A \neg(\neg\Phi \wedge \Phi')$;

$s \models_A \bigvee\{\Phi_i \mid i \in \{i_1, i_2, \ldots, i_n\}\}$ for $s \models_A \Phi_{i_1} \vee \Phi_{i_2}, \ldots, \vee\Phi_{i_n}$;

$s \models_A \Phi \rightarrow \Phi'$ for $s \models_A \neg\Phi \vee \Phi'$;

$\rho \models_A \Diamond\Psi$ for $\rho \models_A \mathbf{true}\mathbf{U}\Psi$;

$\rho \models_A \Box\Psi$ for $\rho \models_A \neg\Diamond\neg\Psi$; and

$\rho \models_A \mathbf{X}_\tau\Psi$ for $\rho \models_A \mathbf{X}\Psi \wedge \neg(\bigvee\{\mathbf{X}_a\Psi \mid a \in Act \setminus \{\tau\}\})$.

The last operator $\mathbf{X}_\tau\Psi$ refers to run(s) which execute an invisible action τ as the first transition and the subsequent run satisfies Ψ. Next, we introduce an auxiliary logic of actions which can be used to express more interesting modalities in the action-based setting. The collection $ActFor$ of action formulae is defined as:

$$\chi ::= a \mid \neg\chi \mid \chi_1 \vee \chi_2 \text{ where } a \in Act \setminus \{\tau\}$$

where χ, χ_1, χ_2 range over action formulae. We write $\mathbf{true}$ as $a \vee \neg a$, where a is some arbitrarily chosen action, and $\mathbf{false}$ as $\neg\mathbf{true}$. An action formula allows us to express constraints on actions which are visible (along a run or after the next step), e.g. $a \vee b$ implies that only actions a and b are allowed in the run. Here, $\mathbf{true}$ stands for 'all visible actions are allowed'. The satisfaction relation for $ActFor$ is given by:

- $a \models_A b \Leftrightarrow a = b$;
- $a \models_A \neg\chi \Leftrightarrow a \not\models_A \chi$; and
- $a \models_A \chi_1 \vee \chi_2 \Leftrightarrow a \models_A \chi_1$ or $a \models_A \chi_2$.

With the help of action formulae, we now look at some more interesting modalities and how APCTL* can express them:

$\rho \models_A \mathbf{X}_\chi\Psi$ for $\rho \models_A \bigvee\{\mathbf{X}_a\Psi \mid a \in Act \setminus \{\tau\} \wedge a \models_A \chi\}$;

$\rho \models_A \Psi_\chi\mathbf{U}_{\chi'}\Psi'$ for $\rho \models_A (\Psi \wedge (\mathbf{X}_\tau\mathbf{true} \vee \mathbf{X}_\chi\mathbf{true}))\mathbf{U}(\Psi \wedge (\mathbf{X}_{\chi'}\Psi'))$;

$\rho \models_A \Psi_\chi\mathbf{U}\Psi'$ for $\rho \models_A (\Psi \wedge (\mathbf{X}_\tau\mathbf{true} \vee \mathbf{X}_\chi\mathbf{true}))\mathbf{U}\Psi'$;

$s \models_A \Psi <a>_p \Psi'$ for $s \models_A P_{[0,p]}(\Psi_\mathbf{false}\mathbf{U}_a\Psi')$; and

$s \models_A \Psi <\epsilon>_p \Psi'$ for $s \models_A P_{[0,p]}(\Psi_\mathbf{false}\mathbf{U}\Psi')$.

The formula $\mathbf{X}_\chi\Psi$ will be satisfied by a run if the run begins with a visible action, say a, s.t. $a \models_A \chi$ and the subsequent run satisfies Ψ. A run satisfies $\Psi_\chi\mathbf{U}\Psi'$ if there exists a suffix which satisfies Ψ' and prior to that, at any point, the run satisfies Ψ and all the visible transitions satisfy χ. Intuitively, a run satisfies $\Psi_\chi\mathbf{U}_{\chi'}\Psi'$ if it reaches a point where the visible transition satisfies χ' and the subsequent run satisfies Ψ', whereas at any instance before this event, Ψ holds and all visible transitions satisfy χ. The formula $\Psi <a>_p \Psi'$ will hold true for a state

s if the cumulative probabilities of all the run(s) ρ such that $\rho[0] = s$ and Ψ holds along ρ as it continues to perform invisible action (τ) until a point where visible a labeled transition is performed following which Ψ' is true is bounded by p. $\Psi <\epsilon>_p \Psi'$ holds true for a state s if the cumulative probabilities of all the run(s) in which Ψ' holds after zero or more invisible τ labeled transitions and Ψ is true at any moment before the previous event along the run is bounded by p. Next, we compare the expressive power of APCTL* and prHML.

Proposition 2. *The logic APCTL* is strictly more expressive than prHML.*

5.3   APCTL

In order to define APCTL, we make use of the same auxiliary logic of actions *ActFor* as defined in the case of APCTL*. This helps us to shorten the notations significantly. The syntax of APCTL is defined by the state formulae generated by the following grammar, where $\Phi, \Phi', \ldots$ range over state formulae, Ψ over run formulae and χ and χ' are action formulae (generated by *ActFor*):

- State Formulae:
  $$\Phi ::== \textbf{true} \mid \neg\Phi \mid \Phi \wedge \Phi' \mid \mathcal{P}_J(\Psi)$$
- Run Formulae:
  $$\Psi ::== \mathbf{X}_\chi \Phi \mid \mathbf{X}_\tau \Phi \mid \Phi_\chi \mathbf{U} \Phi' \mid \Phi_\chi \mathbf{U}_{\chi'} \Phi'$$

where $J \subseteq [0,1] \subset \mathbb{R}$ is an interval. The satisfaction relation for APCTL formulae is given below. Let $\mathcal{D}$ be an ADTMC. Satisfaction of a APCTL state formula Φ by a state s or a run formula Ψ by a run ρ, notation, $\mathcal{D}, s \models_A \Phi$ ($\mathcal{D}, \rho \models_A \Phi$ resp.) or just $s \models_A \Phi$ ($\rho \models_A \Phi$ resp.) is defined inductively by:

$s \models_A \textbf{true}$ always;

$s \models_A \neg\Phi$ iff $s \not\models_A \Phi$;

$s \models_A \Phi \wedge \Phi'$ iff $s \models_A \Phi$ and $s \models_A \Phi'$;

$s \models_A \mathcal{P}_J(\Psi)$ iff $\displaystyle\sum_{\rho \in Runs(s), \rho \models_A \Psi} Pr(\rho) \in J$;

$\rho \models_A \mathbf{X}_\chi \Phi$ iff $\rho_\alpha[1] \models_A \chi$ and $\rho[1] \models_A \Phi$;

$\rho \models_A \mathbf{X}_\tau \Phi$ iff $\rho_\alpha[1] = \tau$ and $\rho[1] \models_A \Phi$;

$\rho \models_A \Phi_\chi \mathbf{U} \Phi'$ iff $\exists k \geq 0$, s.t. $\rho[k] \models_A \Phi'$ and $\forall 0 \leq i < k, \rho[i] \models_A \Phi$ and
     $\forall 1 \leq j \leq k, \rho_\alpha[j] \models_A \chi$ or $\rho_\alpha[j] = \tau$; and

$\rho \models_A \Phi_\chi \mathbf{U}_{\chi'} \Phi'$ iff $\exists k > 0$, s.t. $\rho[k] \models_A \Phi'$ and $\rho_\alpha[k] \models_A \chi'$, and $\forall 0 \leq i < k$,
     $\rho[i] \models_A \Phi$ and $\forall 1 \leq j < k, \rho_\alpha[j] \models_A \chi$ or $\rho_\alpha[j] = \tau$.

Other derived modalities with their interpretations as defined before are as follows:

- $\Phi <a>_p \Phi' \Leftrightarrow P_{[0,p]} \Phi_{\textbf{false}} \mathbf{U}_a \Phi'$; and
- $\Phi <\epsilon>_p \Phi' \Leftrightarrow P_{[0,p]} \Phi_{\textbf{false}} \mathbf{U} \Phi'$.

Proposition 3. *The logic APCTL* is strictly more expressive than APCTL.*

Proposition 4. *The logic APCTL is strictly more expressive than prHML.*

Intuitively, this proposition claims that any prHML formula can be expressed using APCTL but not vice versa.

6   Embeddings for Probabilistic Logics

This section defines the logical embeddings which enable us to move between probabilistic state-based and action-based specifications. This leads to leveraging the prowess of model checking and minimization in either setting. To relate the two logics and their branching time variants presented in the previous section, we will require the modified embeddings which allow us to relate the two types of DTMC models over which these logics have been interpreted. Based on the modified model embeddings, we will define the logical embeddings and prove that the truth is preserved.

6.1   PCTL* → APCTL*

Given an SDTMC $\mathcal{D}$ and a PCTL* formula Φ, the logical embedding *aldl* constructs an equivalent APCTL* formula which can be interpreted over the embedded ADTMC, i.e. $ald(\mathcal{D})$. We define the embedding $aldl : PCTL^* \rightarrow APCTL^*$ inductively as follows:

$$
\begin{aligned}
aldl(\mathbf{true}) &= \mathbf{true}; \\
aldl(a) &= P_{[1,1]}(\mathbf{X}_\chi \mathbf{true}), \text{ where } a \in \chi \subseteq AP; \\
aldl(\neg \Phi) &= \neg aldl(\Phi); \\
aldl(\Phi \wedge \Phi') &= aldl(\Phi) \wedge aldl(\Phi'); \\
aldl(P_J(\Psi)) &= P_J(aldl(\Psi)); \\
aldl(\Psi \wedge \Psi') &= aldl(\Psi) \wedge aldl(\Psi'); \\
aldl(\neg \Psi) &= \neg aldl(\Psi); \\
aldl(\mathbf{X}\Psi) &= \left(\mathbf{X_{true}} \left(\mathbf{X}_{\perp \vee \tau} \left(aldl(\Psi) \right) \right) \right); \text{ and} \\
aldl(\Psi \mathbf{U} \Psi') &= \left(\left(aldl(\Psi) \wedge \mathbf{X}_{\neg(\perp \vee \tau)} \mathbf{true} \right) \vee \left(\mathbf{X}_{\perp \vee \tau} \mathbf{true} \right) \right) \mathbf{U} \\
&\qquad\qquad\qquad\qquad \left(aldl(\Psi') \wedge \mathbf{X}_{\neg(\perp \vee \tau)} \mathbf{true} \right).
\end{aligned}
$$

Theorem 1. *Let $\mathcal{D}$ be an SDTMC, s be a state in $\mathcal{D}$ and Φ be a PCTL* formula. Then: $\mathcal{D}, s \models_S \Phi \Leftrightarrow ald(\mathcal{D}), s \models_A aldl(\Phi)$.*

Intuitively, this theorem says that if an SDTMC satisfies a PCTL* formula, then its corresponding ADTMC would satisfy the APCTL* formula obtained by applying the *aldl* embedding.

6.2 APCTL* → PCTL*

Given an ADTMC $\mathcal{D}$ and a APCTL* formula Φ, the logical embedding $sldl$ constructs an equivalent PCTL* formula which can be interpreted over the embedded SDTMC, i.e. $sld(\mathcal{D})$. We define the embedding $sldl : APCTL^* \to PCTL^*$ inductively as follows:

$$
\begin{aligned}
sldl(\textbf{true}) &= \textbf{true}; \\
sldl(\neg\Phi) &= \neg sldl(\Phi); \\
sldl(\Phi \wedge \Phi') &= sldl(\Phi) \wedge sldl(\Phi'); \\
sldl(P_J(\Psi)) &= P_J(sldl(\Psi)); \\
sldl(\Psi \wedge \Psi') &= sldl(\Psi) \wedge sldl(\Psi'); \\
sldl(\neg\Psi) &= \neg sldl(\Psi); \\
sldl(\mathbf{X}\Psi) &= \mathbf{X}\Big(\big(\bot \wedge sldl(\Psi) \big) \vee \big(\neg\bot \wedge \mathbf{X}(sldl(\Psi)) \big) \Big); \\
sldl(\mathbf{X}_a\Psi) &= \mathbf{X}\Big(a \wedge \mathbf{X}(sldl(\Psi)) \Big), \text{ and} \\
sldl(\Psi\mathbf{U}\Psi') &= \Big(\bot \to sldl(\Psi) \Big)\mathbf{U}\Big(\bot \wedge sldl(\Psi') \Big).
\end{aligned}
$$

Theorem 2. *Let $\mathcal{D}$ be an ADTMC, let s be a state in $\mathcal{D}$ and Φ be an APCTL* formula. Then: $\mathcal{D}, s \models_A \Phi \Leftrightarrow sld(\mathcal{D}), s \models_S sldl(\Phi)$.*

Intuitively, this theorem says that if an ADTMC satisfies an APCTL* formula, then its corresponding SDTMC would satisfy the PCTL* formula obtained by applying the $sldl$ embedding.

Theorem 3. *PCTL* and APCTL* are equally expressive.*

From these theorems we know that APCTL* model checking can be reduced to PCTL* model checking and vice versa. Since the verification of branching time subset of PCTL*, i.e. PCTL is more efficient (polynomial time [6]), our next focus is to propose a transformation function which can be used to convert APCTL into PCTL. Note that the embeddings $sldl$ and $aldl$ defined in the previous section cannot be directly lifted for APCTL to PCTL (PCTL to APCTL resp.) transformations. To understand this issue, let us see an example property given below.

Example 2. Consider the APCTL formula $P_{[1,1]}(\mathbf{X}_\chi\Phi)$. Since the notion of an *Action formula* is not an integral part of APCTL*, the above formula will be expressed in APCTL* as follows:

$$
P_{[1,1]}(\bigvee\{\mathbf{X}_a\Phi \mid a \in Act \text{ and } a \models_A \chi\})
$$

Now using the definition of $sldl$ proposed in the previous section, the corresponding PCTL* formula will be as follows:

$$
P_{[1,1]}\Big(\bigvee\{\mathbf{X}\big(a \wedge \mathbf{X}(sldl(\Phi)) \big) \mid a \in Act \text{ and } a \models_A \chi\} \Big)
$$

Note that this formula is not a PCTL formula as PCTL doesn't allow disjunction (or conjunction) of the **X** (next) operator.

6.3 PCTL → APCTL

Given an SDTMC $\mathcal{D}$ and a PCTL formula Φ, the logical embedding $aldl'$ constructs an equivalent APCTL formula which can be interpreted over the embedded ADTMC, i.e. $ald(\mathcal{D})$. We define the embedding $aldl' : PCTL \rightarrow APCTL$ inductively as follows:

$$aldl'(\textbf{true}) \qquad = \textbf{true};$$
$$aldl'(a) \qquad\qquad = P_{[1,1]}(\mathbf{X}_\chi \textbf{true}), \text{ where } a \in \chi \subseteq AP;$$
$$aldl'(\neg\Phi) \qquad\quad = \neg aldl'(\Phi);$$
$$aldl'(\Phi \wedge \Phi') \qquad = aldl'(\Phi) \wedge aldl'(\Phi');$$

$$aldl'\Big(P_J(\mathbf{X}\Phi)\Big) \quad = P_{[1,1]}\left(\mathbf{X_{true}}\left(P_J\Big(\mathbf{X}_{\perp\vee\tau}\big(aldl'(\Phi)\big)\Big)\right)\right); \text{ and}$$

$$aldl'\Big(P_J(\Phi\mathbf{U}\Phi')\Big) \; = P_J\left(\left(\Big(aldl'(\Phi) \wedge P_{[1,1]}(\mathbf{X}_{\neg(\perp\vee\tau)}\textbf{true})\Big)\vee\right.\right.$$
$$\left.\left.\Big(P_{(0,1]}(\mathbf{X}_{\perp\vee\tau}\textbf{true})\Big)\right)_{\textbf{true}}\mathbf{U}_{\perp\vee\tau}\Big(aldl'(\Phi')\Big)\right).$$

The next theorem proves that $aldl'$ combined with ald can be used for property preservation.

Theorem 4. *Let $\mathcal{D}$ be an SDTMC, let s be a state in $\mathcal{D}$ and Φ be an PCTL formula. Then: $\mathcal{D}, s \models_S \Phi \Leftrightarrow ald(\mathcal{D}), s \models_A aldl'(\Phi)$.*

Example 3. Consider the PCTL formula $\Phi = P_{(0,\frac{1}{2}]}(a\mathbf{U}b)$ and the SDTMC $\mathcal{D}$ given in Fig. 1. We can check that $s \models_S \Phi$ as there are two paths starting from s (s,t,u,v and s,t,x,v) which satisfy $(a\mathbf{U}b)$ and the combined probability of both these paths is $(\frac{1}{2} \cdot \frac{1}{3} \cdot 1) + (\frac{1}{2} \cdot \frac{1}{3} \cdot 1) = \frac{1}{6} + \frac{1}{6} = \frac{1}{3} \in (0, \frac{1}{2}]$. From the definition of $aldl'$ we have that $aldl'(\Phi) = P_{(0,\frac{1}{2}]}\left(\left(\Big(aldl'(a) \wedge P_{[1,1]}(\mathbf{X}_{\neg(\perp\vee\tau)}\textbf{true})\Big) \vee\right.\right.$

$\left.\left.\Big(P_{(0,1]}(\mathbf{X}_{\perp\vee\tau}\textbf{true})\Big)\right)\mathbf{U}_{\perp\vee\tau}\Big(aldl'(b)\Big)\right)$. From Fig. 3, it can be observed that in the embedded model only those runs satisfy the embedded formula (i.e. $s\{a\}, \overline{s}\tau, t\{a\}, \overline{t}\tau, u\{a\}, \overline{u}\perp, v$ and $s\{a\}, \overline{s}\tau, t\{a\}, \overline{t}\tau, x\{a\}, \overline{x}\perp, v$) for which there is a corresponding path satisfying the until formula in the original model. All the other run(s) do not satisfy the formula. As the probability of all the new additional transitions introduced under the embedding is 1; the probability of each run in the embedded model corresponding to each accepting path of the original model remains preserved, i.e. $(1 \cdot \frac{1}{2} \cdot 1 \cdot \frac{1}{3} \cdot 1 \cdot 1) + (1 \cdot \frac{1}{2} \cdot 1 \cdot \frac{1}{3} \cdot 1 \cdot 1) = \frac{1}{6} + \frac{1}{6} = \frac{1}{3} \in (0, \frac{1}{2}]$. Therefore the probability of satisfying the APCTL formula $aldl'(\Phi)$ is also preserved.

6.4   APCTL → PCTL

Given an ADTMC $\mathcal{D}$ and a APCTL formula Φ, the logical embedding $sldl'$ constructs an equivalent PCTL property which can be interpreted over the embedded SDTMC, i.e. $sld(\mathcal{D})$. We define the embedding $sldl' : APCTL \rightarrow PCTL$ inductively as follows:

$$sldl'(\mathbf{true}) \qquad\qquad = \mathbf{true};$$
$$sldl'(\neg\Phi) \qquad\qquad = \neg sldl'(\Phi);$$
$$sldl'(\Phi \wedge \Phi') \qquad\qquad = sldl'(\Phi) \wedge sldl'(\Phi');$$
$$sldl'\Big(P_J(\mathbf{X}_\chi\Phi)\Big) \qquad = P_J\Big(\mathbf{X}\Big(\neg\perp \wedge \chi \wedge P_{[1,1]}\Big(\mathbf{X}(\perp \wedge sldl'(\Phi))\Big)\Big)\Big);$$
$$sldl'\Big(P_J(\mathbf{X}_\tau\Phi)\Big) \qquad = \Big(\perp \wedge P_J\Big(\mathbf{X}(\perp \wedge sldl'(\Phi))\Big)\Big);$$
$$sldl'\Big(P_J(\Phi_\chi\mathbf{U}\Phi')\Big) \quad = P_J\Big(\big((\perp \wedge sldl'(\Phi)) \vee (\neg\perp \wedge \chi)\big)\mathbf{U}\Big(\perp \wedge sldl'(\Phi')\Big)\Big); \text{ and}$$
$$sldl'\Big(P_J(\Phi_\chi\mathbf{U}_{\chi'}\Phi')\Big) \; = P_J\Big(\big((\perp \wedge sldl'(\Phi)) \vee (\neg\perp \wedge \chi)\big)\mathbf{U}\Big((\neg\perp \wedge \chi') \wedge$$
$$P_{[1,1]}\Big(\mathbf{X}(\perp \wedge sldl'(\Phi'))\Big)\Big)\Big).$$

Theorem 5. *Let $\mathcal{D}$ be an ADTMC, s be a state in $\mathcal{D}$ and Φ be an APCTL formula. Then: $\mathcal{D}, s \models_A \Phi \Leftrightarrow sld(\mathcal{D}), s \models_S sldl'(\Phi)$.*

Intuitively, this theorem says that model checking APCTL formulas on ADTMCs can be reduced to model checking PCTL formulas on SDTMCs using the model and logical embeddings. In other words, existing machinery for PCTL model checking can be used to verify action-labeled probabilistic systems (in polynomial time).

Example 4. Consider the APCTL formula $\Phi = P_{[\frac{1}{2},\frac{1}{2}]}\mathbf{X}_b\mathbf{true}$ and the ADTMC $\mathcal{D}$ in Fig. 2 (left). Note that, $s \models_A \Phi$ which is realized by the finite run sb, u in Fig. 2 (left). The corresponding embedded SDTMC $sld(\mathcal{D})$ is shown in Fig. 2 (right). From the definition of $sldl'$, we have that, $sldl'(\Phi) = P_{[\frac{1}{2},\frac{1}{2}]}\Big(\mathbf{X}\Big(\neg\perp \wedge b \wedge$

$P_{[1,1]}\Big(\mathbf{X}(\perp \wedge \mathbf{true})\Big)\Big)\Big)$. Here, $s \models_S sldl'(\Phi)$ in $sld(\mathcal{D})$ (see the corresponding finite path $s, (b, u), u$ in Fig. 2 (right)). No other path(s) satisfy the formula as s is unable to reach a state where $(\neg\perp \wedge b)$ is true other than (b, u).

Next, we compare the expressiveness of PCTL and APCTL.

Theorem 6. *PCTL and APCTL are equally expressive.*

7   Related Work

Model and Logical Embeddings: Transformation functions between Kripke structures (KSs) and labeled transitions systems (LTSs) have been defined in [39,40]. This framework enabled using a process algebra to describe the system behavior and to use Computational Tree Logic (CTL) or CTL* to specify the requirement the system has to comply with. This was achieved by defining transformation functions for CTL* (CTL resp.) and action-based CTL*, i.e. ACTL* (ACTL resp.). The authors have also proved that stuttering equivalence for KSs coincides with divergence-sensitive branching bisimulation for LTSs. These results were extended in [41,42] by defining two additional translations, i.e. inverse embeddings which enable minimization modulo behavioral equivalences. In this paper, authors have also proved that these embeddings can be used for a range of other equivalences of interest, e.g. strong bisimilarity, simulation equivalence, and trace equivalence. A tool was proposed in [38] which takes a process description and a ACTL formula to be verified, and then translates them into a corresponding KS and CTL formula, respectively. For software product lines (SPLs), embeddings have been defined between modal transition systems (MTSs) with variability constraints and featured transition systems (FTSs) [9,10]. Recently, embeddings and inverse embeddings between two types of probabilistic models, i.e. SDTMCs and ADTMCs have been proposed in [14]. These embeddings preserve strong forward bisimulation, strong backward bisimulation and trace equivalence relations. More recently, a formal framework of embeddings has been defined between state and action-labeled continuous-time Markov chains (CTMCs) [15]. In addition to strong bisimulation relations, these embeddings also preserve weak forward and backward bisimulation. Using this framework, a model can be minimized in one setting by minimizing its embedded model in the other setting and applying the inverse of the embedding.

Comparison with Other Probabilistic Logics: In this paper we have proved that APCTL* and APCTL are strictly more expressive than the probabilistic variant of HML [32]. A probabilistic μ-calculus (PMC) for specifying properties on Markov processes has been proposed in [31]. Authors have proved that PMC is incomparable with PCTL and PCTL* as properties with "Until" operator cannot be expressed using this logic. This also means that PMC is incomparable with APCTL and APCTL*. Another interesting probabilistic extension of μ-calculus (PμTL) has been introduced in [33]. This logic is also incomparable with PCTL but the qualitative fragment of PμTL is strictly more expressive than qualitative PCTL. These results also hold true for the case of APCTL and qualitative APCTL, respectively. In [37], probabilistic modal μ-calculus (pLμ) [1,24] has been extended by adding a second conjunction operator called product and its De Morgan dual operator called co-product. This logic is expressive enough to encode the qualitative fragment of PCTL (and therefore qualitative APCTL). This logic is further extended by adding yet another pair of De Morgan dual connectives which allows it to encode full PCTL (and APCTL). In [13], Generalized probabilistic logic (GPL) has been proposed-based on the modal

μ-calculus. This logic is divided into a probabilistic part and a non-probabilistic part involving fixed points and can encode PCTL*. This means APCTL* can also be encoded using GPL.

8   Conclusion

We have defined a formal framework which allows one to move between action-based and state-based probabilistic specifications. We introduced the syntax and semantics of an action-based PCTL* (APCTL*) and an action-based PCTL (APCTL). Both these logics are interpreted over ADTMCs and can be used to specify interesting probabilistic properties. We have proved that our logics are strictly more expressive than the probabilistic variant of HML (prHML). Next, we defined the embeddings between APCTL* (APCTL resp.) and PCTL* (PCTL resp.) and proved that truth is preserved when combined with model embeddings. Compared to probabilistic variants of modal μ-calculus which involve fix-point operators and have limited tool support, our logics can be easily used to specify interesting properties and our logical embeddings reduce the problem of verification of APCTL* (APCTL resp.) to PCTL* (PCTL resp.) model checking. From a practical point of view, our results have enabled researchers to use the state-of-the-art probabilistic model checking tools developed in one setting for model minimization and analysis in the other setting.

This research work can be extended in several interesting directions which are as follows:

– Implement a tool that converts an action-based specification to a state-based specification and vice versa. This tool should also allow moving between SDTMCs and ADTMCs. By doing this it would support both model minimization and probabilistic verification.
– Define embeddings between ADTMCs and SDTMCs which preserve weak equivalence relations, e.g. weak forward and backward bisimulation etc. It would also be interesting to investigate if these new embeddings support APCTL*/PCTL* model checking.
– Extend these embeddings to logics that support stochastic behavior, e.g. [3].

References

1. de Alfaro, L., Majumdar, R.: Quantitative solution of omega-regular games. J. Comput. Syst. Sci. **68**(2), 374–397 (2004)
2. Andova, S., Hermanns, H., Katoen, J.-P.: Discrete-time rewards model-checked. In: Larsen, K.G., Niebert, P. (eds.) FORMATS 2003. LNCS, vol. 2791, pp. 88–104. Springer, Heidelberg (2004). https://doi.org/10.1007/978-3-540-40903-8_8
3. Aziz, A., Sanwal, K., Singhal, V., Brayton, R.K.: Verifying continuous time Markov chains. In: Proceedings of the 8th International Conference on Computer Aided Verification, CAV 1996, New Brunswick, NJ, USA, 31 July–3 August 1996, pp. 269–276 (1996)

4. Aziz, A., Singhal, V., Balarin, F.: It usually works: The temporal logic of stochastic systems. In: Proceedings of the 7th International Conference on Computer Aided Verification, Liège, Belgium, 3–5 July 1995, pp. 155–165 (1995)

5. Baeten, J.C.M., Bergstra, J.A., Smolka, S.A.: Axiomatizing probabilistic processes: ACP with generative probabilities. Inf. Comput. **121**(2), 234–255 (1995)

6. Baier, C., Katoen, J.P.: Principles of Model Checking. MIT Press, Cambridge (2008)

7. Baier, C., Hermanns, H.: Weak bisimulation for fully probabilistic processes. In: Grumberg, O. (ed.) CAV 1997. LNCS, vol. 1254, pp. 119–130. Springer, Heidelberg (1997). https://doi.org/10.1007/3-540-63166-6_14

8. Baier, C., Hermanns, H., Katoen, J., Wolf, V.: Bisimulation and simulation relations for Markov chains. Electron. Notes Theor. Comput. Sci. **162**, 73–78 (2006)

9. ter Beek, M.H., Damiani, F., Gnesi, S., Mazzanti, F., Paolini, L.: From featured transition systems to modal transition systems with variability constraints. In: Calinescu, R., Rumpe, B. (eds.) SEFM 2015. LNCS, vol. 9276, pp. 344–359. Springer, Cham (2015). https://doi.org/10.1007/978-3-319-22969-0_24

10. ter Beek, M.H., Damiani, F., Gnesi, S., Mazzanti, F., Paolini, L.: On the expressiveness of modal transition systems with variability constraints. Sci. Comput. Program. **169**, 1–17 (2019)

11. Buchholz, P.: Exact and ordinary lumpability in finite Markov chains. J. Appl. Prob. **31**, 59–75 (1994)

12. Bunte, O., et al.: The mCRL2 toolset for analysing concurrent systems - improvements in expressivity and usability. In: Vojnar, T., Zhang, L. (eds.) TACAS 2019. LNCS, vol. 11428, pp. 21–39. Springer, Cham (2019). https://doi.org/10.1007/978-3-030-17465-1_2

13. Cleaveland, R., Iyer, S.P., Narasimha, M.: Probabilistic temporal logics via the modal mu-calculus. Theor. Comput. Sci. **342**(2–3), 316–350 (2005)

14. Das, S., Sharma, A.: Embeddings between state and action labeled probabilistic systems. In: SAC 2021: The 36th ACM/SIGAPP Symposium on Applied Computing, Virtual Event, Republic of Korea, 22–26 March 2021, pp. 1759–1767. ACM (2021)

15. Das, S., Sharma, A.: State space minimization preserving embeddings for continuous-time Markov chains. In: Ballarini, P., Castel, H., Dimitriou, I., Iacono, M., Phung-Duc, T., Walraevens, J. (eds.) EPEW/ASMTA 2021. LNCS, vol. 13104, pp. 44–61. Springer, Cham (2021). https://doi.org/10.1007/978-3-030-91825-5_3

16. Dehnert, C., Junges, S., Katoen, J.-P., Volk, M.: A *storm* is coming: a modern probabilistic model checker. In: Majumdar, R., Kunčak, V. (eds.) CAV 2017. LNCS, vol. 10427, pp. 592–600. Springer, Cham (2017). https://doi.org/10.1007/978-3-319-63390-9_31

17. Desharnais, J.: Labelled Markov processes. Ph.D. thesis, McGill University (1999)

18. Garavel, H., Lang, F., Mateescu, R., Serwe, W.: CADP 2010: a toolbox for the construction and analysis of distributed processes. In: Abdulla, P.A., Leino, K.R.M. (eds.) TACAS 2011. LNCS, vol. 6605, pp. 372–387. Springer, Heidelberg (2011). https://doi.org/10.1007/978-3-642-19835-9_33

19. van Glabbeek, R.J., Smolka, S.A., Steffen, B.: Reactive, generative and stratified models of probabilistic processes. Inf. Comput. **121**(1), 59–80 (1995)

20. Hansson, H., Jonsson, B.: A calculus for communicating systems with time and probabilities. In: RTSS, pp. 278–287. IEEE Computer Society (1990)

21. Hansson, H., Jonsson, B.: A logic for reasoning about time and reliability. Formal Asp. Comput. **6**(5), 512–535 (1994)

22. Hennessy, M., Milner, R.: Algebraic laws for nondeterminism and concurrency. J. ACM **32**(1), 137–161 (1985)
23. Hillston, J.: A Compositional Approach to Performance Modelling. Cambridge University Press, Cambridge (1996)
24. Huth, M., Kwiatkowska, M.Z.: Quantitative analysis and model checking. In: Proceedings of 12th Annual IEEE Symposium on Logic in Computer Science, Warsaw, Poland, 29 June–2 July 1997, pp. 111–122. IEEE Computer Society (1997)
25. Jonsson, B., Larsen, K.G.: Specification and refinement of probabilistic processes. In: LICS, pp. 266–277. IEEE Computer Society (1991)
26. Jou, C.-C., Smolka, S.A.: Equivalences, congruences, and complete axiomatizations for probabilistic processes. In: Baeten, J.C.M., Klop, J.W. (eds.) CONCUR 1990. LNCS, vol. 458, pp. 367–383. Springer, Heidelberg (1990). https://doi.org/10.1007/BFb0039071
27. Katoen, J., Khattri, M., Zapreev, I.S.: A Markov reward model checker. In: QEST, pp. 243–244. IEEE Computer Society (2005)
28. Kemeny, J.G., Snell, J.L.: Denumerable Markov Chains. Springer, New York (1976). https://doi.org/10.1007/978-1-4684-9455-6
29. Kemeny, J.G., Snell, J.L., et al.: Finite Markov chains, vol. 356. van Nostrand, Princeton (1960)
30. Kwiatkowska, M., Norman, G., Parker, D.: PRISM 4.0: verification of probabilistic real-time systems. In: Gopalakrishnan, G., Qadeer, S. (eds.) CAV 2011. LNCS, vol. 6806, pp. 585–591. Springer, Heidelberg (2011). https://doi.org/10.1007/978-3-642-22110-1_47
31. Larsen, K.G., Mardare, R., Xue, B.: Probabilistic mu-calculus: decidability and complete axiomatization. In: 36th IARCS Annual Conference on Foundations of Software Technology and Theoretical Computer Science, FSTTCS 2016, 13–15 December 2016, Chennai, India. LIPIcs, vol. 65, pp. 25:1–25:18. Schloss Dagstuhl - Leibniz-Zentrum für Informatik (2016)
32. Larsen, K.G., Skou, A.: Bisimulation through probabilistic testing. Inf. Comput. **94**(1), 1–28 (1991)
33. Liu, W., Song, L., Wang, J., Zhang, L.: A simple probabilistic extension of modal mu-calculus. In: Proceedings of the Twenty-Fourth International Joint Conference on Artificial Intelligence, IJCAI 2015, Buenos Aires, Argentina, 25–31 July 2015, pp. 882–888 (2015)
34. Lowe, G.: Probabilistic and prioritized models of timed CSP. Theor. Comput. Sci. **138**(2), 315–352 (1995)
35. Marsan, M.A., Conte, G., Balbo, G.: A class of generalized stochastic Petri nets for the performance evaluation of multiprocessor systems. ACM Trans. Comput. Syst. **2**(2), 93–122 (1984)
36. Miguel, C., Fernández, A., Vidaller, L.: LOTOS extended with probablistic behaviours. Formal Asp. Comput. **5**(3), 253–281 (1993)
37. Mio, M.: Probabilistic modal mu-calculus with independent product. Log. Methods Comput. Sci. **8**(4) (2012)
38. Nicola, R.D., Fantechi, A., Gnesi, S., Ristori, G.: An action-based framework for verifying logical and behavioural properties of concurrent systems. Comput. Networks ISDN Syst. **25**(7), 761–778 (1993)
39. De Nicola, R., Vaandrager, F.: Action versus state based logics for transition systems. In: Guessarian, I. (ed.) LITP 1990. LNCS, vol. 469, pp. 407–419. Springer, Heidelberg (1990). https://doi.org/10.1007/3-540-53479-2_17
40. Nicola, R.D., Vaandrager, F.W.: Three logics for branching bisimulation. J. ACM **42**(2), 458–487 (1995)

41. Reniers, M.A., Schoren, R., Willemse, T.A.C.: Results on embeddings between state-based and event-based systems. Comput. J. **57**(1), 73–92 (2014)
42. Reniers, M.A., Willemse, T.A.C.: Folk theorems on the correspondence between state-based and event-based systems. In: Černá, I., Gyimóthy, T., Hromkovič, J., Jefferey, K., Královič, R., Vukolić, M., Wolf, S. (eds.) SOFSEM 2011. LNCS, vol. 6543, pp. 494–505. Springer, Heidelberg (2011). https://doi.org/10.1007/978-3-642-18381-2_41
43. Sharma, A.: Weighted probabilistic equivalence preserves ω-regular properties. In: Schmitt, J.B. (ed.) MMB&DFT 2012. LNCS, vol. 7201, pp. 121–135. Springer, Heidelberg (2012). https://doi.org/10.1007/978-3-642-28540-0_9

Footprint Logic for Object-Oriented Components

Frank S. de Boer[1,2(✉)], Stijn de Gouw[1,3(✉)], Hans-Dieter A. Hiep[1,2(✉)][iD],
and Jinting Bian[1,2][iD]

[1] Centrum Wiskunde & Informatica, Amsterdam, The Netherlands
{frb,hdh,jinting.bian}@cwi.nl
[2] Leiden Institute of Advanced Computer Science (LIACS),
Leiden, The Netherlands
[3] Open Universiteit, Heerlen, The Netherlands
sdg@ou.nl

Abstract. We introduce a new way of reasoning about invariance in terms of *footprints* in a program logic for object-oriented components. A footprint of an object-oriented component is formalized as a monadic predicate that describes which objects on the heap can be affected by the execution of the component. Assuming encapsulation, this amounts to specifying which objects of the component can be called. Adaptation of local specifications into global specifications amounts to showing invariance of assertions, which is ensured by means of a form of *bounded quantification* which excludes references to a given footprint.

Keywords: Hoare logic · Invariance · Strong partial correctness

1 Introduction

A major and important challenge in theoretical computer science is finding practical and efficient verification techniques for showing functional correctness properties of object-oriented programs. A common approach to the verification task is divide and conquer: to split a program into separate components, where the behavior of each component is locally formally specified. The benefits of employing local specifications are: simplification of the verification task of each component, reusability of verified components, and distribution of the verification task among verifiers. Verified components can then be composed into a larger program, and the correctness of the resulting composed program with respect to a global specification can be established by adapting the local specifications to the larger context in which the components are used, abstracting from their internal implementation.

But what exactly distinguishes a local specification from a global specification? In this paper, we introduce a novel concept of *footprints*. We define the

S. L. Tapia Tarifa and J. Proença (Eds.): FACS 2022, LNCS 13712, pp. 141–160, 2022.
https://doi.org/10.1007/978-3-031-20872-0_9

footprint of a component as the set of objects which that component may call: if one assumes encapsulation of objects, as is often the case in component-based software, the state of objects outside of the footprint of a component thus remains unaffected by the execution of that component. Furthermore, we interpret the semantics of local component specifications with respect to such a footprint. Verifying the local correctness of a component then requires to show that the execution of a component is restricted to method calls to objects belonging to its footprint. As such, we can formulate precisely which properties of objects remain invariant within a larger context than that of the component alone.

Traditionally, in program logics, such as Hoare logic, the so-called *adaptation* rules are used to adapt a local correctness specification of a component to the larger context in which the component is used. For example this can be achieved by adding to the pre- and postcondition an invariant. Hoare introduced in [10] one rule, *the* adaptation rule, which generalizes all adaptation rules. In his seminal paper [13], Olderog studied the expressiveness and the completeness of the adaptation rule. These adaptation rules form the basis for reasoning about program components, abstracting from their internal implementation. The adaptation rules, including Hoare's adaptation rule [10], however, are of limited use in the presence of *aliasing* in object-oriented programs. Aliasing arises when syntactically different expressions refer to the same memory location. Well-known data structures which give rise to aliasing are arrays and pointer structures. In the presence of aliasing we can no longer syntactically determine general invariant properties, but the standard adaptation rules are based on purely syntactic conditions, e.g. whether a given formula contains free variables which also appear in a given statement.

In this paper we introduce a novel Hoare logic for reasoning about invariant properties using footprints as they arise in object-oriented components, viz. components of object-oriented programs. One of the main challenges addressed is a formalization of footprints at an abstraction level that coincides with the programming language. For example, in object-oriented programming languages such as Java, we can only refer to objects that *exist*, i.e. that have been dynamically created. Thus quantifiers only range over objects which actually exist.

We therefore generalize the weakest precondition calculus underlying the Hoare logic for object-orientation introduced in [7], to the specification and verification of footprints as sets of objects. To represent and reason about such sets at an appropriate abstraction level, we formalize the assertion language in a *second-order monadic logic*. To reason about footprints in a modular manner we introduce in this paper a new *hybrid* Hoare logic which combines two different interpretations of correctness specifications. One interpretation requires absence of so-called 'null-pointer exceptions', e.g., calling a method on 'null'. The other interpretation generalizes such absence of failures to footprints, requiring that only objects included in the footprint can be called. That is, semantically, calling an object which does not belong to the footprint generates a failure. In the context of this latter interpretation we introduce a new invariance (or 'frame') rule. This rule enforces invariance by a form of *bounded quantification* which restricts the description to that part of the heap disjoint from the footprint.

We compare our approach to the two main existing approaches to reasoning about invariance: separation logic [15] and dynamic frames [12], showing the differences and commonalities between the three approaches by proving a main invariant property of the push operation on a stack data structure.

2   The Programming Language

To focus on the main features of our approach, we consider a basic object-oriented programming language. This model features the following main characteristics of component-based software: objects *encapsulate* their own local state, that is, objects only interact via *method calls*, and objects can be dynamically *created*. For technical convenience, we restrict the data types of our language to the basic value types of **integer** and **Boolean**, and the reference type **object**, which denotes the set of object identities. It should be emphasized here that only for notational convenience we abstract from classes in the presentation of the footprint logic (it is straightforward to generalize our method to classes, and integrate standard approaches for reasoning about other object-oriented features like inheritance, polymorphism, dynamic dispatch, etc.). In our model, the only built-in operation which involves the type **object** is reference equality. The constant **null** of type **object** represents the *invalid reference*.

The set of program variables with typical elements $x, y, \ldots$ is denoted by *Var*. These include the formal parameters of methods. The variable **this** is a distinguished variable of type **object**. By *Field* we denote a finite set of field identifiers, with typical element f. We assume variables and field are implicitly typed. Expressions $t, t_0, \ldots$ are constructed from program variables, fields, and built-in operations. For notational convenience we denote by $u, v, \ldots$ a variable x or a field f.

We have the following abstract grammar of statements, that are used for describing component behavior, including the method bodies (we again leave typing information implicit).

$$S ::= u := t \mid x := \mathbf{new} \mid y.m(\bar{t}) \mid S_1;\ S_2 \mid$$
$$\mathbf{if}\ B\ \mathbf{then}\ S_1\ \mathbf{else}\ S_2\ \mathbf{fi} \mid \mathbf{while}\ B\ \mathbf{do}\ S_1\ \mathbf{od}$$

In the assignment $u := t$, the left-hand side u denotes a variable or a field and t is an expression of the same type. For technical convenience (and without loss of generality), we restrict to object creation statements $x := \mathbf{new}$, where the left-hand side is a variable x (we thus exclude direct assignments of new object references to fields). Furthermore, we restrict to method calls $y.m(\bar{t})$, where the callee is denoted by a variable y ($\bar{t}$ are the actual parameters). Returning a value can be modeled by the use of a global variable which temporarily stores the return value. A program consists of a main statement S and a set of method definitions $m(x_1, \ldots, x_n) :: S$, with formal parameters $x_1, \ldots, x_n \in Var$ (the formal parameters of a method are considered local to the method body) and body S where **this** does not occur on the left-hand side of any assignment.

The formal semantics assumes an infinite set O of object identities with typical elements $o, o', \ldots$, and a set of values V such that $O \subseteq V$, that is, object identities are values. Furthermore, we assume that **null** $\in O$. Each object has its own local state which assigns values to its fields. A *heap* $\sigma \in \Sigma$ is a *partial function*, i.e. $\Sigma = O \rightharpoonup (\textit{Field} \rightarrow V)$, which assigns to an object o its local state. A local state $\sigma(o) \in \textit{Field} \rightarrow V$ assigns values to fields. By $\sigma(o) = \perp$ we denote that $\sigma(o)$ is undefined, in other words, o does not 'exist' in the heap σ. We always have $\sigma(\textbf{null}) = \perp$. By $dom(\sigma)$ we denote the set of existing objects, that is, the domain of σ consists of the objects o for which $\sigma(o)$ is defined. A *store* is a function $\tau \in T$ that assigns values to variables, i.e. $T = \textit{Var} \rightarrow V$. We restrict to stores and local states that are *type safe*, meaning that variables and fields are assigned to values of their type. A *configuration* $(\sigma, \tau) \in \Sigma \times T$ consists of a heap σ and a store τ. Both in the program semantics, and later in the semantics of assertions, we restrict to configurations that are *consistent*, i.e. fields of type **object** of existing objects only refer to existing objects or to **null**, and the store assigns variables of type **object** only to existing objects or **null**.

Given a set of method definitions, the basic input/output semantics $\mathcal{M}(S)$ of a statement S can be defined as a *partial* function $\Sigma \times T \rightharpoonup \Sigma \times T$. Note that the programming language is *deterministic*, so $\mathcal{M}(S)(\sigma, \tau)$ is either undefined, indicating that S does not terminate or it does not terminate properly, or it denotes a single final configuration of a properly terminating computation.

Following [2], we have the following partial correctness semantics of method calls: if $\tau(y) = \textbf{null}$ then $\mathcal{M}(y.m(\bar{t}))(\sigma, \tau)$ is undefined, and, otherwise, it is defined in terms of the (run-time) statement

$$\textbf{this}, \bar{x} := y, \bar{t}; S; \textbf{this}, \bar{x} := \tau(\textbf{this}), \tau(\bar{x})$$

where S denotes the body of m. The *parallel assignment* $\textbf{this}, \bar{x} := y, \bar{t}$ initializes the variables $\textbf{this}, \bar{x}$ of S, and the parallel assignment $\textbf{this}, \bar{x} := \tau(\textbf{this}), \tau(\bar{x})$ restores the initial values of the variables $\textbf{this}, \bar{x}$. For further details of the program semantics we refer to [2, p.206].

To observe *failures* due to calling a method on **null** (which generates a 'null-pointer exception'), we introduce a so-called strong partial correctness semantics $\mathcal{M}_{\textbf{null}}(S)(\sigma, \tau) \in (\Sigma \times T) \cup \{\textbf{fail}\}$ where $\mathcal{M}_{\textbf{null}}(y.m(\bar{t}))(\sigma, \tau) = \textbf{fail}$, if $\tau(y) = \textbf{null}$. We extend this semantics to additionally model the execution of a statement restricted by a footprint. We restrict to a *coarse-grained* notion of footprints which specifies the objects that may be called (either directly or indirectly through called methods). Note that this includes all the objects that may be changed, since we assume encapsulation and objects can only modify their own state (field assignments are relative to **this**). Our model then is based on extending configurations with an additional set of objects $\phi \subseteq O$ and the definition of a semantics $\mathcal{M}_{\textbf{F}}(S)(\sigma, \tau, \phi) \in (\Sigma \times T \times \mathcal{P}(O)) \cup \{\textbf{fail}\}$ where **fail** indicates failure which arises from a call of a method of an object that is not in the footprint ϕ. More specifically, $\mathcal{M}_{\textbf{F}}(y.m(\bar{t}))(\sigma, \tau, \phi)$ requires that the object $\tau(y)$ is callable, that is, it belongs to the set ϕ, otherwise $\mathcal{M}_{\textbf{F}}(y.m(\bar{t}))(\sigma, \tau, \phi) = \textbf{fail}$.

We restrict to footprints ϕ including only objects that exist in the given σ (which thus means failures arise from calling a method on **null**). Another crucial characteristic of our model is that newly created objects are naturally considered callable, and therefore new objects are added to the footprint ϕ, that is, $\mathcal{M}_\mathbf{F}(x := \mathbf{new})(\sigma, \tau, \phi) = (\sigma', \tau[x := o], \phi \cup \{o\})$ where o is a fresh object identity, i.e. $\sigma(o) = \bot$ and $o \neq \mathbf{null}$, and σ' results from σ by initializing the local state of the newly created object o.

3   The Specification Language

Given a set of method definitions, we specify the correctness of a statement by a pre- and postcondition which are formally specified by logical formulas $p, q, \ldots$, also called assertions. In order to express global properties of the heap, logical expressions e extend the expressions of our programming language with the *dereferencing* operator: $x.f$ denotes the value of field f of the object denoted by the variable x. As a special case, we abbreviate $\mathbf{this}.f$ by f.

The assertion language further features *logical* variables (which are assumed not to appear in the programs that describe the component behavior). A logical variable can either be a variable x or a *monadic predicate* X. Second-order quantification of such predicates allow for the specification of properties like *reachability*: $\forall Y\big((Y(e) \wedge \forall x(Y(x) \rightarrow Y(x.next))) \rightarrow Y(e')\big)$ states that the object denoted by e' can be reached from the one denoted by e via a chain of *next* fields (see Sect. 5 for more details). We represent the footprint by the distinguished monadic predicate F. The assertion $F(e)$ then expresses that the object denoted by e belongs to the set of objects denoted by the footprint F.

We omit the straightforward details of the definition $\mathcal{E}(e)(\sigma, \tau, \omega)$ of the semantics of logical expressions e, and the standard Tarski inductive truth definition for $\sigma, \tau, \phi, \omega \models p$, where ω assigns values to the logical variables (assuming that the sets of logical variables and program variables are disjoint). As a particular case, $\sigma, \tau, \phi, \omega \models F(e)$ indicates that $o \in \phi$, where $o = \mathcal{E}(e)(\sigma, \tau, \omega)$. For any other monadic predicate X, $\sigma, \tau, \phi, \omega \models X(e)$ indicates that $o \in \omega(X)$, where, as above, $o = \mathcal{E}(e)(\sigma, \tau, \omega)$. We restrict this truth definition to $\sigma, \tau, \phi, \omega$ which are consistent: $\tau(x)$ and $\omega(x)$ denote an object that exists in σ for every object variable x, and ϕ, and $\omega(X)$ for every monadic predicate X, denote a set of existing objects. Quantification over (sets of) objects ranges over (sets of) *existing* objects. Objects other than **null** that do not exist cannot be referred to—neither in the programming language nor in the assertion language. So $\sigma, \tau, \phi, \omega \models \exists x(p)$, where the variable x is of type **object**, holds precisely when there is an object $o \in dom(\sigma)$ such that $\sigma, \tau, \phi, \omega[x := o] \models p$. Here $\omega[x := o]$ results from ω by assigning the object o to the variable x.

By $\{p\}\ S\ \{q\}$ we denote the usual correctness formula in program logics. We distinguish two different interpretations corresponding to the two semantics $\mathcal{M}_{\mathbf{null}}$ and $\mathcal{M}_\mathbf{F}$. By $\models_{\mathbf{null}} \{p\}\ S\ \{q\}$ we denote the interpretation: for any $\sigma, \tau, \phi, \omega$ which are consistent, if $\sigma, \tau, \phi, \omega \models p$ then $\mathcal{M}_{\mathbf{null}}(S)(\sigma, \tau) \neq \mathbf{fail}$

and $\sigma', \tau', \phi, \omega \models q$ in case $\mathcal{M}_{null}(S)(\sigma, \tau) = (\sigma', \tau')$. Note that in this interpretation the predicate F does not have a special meaning. The interpretation $\models_F \{p\} \, S \, \{q\}$ is defined similarly, but with respect to the semantics $\mathcal{M}_F$, which thus requires reasoning about the footprint (in preconditions, postconditions, and loop invariants) to ensure that the statement does not lead to failure. In these specifications one formalizes facts about the contents of the footprint so as to ensure that components do not fail with respect to the strong partial correctness semantics. Note that both interpretations do *not* require reasoning about termination (i.e., absence of divergence).

4   The Hoare Logic of Footprints

We introduce a *hybrid* proof system which integrates reasoning about the two different interpretations of correctness specifications. We prefix Hoare triples by $\vdash_{null}$ and $\vdash_F$ to distinguish between the correctness specifications interpreted with respect to the semantics $\mathcal{M}_{null}$ and $\mathcal{M}_F$, respectively. By $\vdash$ we denote either of these two.

We have the standard axiom for (parallel) assignments to program variables, the usual consequence rule, and the standard rules for sequential composition, if-then-else and while statements, for both interpretations of correctness formulas. To axiomatize field assignments for both the interpretations, we introduce the substitution $p[f := t]$ which caters for *aliasing*: it replaces every expression of the form $x.f$ by the *conditional expression* **if** $x =$ **this then** t **else** $x.f$ **fi**.

Axiom 1 (FIELD ASSIGNMENT). $\vdash \{p[f := t]\} \, f := t \, \{p\}$

As a simple example, we have the following instantiation

$$\{\textbf{if } x = \textbf{this then } 0 \textbf{ else } x.f \textbf{ fi} = 1\} \, f := 0 \, \{x.f = 1\}$$

of this axiom, where the precondition is logically equivalent to $x \neq \textbf{this} \wedge x.f = 1$.

To axiomatize object creation for both interpretations of correctness formulas we introduce a substitution operator $p[x := \textbf{new}, \Phi]$, where Φ denotes a set of monadic predicates with the intention that X holds for the new object (that is, the set denoted by X includes the new object) if and only if $X \in \Phi$. This substitution involves an extension to second-order quantification of the *contextual* analysis of the occurrences of the variable x, as described in [7], assuming that in assertions an object variable x can only be dereferenced, tested for equality or appear as argument of a monadic predicate. Without loss of generality we restrict the definition of $p[x := \textbf{new}, \Phi]$ to assertions p which do not contain conditional expressions which have x as argument, since such conditional expressions (that have been introduced by a prior operation to cater for aliasing) can be systematically removed before applying this substitution.

Definition 1. *We have the following main clauses.*

$$x[x := \mathbf{new}, \Phi] \text{ is left undefined}$$
$$(x.f)[x := \mathbf{new}, \Phi] \text{ has a fixed default value}$$
$$(x = x)[x := \mathbf{new}, \Phi] = \mathbf{true}$$
$$(e = x)[x := \mathbf{new}, \Phi] = \mathbf{false} \text{ if } x \text{ and } e \text{ are distinct}$$
$$op(e_1, \ldots, e_n)[x := \mathbf{new}, \Phi] = op(e_1[x := \mathbf{new}, \Phi], \ldots, e_n[x := \mathbf{new}, \Phi])$$
$$X(x)[x := \mathbf{new}, \Phi] = \begin{cases} \mathbf{true} & \text{if } X \in \Phi \\ \mathbf{false} & \text{if } X \notin \Phi \end{cases}$$
$$X(e)[x := \mathbf{new}, \Phi] = X(e[x := \mathbf{new}, \Phi])$$
$$(\exists y(p))[x := \mathbf{new}, \Phi] = p[y := x][x := \mathbf{new}, \Phi] \ \lor \exists y(p[x := \mathbf{new}, \Phi])$$
$$(\exists X(p))[x := \mathbf{new}, \Phi] = \exists X(p[x := \mathbf{new}, \Phi \cup \{X\}]) \ \lor \exists X(p[x := \mathbf{new}, \Phi])$$

Definition 1 lists a few main cases. It should be noted that despite that $x[x := \mathbf{new}, \Phi]$ is undefined, $p[x := \mathbf{new}, \Phi]$ *is* defined for every *assertion*. Furthermore note that $(x = e)[x := \mathbf{new}, \Phi]$, for any logical expression e (syntactically) different from x, reduces to **false** because the value of e is not affected by the execution of $x := \mathbf{new}$ and as such refers after its execution to an 'old' object.

Because of the restriction that quantification over objects ranges over *existing* objects, we have a *changing scope of quantification* when applying the substitution $[x := \mathbf{new}, \Phi]$ to a formula $\exists y(p)$ or $\exists X(p)$. The resulting assertion namely is evaluated *prior* to the object creation, that is, in the heap where the newly created object does not exist yet. The case $(\exists y(p))[x := \mathbf{new}, \Phi]$ handles this changing scope of quantification by distinguishing two cases, as described initially in [7], namely: p is true for the new object, i.e. $p[y := x][x := \mathbf{new}, \Phi]$ holds; or there exists an 'old' object for which $p[x := \mathbf{new}, \Phi]$ holds, i.e. $\exists y(p[x := \mathbf{new}, \Phi])$ holds.

Similarly, $(\exists X(p))[x := \mathbf{new}, \Phi]$ is handled by distinguishing between whether or not X contains the newly created object. Without loss of generality we may assume that $X \notin \Phi$, since otherwise we apply the substitution to an alphabetic variant of $\exists X(p)$. The assertion $p[x := \mathbf{new}, \Phi \cup \{X\}]$ then describes the case that $X(x)$ is assumed to hold for the newly created object x. On the other hand, $p[x := \mathbf{new}, \Phi]$ describes the case that $X(x)$ is assumed not to hold.

We have the following axiomatization of object creation.

Axiom 2 (OBJECT CREATION).

$$\vdash_{\mathbf{null}} \{q[x := \mathbf{new}, \emptyset]\} \ x := \mathbf{new} \ \{q\} \ and \ \vdash_{\mathbf{F}} \{q[x := \mathbf{new}, \{F\}]\} \ x := \mathbf{new} \ \{q\}$$

For reasoning about the footprints of (non-recursive) method calls, we have the following rule which guarantees absence of failure in calling an object that does not belong to the footprint.

Rule 3 (METHOD CALL). Given the method definition $m(\bar{x}) :: S$, we have

$$\frac{\vdash_{\mathbf{F}} \{p \wedge F(y)\} \textbf{ this}, \bar{x} := y, \bar{t};\; S \{q\}}{\vdash_{\mathbf{F}} \{p \wedge F(y)\}\; y.m(\bar{t})\; \{q\}}$$

where, as above, the formal parameters (which include the variable **this**) do not occur free in the postcondition q.

In order to reason *semantically* about invariance in terms of footprints we introduce the *restriction* $p \downarrow$ of a formula p which restricts the quantification over objects in p to the invariant part of the state, namely that part that is *disjoint* from the footprint as denoted by F. For notational convenience, we introduce the complement F^c of F, e.g., $F^c(x)$ then denotes the assertion $\neg F(x)$. This restriction *bounds* every quantification involving objects to those objects which do *not* belong to the footprint. More specifically, $(\exists x(p)) \downarrow$ reduces to $\exists x(F^c(x) \wedge (p \downarrow))$, and similarly $(\forall x(p)) \downarrow$ reduces to $\forall x(F^c(x) \rightarrow (p \downarrow))$. Quantification over monadic predicates is treated similarly, i.e., $(\exists X(p)) \downarrow$ reduces to $\exists X(\forall x(X(x) \rightarrow F^c(x)) \wedge (p \downarrow))$.

We have the following main semantic invariance property of an assertion $p \downarrow$.

Theorem 1. *Let p be an assertion which does not contain free occurrences of dereferenced variables. We have $\sigma, \tau, \phi, \omega \models p \downarrow$ iff $\sigma', \tau', \phi', \omega \models p \downarrow$ where*

- $dom(\sigma) \subseteq dom(\sigma')$,
- $\sigma(o) = \sigma'(o)$, *for any* $o \in dom(\sigma) \setminus \phi$,
- $\phi \subseteq \phi'$, $\phi' \cap dom(\sigma) \subseteq \phi$, *and* $dom(\sigma') \setminus dom(\sigma) \subseteq \phi'$,
- $\tau(x) = \tau'(x)$, *for any variable x occurring free in p.*

Note that any assertion $p(x)$ which contains free occurrences of the variable x is logically equivalent to $\exists y(y = x \wedge p(y))$, where $p(y)$ results from replacing the free occurrences of x by y. The first three conditions of this theorem describe general properties of the semantics $\mathcal{M}_{\mathbf{F}}(S)(\sigma, \tau, \phi) = (\sigma', \tau', \phi')$. The last condition is true whenever the variables x that occur free in p do not occur in the statement S nor in any of the method bodies of the (indirectly) called methods.

Proof. We prove this theorem for assertions p which may contain free occurrences of dereferenced variables, requiring that $\tau(x) \notin \phi$, for any such variable. We proceed by induction on the assertion p. We highlight the main case of $\exists x(p)$.

$$\sigma, \tau, \phi, \omega \models (\exists x(p)) \downarrow \text{ iff (definition } (\exists x(p)) \downarrow)$$
$$\sigma, \tau, \phi, \omega \models \exists x(F^c(x) \wedge (p \downarrow)) \text{ iff } (o \in dom(\sigma) \setminus \phi)$$
$$\sigma, \tau, \phi, \omega[x := o] \models p \downarrow \text{ iff (induction hypothesis)}$$
$$\sigma', \tau', \phi', \omega[x := o] \models p \downarrow \text{ iff } (o \in dom(\sigma') \setminus \phi')$$
$$\sigma', \tau', \phi', \omega \models \exists x(F^c(x) \wedge (p \downarrow)) \text{ iff (definition } (\exists x(p)) \downarrow)$$
$$\sigma', \tau', \phi', \omega \models (\exists x(p)) \downarrow$$

Note that $o \in dom(\sigma) \setminus \phi$ implies that $o \in dom(\sigma') \setminus \phi'$, since $dom(\sigma) \subseteq dom(\sigma')$ and $\phi' \cap dom(\sigma) \subseteq \phi$. On the other hand, $o \in dom(\sigma') \setminus \phi'$ implies $o \in dom(\sigma) \setminus \phi$, because $dom(\sigma') \setminus dom(\sigma) \subseteq \phi'$ and $\phi \subseteq \phi'$. □

Given the above we can now introduce the following rule for reasoning about invariance.

Rule 4 (SEMANTIC INVARIANCE).

$$\frac{\vdash_{\mathbf{F}} \{p\}\ S\ \{q\}}{\vdash_{\mathbf{F}} \{p \wedge r\!\downarrow\}\ S\ \{q \wedge r\!\downarrow\}}$$

where the assertion r does not contain free occurrences of dereferenced variables and its program variables do not occur in the implicitly given set of method definitions and the statement S.

Soundness of the above semantic invariance rule follows directly from Theorem 1. We conclude with the following meta-rule for *eliminating* footprints.

Rule 5 (FOOTPRINT ELIMINATION).

$$\frac{\vdash_{\mathbf{F}} \{p\}\ S\ \{q\}}{\vdash_{\mathbf{null}} \{\exists F(p)\}\ S\ \{q\}}$$

where the monadic predicate F does not occur in the postcondition q.

Note that this rule allows a modular reasoning about footprints, that is, it caters for the introduction and elimination of footprints at the level of individual statements. Soundness of this rule follows from the main observation that, since F does not appear in the postcondition q, we have that $\models_{\mathbf{F}} \{p\}\ S\ \{q\}$ implies $\models_{\mathbf{null}} \{p\}\ S\ \{q\}$.

5   A Comparison Between Related Approaches

The two other main approaches to reasoning about invariant properties of unbounded heaps are that of separation logic [15] and dynamic frames [12]. In this section we illustrate the main differences between our approach and that of separation logic and dynamic frames by means of respective correctness proofs of a stack data structure where items can be added at the top.

To illustrate our approach, we model the stack structure by the two classes *LinkedList* and *Node*. In our framework, classes can be introduced by partitioning the type **object** into disjoint classes of objects. An instance of the class *LinkedList* holds a pointer, stored in field *first*, to the first node of the linked list. Each instance of class *Node* stores a value in field *val* and a pointer stored in field *next* to the next node. The implementation of the *push* method of the class *LinkedList* is shown in Listing 1.1. It uses as constructor the method *setAttributes*, which stores the pointer to the next node and the desired value. This refactoring of *first* := **new** *Node(first,v)* enables the separation of concerns between the creation of an object and calling its constructor in the proof theory.

```
push(v) ::
    u := new Node; u.setAttributes(first, v); first := u
setAttributes(node, v) ::
    next := node; val := v
```

Listing 1.1. Stack implementation for pushing items

We will focus on proving a natural global reachability property about the nodes in the stack. In particular, we will apply our approach and that of separation logic and dynamic frames to proving that all nodes in the stack that were reachable (starting from the *first* node, by repeatedly following *next* links) before a call to *push*, are still reachable after the call to *push*. It should be noted that this classic example serves primarily as an illustration of the different approaches. As such it provides a typical example of reasoning about invariant properties of unbounded heaps, and lends itself well for this purpose.

5.1   Footprints

Since we assume an implicit strongly typed assertion language, we introduce for the *LinkedList* and *Node* classes the distinct footprint predicates F_L and F_N which apply to *LinkedList* objects and *Node* objects, respectively. We omit the straightforward details of refining the definition of the proof system to a class-based programming language and an assertion language which features for each class a corresponding footprint predicate. Note that the restriction operator introduces bounded quantification over F_L and F_N, respectively.

We denote by $r'(Y, e, e')$ the assertion

$$[Y(e) \land \forall x(Y(x) \rightarrow Y(x.next))] \rightarrow Y(e')$$

and by $r(e, e')$ the assertion $\forall Y(r'(Y, e, e'))$ which formalizes reachability of the *Node* object denoted by e' from the one denoted by e via a chain of *next* fields (the predicate Y thus is assumed to range over *Node* objects).

We first show how to derive the global reachability specification

$$\vdash_{\textbf{null}} \{r(\textit{first}, y)\} \ \textbf{this}.push(v) \ \{r(\textit{first}, y)\} \tag{1}$$

from the local specification

$$\vdash_{\textbf{F}} \quad \begin{array}{c} \{F_L = \{\textbf{this}\} \land F_N = \emptyset \land \textit{first} = z\} \\ \textbf{this}.push(v) \\ \{F_N = \{\textit{first}\} \land \textit{first} \neq z \land \textit{first}.next = z \land \forall x(x.next \neq \textit{first})\} \end{array} \tag{2}$$

and, consequently, we prove the correctness specification 2. Recall that **this**.*first* and *first* denote the same value, and we use basic set-theoretic notations to abbreviate assertions about monadic predicates, e.g. $F_L = \{\textbf{this}\}$ abbreviates the assertion $F_L(\textbf{this}) \land \forall x(F_L(x) \rightarrow x = \textbf{this})$. The variables y and z in the correctness specifications (1) and (2) are assumed to be logical variables which do not appear in the definitions of the 'push' and the 'setAttributes' methods, and as such are not affected by the execution of the 'push' method. The variable z is used to 'freeze' the initial value of the field $first$. Note that the correctness specification (2) only describes the local changes affected by execution of the 'push' method in terms of its footprint.

Since the only variable that is dereferenced in the assertion $r(z, y)$ is the bound variable x, and $r(z, y)$ does not refer to program variables, we can apply the semantic invariance rule (Rule 4) to the correctness specification (2), and

$$\vdash_{\mathbf{F}} \{p \wedge r(z, y) \downarrow\} \ \mathbf{this}.push(v) \ \{q \wedge r(z, y) \downarrow\}$$

where p denotes the precondition and q denotes the postcondition of the correctness specification (2). By extending its definition to deal with classes, $r(z, y) \downarrow$ reduces to $\forall Y(Y \subseteq F_N^c \rightarrow r'(Y, z, y) \downarrow)$ where $Y \subseteq F_N^c$ abbreviates $\forall x(Y(x) \rightarrow F_N^c(x))$ and F_N^c denotes the complement of F_N, and $r'(Y, z, y) \downarrow$ reduces to $[Y(z) \wedge \forall x(F_N^c(x) \rightarrow (Y(x) \rightarrow Y(x.next)))] \rightarrow Y(y)$.

First we observe that $F_N = \emptyset$ implies $\forall x(F_N^c(x))$ (all existing objects belong to the complement of the footprint), and so the restriction operator has no effect, that is, $r(z, y) \downarrow$ is equivalent to $r(z, y)$.

Furthermore, we show that the assertion $r(z, y) \downarrow$ implies $r(z, y)$, assuming that $F_N = \{first\}$ and $first \neq z \wedge \forall x(x.next \neq first)$. The argument is a straightforward argument in second-order logic: Under the assumption $F_N = \{first\}$, $r(z, y) \downarrow$ is equivalent to

$$\forall Y(Y^c(first) \rightarrow r'(Y, z, y) \downarrow)$$

where $r'(Y, z, y) \downarrow$ reduces to

$$[Y(z) \wedge \forall x(x \neq first \rightarrow (Y(x) \rightarrow Y(x.next)))] \rightarrow Y(y).$$

As above, Y^c denotes the complement of Y: $Y^c(first)$ if and only if $\neg Y(first)$. Let Y be such that $Y(z) \wedge \forall x(Y(x) \rightarrow Y(x.next))$. We show that $Y(y)$. Let $Y' = Y \setminus \{first\}$. Since $first \neq z \wedge \forall x(x.next \neq first)$, it follows that $Y'(z) \wedge \forall x(Y'(x) \rightarrow Y'(x.next))$. From $r(z, y) \downarrow$ it follows that $r'(Y', z, y) \downarrow$, which is equivalent to $r'(Y', z, y)$ (note that the bounded quantification here has no effect). So we have $Y'(y)$, and by definition of Y', we conclude that $Y(y)$.

By the consequence rule the correctness formula for **push** then reduces to

$$\vdash_{\mathbf{F}} \{F_L = \{\mathbf{this}\} \wedge F_N = \emptyset \wedge first = z \wedge r(z, y)\} \ \mathbf{this}.push(v) \ \{first.next = z \wedge r(z, y)\}.$$

We next observe that $first.next = z \wedge r(z, y)$ implies $r(first, y)$: Let $Y(first)$ and $\forall x(Y(x) \rightarrow Y(x.next))$. From $first.next = z$ it then follows that $Y(z)$. From $r(z, y)$ we then conclude that $Y(y)$.

Thus, we obtain

$$\vdash_{\mathbf{F}} \{F_L = \{\mathbf{this}\} \wedge F_N = \emptyset \wedge first = z \wedge r(z, y)\} \ \mathbf{this}.push(v) \ \{r(first, y)\}$$

by another application of the consequence rule. Finally, substituting $first$ for the variable z in the precondition and subsequently eliminating the footprints predicates F_L and F_N, we obtain by a trivial application of the consequence rule the global correctness specification (1).

We now have to give a proof of the local correctness specification (2), and let p denote the precondition

$$F_l = \{\mathbf{this}\} \wedge F_n = \emptyset \wedge \textit{first} = z$$

and q denote the postcondition

$$F_n = \{\textit{first}\} \wedge \textit{first} \neq z \wedge \textit{first.next} = z \wedge \forall x(x.\textit{next} \neq \textit{first}).$$

In this proof we omit the prefix $\vdash_F$. By the standard axiom for assignments to variables we have

$$\{F_n = \{u\} \wedge u \neq z \wedge u.\textit{next} = z \wedge \forall x(x.\textit{next} \neq u)\}$$
$$\textit{first} := u$$
$$\{q\}.$$

Next we observe that

$$\{F_n = \{u\} \wedge u \neq z \wedge$$
$$\mathbf{if}\ u = \mathbf{this}\ \mathbf{then}\ \textit{node}\ \mathbf{else}\ u.\textit{next}\ \mathbf{fi} = z \wedge$$
$$\forall x(\mathbf{if}\ x = \mathbf{this}\ \mathbf{then}\ \textit{node}\ \mathbf{else}\ x.\textit{next}\ \mathbf{fi} \neq u)\}$$
$$\textit{next} := \textit{node}$$
$$\{F_n = \{u\} \wedge u \neq z \wedge u.\textit{next} = z \wedge \forall x(x.\textit{next} \neq u)\}$$

by Axiom 1 for field assignments. Towards deriving the correctness formula for the method call, we work with the assignment of actual parameters to formal parameters. By the standard axiom for assignments to variables again, we then derive

$$\{F_n = \{u\} \wedge u \neq z \wedge$$
$$\mathbf{if}\ u = \mathbf{this}\ \mathbf{then}\ \textit{first}\ \mathbf{else}\ u.\textit{next}\ \mathbf{fi} = z \wedge$$
$$\forall x(\mathbf{if}\ x = \mathbf{this}\ \mathbf{then}\ \textit{first}\ \mathbf{else}\ x.\textit{next}\ \mathbf{fi} \neq u)\}$$
$$\textit{node} := \textit{first}$$
$$\{F_n = \{u\} \wedge u \neq z \wedge$$
$$\mathbf{if}\ u = \mathbf{this}\ \mathbf{then}\ \textit{node}\ \mathbf{else}\ u.\textit{next}\ \mathbf{fi} = z \wedge$$
$$\forall x(\mathbf{if}\ x = \mathbf{this}\ \mathbf{then}\ \textit{node}\ \mathbf{else}\ x.\textit{next}\ \mathbf{fi} \neq u)\}.$$

Substituting $\mathbf{this}$ by u and a trivial application of the consequence rule we obtain

$$\{F_n = \{u\} \wedge u \neq z \wedge \textit{first} = z \wedge$$
$$\forall x(\mathbf{if}\ x = u\ \mathbf{then}\ \textit{first}\ \mathbf{else}\ x.\textit{next}\ \mathbf{fi} \neq u)\}$$
$$\mathbf{this} := u$$
$$\{F_n = \{u\} \wedge u \neq z \wedge$$
$$\mathbf{if}\ u = \mathbf{this}\ \mathbf{then}\ \textit{first}\ \mathbf{else}\ u.\textit{next}\ \mathbf{fi} = z \wedge$$
$$\forall x(\mathbf{if}\ x = \mathbf{this}\ \mathbf{then}\ \textit{first}\ \mathbf{else}\ x.\textit{next}\ \mathbf{fi} \neq u)\}.$$

Putting the above together using the rule for sequential composition and Rule 3 for method calls, we derive

$$\{F_n = \{u\} \wedge u \neq z \wedge \textit{first} = z \wedge$$
$$\forall x(\mathbf{if}\ x = u\ \mathbf{then}\ \textit{first}\ \mathbf{else}\ x.\textit{next}\ \mathbf{fi} \neq u)\}$$
$$u.\textit{setAttributes}(\textit{first}, v)$$
$$\{q\}.$$

Applying Axiom 2, we have next have to prove

$$\{F_n = \emptyset \wedge \mathit{first} = z\}$$
$$u := \mathbf{new}\ \mathit{Node}$$
$$\{F_n = \{u\} \wedge u \neq z \wedge \mathit{first} = z \wedge$$
$$\forall x(\mathbf{if}\ x = u\ \mathbf{then}\ \mathit{first}\ \mathbf{else}\ x.next\ \mathbf{fi} \neq u)\}.$$

The calculation of $(F_n = \{u\})[x := \mathbf{new}, \{F_n\}]$ follows the same steps as that of $(\forall y(F(y) \leftrightarrow y = x))[x := \mathbf{new}, \{F\}]$, already shown above as an example. We then calculate

$$(\forall x(\mathbf{if}\ x = u\ \mathbf{then}\ \mathit{first}\ \mathbf{else}\ x.next\ \mathbf{fi} \neq u))[u := \mathbf{new}, \{F_n\}],$$

and skip the straightforward details of the application of the substitution to the remaining conjuncts. Before applying the substitution we first have to remove the conditional expression from the inequality. Here we go:

$$(\forall x(\mathbf{if}\ x = u\ \mathbf{then}\ \mathit{first} \neq u\ \mathbf{else}\ x.next \neq u\ \mathbf{fi}))[u := \mathbf{new}, \{F_n\}]$$
$$=$$
$$\mathbf{if}\ u = u\ \mathbf{then}\ \mathit{first} \neq u\ \mathbf{else}\ u.next \neq u\ \mathbf{fi}[u := \mathbf{new}, \{F_n\}]\ \wedge$$
$$\forall x(\mathbf{if}\ x = u\ \mathbf{then}\ \mathit{first} \neq u\ \mathbf{else}\ x.next \neq u\ \mathbf{fi}[u := \mathbf{new}, \{F_n\}])$$
$$=$$
$$\mathbf{if}\ (u = u)[u := \mathbf{new}, \{F_n\}]$$
$$\mathbf{then}\ (\mathit{first} \neq u)[u := \mathbf{new}, \{F_n\}]$$
$$\mathbf{else}\ (u.next \neq u)[u := \mathbf{new}, \{F_n\}]\ \mathbf{fi}\ \wedge$$
$$\forall x(\mathbf{if}\ (x = u)[u := \mathbf{new}, \{F_n\}]$$
$$\mathbf{then}\ (\mathit{first} \neq u)[u := \mathbf{new}, \{F_n\}]$$
$$\mathbf{else}\ (x.next \neq u)[u := \mathbf{new}, \{F_n\}]\ \mathbf{fi})$$
$$=$$
$$\mathbf{if\ true\ then\ true\ else\ true\ fi}\ \wedge$$
$$\forall x(\mathbf{if\ false\ then\ true\ else\ true\ fi})$$

The resulting assertion is clearly logically equivalent to **true**.

Putting the above together using the rule for sequential composition and Rule 3 for method calls, we finally derive the above correctness specification.

5.2   Dynamic Frames

In this section we discuss a proof of the push operation on our stack-like data structure using the approach of dynamic frames [12] as it is implemented in the KeY tool [16]. The KeY proof of our case study can be found in the artifact [4] accompanying this paper, which includes user-defined taclets (describing inference rules in the KeY system) that we used to define the reachability predicate. Also, a video recording [3] shows the steps for reproducing the proof of invariance of reachability over the push method using KeY. This section describes on a more abstract level the basic ideas underlying dynamic frames, in an extension of Hoare logic (instead of dynamic logic) which allows for a better comparison with our footprints approach. As such, we restrict to coarse-grained dynamic frames, similar to our footprints (KeY supports dynamic frames on the finer granularity of fields, see Sect. 7 for a brief discussion on extending the granularity of footprints in our approach).

In the dynamic frames approach footprints are introduced by extending contracts with an assignable clause. For example, the clause **assignable {this}** states that only the fields of the object **this** can be modified. The corresponding proof obligation then requires to prove that the initial and final heap (of an execution of **this**.$push(v)$) differ at most with respect the values of field $first$. To formally express this proof obligation in our assertion language (introduced in Sect. 3), we take a snapshot of the entire initial heap and store it in a logical variable: this allows one to refer in the postcondition to the initial heap and express its relation with the final heap.

Thus, we extend the assertion language with heap variables, with typical element h. Given a current heap σ and environment ω, for any heap variable h, $\omega(h)$ denotes a heap (as defined in Sect. 2) such that its domain is included in that of σ, i.e. $dom(\omega(h)) \subseteq dom(\sigma)$. Given a heap variable h the expression $h(x.f)$ denotes the result of the lookup of field f of object x in heap h, and $h(f)$ abbreviates $h(\mathbf{this}.f)$. Furthermore, we introduce the binary predicate $x \in dom(h)$ which means that the object denoted by x is in the domain of the heap denoted by h. If the object denoted by x is not in the domain of the heap denoted by h, then $h(x.f)$ refers to the value of f in the current heap, denoted by $x.f$.

Given the above, we define the assertion $init(h)$ as $\forall x(x \in dom(h) \wedge \bigwedge_f x.f = h(x.f))$ which expresses that heap variable h stores the current heap, that is, $\sigma, \tau, \phi, \omega \models init(h)$ if and only if $\omega(h) = \sigma$. We note that $\bigwedge_f$ ranges over finitely many field names. We can now present a high-level proof in Hoare logic of our case study, using the above extension of the assertion language with heap variables. We again denote by $r(e, e')$ the reachability assertion defined above. Our goal is

$$\{r(first, y)\} \ \mathbf{this}.push(v) \ \{r(first, y)\}.$$

To derive our goal, we first derive the local specification

$$\{init(h)\}$$
$$\mathbf{this}.push(v) : \mathbf{assignable} \ \{\mathbf{this}\} \qquad (3)$$
$$\{first.next = h(\mathbf{this}.first) \wedge first \notin \mathrm{dom}(h)\}.$$

This specification makes use of an assignable clause which requires a proof of

$$\{init(h)\} \ \mathbf{this}.push(v) \ \{q \wedge \forall x(x \neq \mathbf{this} \rightarrow \bigwedge_f x.f = h(x.f))\}$$

where q denotes the postcondition of the specification (3). The proof of the latter specification is easily derived by inlining the method bodies (in KeY, this proof obligation can even be done automatically).

Next we apply a new invariance rule for heap variables which states the invariance of assertions which are independent of the current heap. Let $r_h(\mathit{first}, y)$ denote the reachability assertion

$$\forall Y\left([Y(h(\mathit{first})) \wedge \forall x(x \in \mathit{dom}(h) \rightarrow (Y(x) \rightarrow Y(h(x.\mathit{next}))))] \rightarrow Y(y)\right)$$

This assertion does not depend on the current heap and thus is invariant. By the conjunction rule, we then obtain

$$\{r_h(\mathit{first}, y) \wedge \mathit{init}(h)\}$$
$$\textbf{this}.\mathit{push}(v) \tag{4}$$
$$\{r_h(\mathit{first}, y) \wedge q \wedge \forall x(x \neq \textbf{this} \rightarrow \textstyle\bigwedge_f x.f = h(x.f))\}$$

That the postcondition of (4) implies $r(\mathit{first}, y)$ can be established as in the previous proof, and similarly that $r(\mathit{first}, y)$ implies the precondition of (4) with the heap variable h existentially quantified. Thus an application of the consequence rule, and the existential elimination rule of variable h, concludes the proof.

Let us remark one difference with the presentation in this section and the one in the KeY system. Instead of the conjunction over all fields, KeY uses heap updates in the formalization of the assignable clause by so-called 'anonymizing updates' [1, Sect. 9.4.1]. For example, consider a method with an assignable clause $\{\textbf{this}\}$ and let heap be the heap before the method executes. In KeY's logic, the heap resulting after the method call is expressed as $\mathit{heap}[\mathit{anon}(\mathit{self.*}, \mathit{anon\_heap})]$. When evaluating a field access $x.f$ in the heap $\mathit{heap}[\mathit{anon}(\mathit{self.*}, \mathit{anon\_heap})]$, the evaluation rules for heaps in KeY require to perform what is essentially an aliasing analysis, e.g. checking whether $x = \textbf{this}$. As a program (fragment) is symbolically executed in KeY, updates to the heap are accumulated, resulting in new, larger heap terms. For real-world programs, the heap terms tend to get large (in our case study, see e.g. [3, 0:41]). Consequently, it may require many proof steps to reason about the value of fields in such heaps.

5.3 Separation Logic

In this section we present a proof in separation logic of the push operation of our stack data structure. For various classes of programs, there are different corresponding separation logics. We shall use the proof system as presented by Reynolds in [14]. It contains a basic programming language with the core rules of separation logic that are shared by many of the separation logic 'versions' that have extended language features, e.g. separation logic suitable for Java [9], and tools for verifying (Java) programs using separation logic such as VeriFast [11] and VerCors [5]. For example, VerCors can be used to prove the correctness of a (concurrent wait-free) push explicitly using permissions[1]. However, for the purposes of this case study we keep the discussion on a more abstract level to

---

[1] https://vercors.ewi.utwente.nl/try_online/example/82.

focus on the basic ideas underlying separation logic which allows for a better comparison with our footprints approach.

Instead of object orientation, Reynolds' uses a fairly low-level programming language with allocation, de-allocation and several pointer operations, such as reading or writing to a location that a pointer variable references. In this setting, object creation can be modeled by allocation, and accessing a field of an object corresponds to accessing a location referenced by a pointer.

Before we perform the proof, we therefore first write the push operation in Reynolds' language, as follows. Note that in our approach we have separated the concerns for object creation and calling its constructor, but here these two happen at the same time by one assignment.

```
push(v) ::
   first := cons(v, first)
```

Listing 1.2. Stack implementation for pushing items in separation logic

The cons operation allocates a new cell in memory where two values are stored: the first component is the value of the item to push (denoted by the formal parameter v), and the second component is a pointer to the 'node' storing the next value. Following Reynolds, reachability can then be expressed in separation logic by the following recursively defined predicates:

$$reach(begin, other) \equiv \exists n(n \geq 0 \wedge reach_n(begin, other))$$
$$reach_0(begin, other) \equiv begin = other$$
$$reach_{n+1}(begin, other) \equiv \exists v, next[(begin \mapsto (v, next)) * reach_n(next, other)]$$

The $*$ operation is the so-called separating conjunction: informally it means that the left and right sub-formulas should be satisfied in disjoint heaps. A heap in this context is a *partial* function from addresses to values. Two heaps are disjoint if their domains are disjoint. Furthermore, the formula $begin \mapsto (v, next)$ asserts that the heap contains precisely one cell (namely, the one at the address stored in the variable $begin$), and that this cell contains the values v and $next$.

The above global correctness specification (1) translates to the following main global correctness property in separation logic

$$\{reach(first, y)\} \ push(v) \ \{reach(first, y)\}. \tag{5}$$

Furthermore, we have the following local specification in separation logic

$$\{first = z \wedge \mathbf{emp}\} \ push(v) \ \{first \mapsto (v, z)\} \tag{6}$$

of the push method corresponding to the above local specification (2) which uses footprints. Here **emp** denotes the empty heap. In fact, at a deeper level, it can be shown that this information corresponds to the assertion $F_N = \emptyset$ (this is further discussed in Sect. 7).

Applying the frame rule of separation logic with the 'invariance' formula $reach(z, y)$ to the local specification then yields:

$$\{first = z \wedge \mathbf{emp} * reach(z, y)\} \ push(v) \ \{first \mapsto (v, z) * reach(z, y)\}.$$

Using the rule for Auxiliary Variable Elimination of separation logic, we infer:

$$\{\exists z(\mathit{first} = z \wedge \mathbf{emp} * \mathit{reach}(z, y))\} \; \mathit{push}(v) \; \{\exists z(\mathit{first} \mapsto (v, z) * \mathit{reach}(z, y))\}.$$

We show that the global correctness formula (5) follows from an application of the consequence rule to the following correctness formula:

$$\{\exists z(\mathit{first} = z \wedge \mathbf{emp} * \mathit{reach}(z, y))\} \; \mathit{push}(v) \; \{\exists z(\mathit{first} \mapsto (v, z) * \mathit{reach}(z, y))\}.$$

- First we show that $\mathit{reach}(\mathit{first}, y)$ (the desired precondition) implies the precondition above. In the precondition above, substituting z for first (equals for equals) the existential quantifier in the precondition can be eliminated so the precondition above is equivalent to $\mathbf{emp} * \mathit{reach}(\mathit{first}, y)$. Further, by the equivalence $\mathbf{emp} * p \leftrightarrow p$ and commutativity of the separating conjunction, the clause with the empty heap can be eliminated. Hence, $\mathit{reach}(\mathit{first}, y)$ implies the precondition above (they are equivalent).
- Next we show that the postcondition implies the assertion $\mathit{reach}(\mathit{first}, y)$. Observe that the postcondition and the definition of $\mathit{reach}(z, y)$ imply that for some z_0 and n_0: $(\mathit{first} \mapsto (v, z_0)) * \mathit{reach}_{n_0}(z_0, y)$. So, by the definition of reach_n we have $\mathit{reach}_{n_0+1}(\mathit{first}, y)$, which by definition of the reach predicate implies the desired postcondition $\mathit{reach}(\mathit{first}, y)$.

We conclude this case study of separation logic with the observation that the above local specification of the push method (6) follows from the Allocation (local) axiom of separation logic:
$$\{\mathit{first} = z \wedge \mathbf{emp}\} \; \mathit{first} := \mathit{cons}(v, \mathit{first}) \; \{\mathit{first} \mapsto (v, z)\}.$$

6   Discussion

All above three correctness proofs are based on the application of some form of frame rule to derive the invariance of the global reachability property from a local specification of the push method. In our approach we express such properties by restricting quantification to objects which do not belong to the footprint which includes all objects that can be affected by the execution. In the approach based on dynamic frames such footprints are used in a general semantic definition of a relation between the initial and final heap, where to prove an invariant property one has to show that it is preserved by this relation. Invariance in separation logic is expressed by a separating conjunction which allows to express invariant properties of disjoint heaps.

Summarizing, in both our approach and that of separation logic, invariant properties are so *by definition*: in separation logic by the use of the separating conjunction, in our approach by bounded quantification. The general semantics of the separating conjunction and bounded quantification *ensures* invariance, which thus does not need to be established for each application of the frame rule separately. In contrast, the approach of dynamic frames *requires* for each property an ad-hoc proof of its invariance.

In separation logic [15] invariant properties are specified and verified with the frame rule which uses the separating conjunction to *split* the heap into two disjoint parts. As a consequence, the assertion language of separation logic by its very nature supports reference to locations that are not allocated (or objects that do not exist). Furthermore, because of the restriction to *finite* heaps, the validity of the first-order language restricted to the so-called 'points to' predicate is non-compact, and as such not recursively enumerable (see [6]).

In contrast, our approach is based on standard (second-order monadic) predicate logic which allows for the use of established theorem proving techniques (e.g. Isabelle/HOL). Furthermore, as discussed above, our approach allows for reasoning at an abstraction level which coincides with the programming language, i.e., it does not require special predicates like the 'points to' predicate or 'dangling pointers' which are fundamental in the definition of the separating conjunction and implication.

Dynamic frames [12] were introduced as an extension of Hoare logic where an explicit heap representation as a function in the assertion language is used to describe the parts of the heap that are modified. As footprints in our approach, a dynamic frame is a specification of a set of objects that can be modified. This specification itself may, in general, involve dynamic properties of the heap. Since it is impossible to capture an *entire* heap structure in a fixed, finite number of first-order freeze variables, heap variables are introduced. Such variables allow to store the entire initial heap in a single freeze variable (which does not occur in the program), and thus describe the modifications as a relation between the initial and the final heap. In this bottom-up approach invariant properties then are verified directly in terms of this relation between the initial and the final heap. This low-level relation, which itself is not part of the program specification, in general complicates reasoning. In contrast, in our approach we incorporate footprints in the Hoare logic of *strong partial correctness*: a footprint simply *constraints* the program execution. This allows the formal verification of invariant properties in a compositional, top-down manner by an extension of the standard invariance rule. As such this rule abstracts from the relation between the initial and the final heap as specified by a footprint: it is only used in proving its *soundness*, and *not* in program verification itself.

7   Conclusion

In this paper, we introduced a program logic extended with footprints and a novel invariance rule for reasoning about invariant properties of unbounded heaps. This invariance rule allows to adapt correctness specifications (contracts), abstracting from the internal implementation details.

We kept the programming language small but expressive, to focus on several core challenges: developing a logic that (1) avoids explicit heaps, (2) uses *abstract* object creation (informally this means: objects that do not *yet* exist cannot be referred to in the assertion language), (3) interprets assertions using (the standard) Tarski's truth definition, and (4) features a frame rule to reason about invariance. Our logic satisfies all four of the above features, as opposed to the main existing approaches of separation logic and dynamic frames. Moreover, our hybrid program logic combines two different forms of strong partial correctness specifications. This combination allows for modular/compositional reasoning about footprints, which can be introduced and eliminated on the level of individual statements. We further discussed the differences between our approach and that of separation logic and dynamic frames in terms of the underlying assertion languages.

Using results from the paper [8], on the verification of object-oriented programs with classes and abstract object creation (but without footprints), the logic in this paper can be further extended in a straightforward manner to cover nearly the entire sequential subset of Java, including other failures and exceptions generated by other language features. That paper also showed how to mechanize proof rules for abstract object creation in the KeY system. Along the same lines, as future work, we aim to extend the KeY system with a mechanized version of the logic in this paper.

Acknowledgments. We are grateful for the constructive feedback that was given by the anonymous reviewers.

References

1. Ahrendt, W., Beckert, B., Bubel, R., Hähnle, R., Schmitt, P.H., Ulbrich, M. (eds.). Deductive Software Verification - The KeY Book - From Theory to Practice, vol. 10001 of Lecture Notes in Computer Science. Springer, Cham (2016). https://doi.org/10.1007/978-3-319-49812-6
2. Apt, K.R., Olderog, E.R., Apt, K.R.: Verification of Sequential and Concurrent Programs. Texts in Computer Science. Springer, London (2009). https://doi.org/10.1007/978-1-84882-745-5
3. Bian, J., Hiep, H.-D.A., de Boer, F.S., de Gouw, S.: Integrating ADTs in KeY and their application to history-based reasoning. In: Huisman, M., Păsăreanu, C., Zhan, N. (eds.) FM 2021. LNCS, vol. 13047, pp. 255–272. Springer, Cham (2021). https://doi.org/10.1007/978-3-030-90870-6_14
4. Bian, J., Hiep, H.-D.A.: Reasoning about invariant properties of object-oriented programs-dynamic frames: Proof files. Zenodo (2021). https://doi.org/10.5281/zenodo.6044345
5. Blom, S., Darabi, S., Huisman, M., Oortwijn, W.: The VerCors tool set: verification of parallel and concurrent software. In: Polikarpova, N., Schneider, S. (eds.) IFM 2017. LNCS, vol. 10510, pp. 102–110. Springer, Cham (2017). https://doi.org/10.1007/978-3-319-66845-1_7
6. Calcagno, C., Yang, H., O'Hearn, P.W.: Computability and complexity results for a spatial assertion language for data structures. In: Hariharan, R., Vinay, V., Mukund, M. (eds.) FSTTCS 2001. LNCS, vol. 2245, pp. 108–119. Springer, Heidelberg (2001). https://doi.org/10.1007/3-540-45294-X_10
7. Thomas, W. (ed.): FoSSaCS 1999. LNCS, vol. 1578. Springer, Heidelberg (1999). https://doi.org/10.1007/3-540-49019-1

8. de Gouw, S., de Boer, F.S., Ahrendt, W., Bubel, R.: Integrating deductive verification and symbolic execution for abstract object creation in dynamic logic. Softw. Syst. Model. **15**(4), 1117–1140 (2016)

9. Distefano, D., Parkinson J, M.J.: jstar: towards practical verification for java. In: Harris, G.E. (eds.), Proceedings of the 23rd Annual ACM SIGPLAN Conference on Object-Oriented Programming, Systems, Languages, and Applications, OOPSLA 2008, 19–23 October 2008, Nashville, TN, USA, pp. 213–226. ACM (2008). https://doi.org/10.1145/1449764.1449782

10. Hoare, C.A.R.: Procedures and parameters: an axiomatic approach. In: Engeler, E. (ed.) Symposium on Semantics of Algorithmic Languages. LNM, vol. 188, pp. 102–116. Springer, Heidelberg (1971). https://doi.org/10.1007/BFb0059696

11. Jacobs, B., Smans, J., Philippaerts, P., Vogels, F., Penninckx, W., Piessens, F.: VeriFast: a powerful, sound, predictable, fast verifier for C and Java. In: Bobaru, M., Havelund, K., Holzmann, G.J., Joshi, R. (eds.) NFM 2011. LNCS, vol. 6617, pp. 41–55. Springer, Heidelberg (2011). https://doi.org/10.1007/978-3-642-20398-5_4

12. Kassios, I.T.: Dynamic frames: support for framing, dependencies and sharing without restrictions. In: Misra, J., Nipkow, T., Sekerinski, E. (eds.) FM 2006. LNCS, vol. 4085, pp. 268–283. Springer, Heidelberg (2006). https://doi.org/10.1007/11813040_19

13. Olderog, E.-R.: On the notion of expressiveness and the rule of adaption. Theor. Comput. Sci. **24**, 337–347 (1983)

14. Reynolds, J.C.: Separation logic: a logic for shared mutable data structures. In: 17th IEEE Symposium on Logic in Computer Science (LICS 2002), 22–25 July 2002, Copenhagen, Denmark, Proceedings, pp. 55–74. IEEE Computer Society (2002). https://doi.org/10.1109/LICS.2002.1029817

15. Reynolds, J.C.: An overview of separation logic. In: Verified Software: Theories, Tools, Experiments, First IFIP TC 2/WG 2.3 Conference, VSTTE 2005, Zurich, Switzerland, 10–13 October 2005, Revised Selected Papers and Discussions, pp. 460–469 (2005)

16. Weiß, B.: Deductive Verification of Object-Oriented Software: Dynamic Frames, Dynamic Logic and Predicate Abstraction. PhD thesis, Karlsruhe Institute of Technology (2011)

Decompositional Branching Bisimulation Minimisation of Monolithic Processes

Mark Bouwman$^{(\boxtimes)}$ (ID), Maurice Laveaux (ID), Bas Luttik (ID), and Tim Willemse (ID)

Eindhoven University of Technology, Eindhoven, The Netherlands
{m.s.bouwman,m.laveaux,s.p.luttik,t.a.c.willemse}@tue.nl

Abstract. One of the biggest challenges in model checking complex systems is the state space explosion problem. A well known technique to reduce the impact of this problem is compositional minimisation. In this technique, first the state spaces of all components are computed and minimised modulo some behavioural equivalence (e.g., some form of bisimilarity). These minimised transition systems are subsequently combined to obtain the final state space.

In earlier work a compositional minimisation technique was presented tailored to mCRL2: it provides support for the multi-action semantics of mCRL2 and allows splitting up a monolithic linear process specification into components. Only strong bisimulation minimisation of components can be used, limiting the effectiveness of the approach. In this paper we propose an extension to support branching bisimulation reduction and prove its correctness. We present a number of benchmarks using mCRL2 models derived from industrial SysML models, showing that a significant reduction can be achieved, also compared to compositional minimisation with strong bisimulation reduction.

Keywords: Model checking · Bisimulation quotienting · Compositional minimisation

1 Introduction

The mCRL2 toolset [3] has been used in various industrial applications [2,12,19]. A recurring challenge is that industrial models are often complex and consist of many concurrent components. The state spaces induced by such models are typically huge, hindering verification efforts. The size of the state space tends to scale exponentially with the number of components; a well known phenomenon often referred to as the state space explosion problem.

The state space induced by a model can be minimised modulo an equivalence relation, such as strong bisimulation [18] or branching bisimulation [8]. For some models the state space induced by the model is too large to generate while the size of the minimised state space is small enough to allow model checking. For such models compositional minimisation is a helpful technique. As the name

S. L. Tapia Tarifa and J. Proença (Eds.): FACS 2022, LNCS 13712, pp. 161–182, 2022.
https://doi.org/10.1007/978-3-031-20872-0_10

suggests, the technique applies minimisation not at the top level but at the level of components. The model is split into several submodels representing the components of the system. The (state spaces of the) components are minimised and then combined to construct a behaviourally equivalent but smaller state space. Compositional minimisation has a history going back to the early 1990s [9]. A complication with compositional minimisation is that the sum of the state spaces can be larger than the state space of the entire system. This issue can be addressed by interface specifications, which model the interfaces between the separated components [6,10,13]. The CADP toolset [5] provides tools for compositional verification, including EXP.OPEN, which computes the parallel composition of a network of state spaces.

We would like to tailor compositional minimisation to the specifics of mCRL2. Any workflow with an mCRL2 model starts with *linearisation*, a symbolic operation which produces a Linear Process Specification (LPS). In linearisation all parallelism is removed and the specification is brought into a normal form. An LPS contains a single process P, parametrised with a number of data parameters. The process contains a number of summands. Each summand in P consists of a guard (referencing the data parameters), an action and a recursion on P with updates to the parameters. Since P has no notion of components, applying compositional minimisation cannot be done using the aforementioned techniques.

Recent additions to the mCRL2 toolset are `lpscleave` and `lpscombine` that, respectively, decompose and recombine a model [16]. Cleaving an LPS P produces two new LPSs, L and R, both containing part of the behaviour. The LPS is cleaved based on a split of the data parameters of P, which the user can specify. Subsequently the state spaces associated to L and R can be computed and minimised with respect to strong bisimilarity, yielding, say L' and R'. The tool `lpscombine` combines L' and R' to the final state space, ensuring that the result is strongly bisimilar to P.

One of the complications these tools have overcome is dealing with the multi-action semantics of mCRL2 [11]. In mCRL2 when L and R can perform some step a and b, then their parallel composition $L \parallel R$ can perform an $a|b$ step. The special unobservable action τ is the identity element for multi-actions, i.e. $a|\tau = a$. As a consequence τ actions arbitrarily communicate with other components in a parallel composition. If components would have τ-transitions, `lpscombine` would be unsound. To prevent this phenomenon, `lpscleave` replaces all occurrences of τs by a visible action `tag` in L and R. The recombination process of `lpscombine` abstracts again from this visible action.

Many behavioural equivalences (such as branching bisimilarity) include a special treatment of τ-transitions [7]. If a process has many τ-transitions, then minimisation with respect to notions of bisimilarity that treat τ-transitions as unobservable will yield a much smaller state space. Since, by construction, L and R do not contain τ-transitions we cannot effectively use minimisation with respect to branching bisimilarity on the level of components.

Contributions. Our first contribution is an extension of the theory of [16] in which we add support for intermediate branching bisimulation minimisation. In short, this is achieved by first hiding the tag action labels in the cleaved processes (turning many transitions into τ-labelled transitions), then minimising the state space and finally reintroducing the tag labels that were hidden in the first step (to avoid communicating τ-actions). In effect we treat tag as unobservable whilst computing the branching bisimulation quotient. The extension is proven correct. The techniques can be generalised to other process algebras using multi-actions.

Our second contribution is illustrating the effectiveness of the updated tools by applying them to mCRL2 models derived from railway SysML [17] specifications. These models have a clear notion of component as they typically consist of the parallel composition of several State Machines. For these models we are able to reduce the state space by several orders of magnitude. For some models verification is improved from being infeasible to being able to check a requirement in a few minutes. For other models, however, verification remains infeasible.

Organisation. This paper is structured as follows. Section 2 provides preliminaries on mCRL2. Section 3 provides the preliminaries of the cleaving and combining processes. Section 4 provides the necessary extension to the theory to support branching bisimulation minimisation. Section 5 details how we concretely implemented support for branching bisimulation minimisation. Benchmarks are presented in Sect. 6. Section 7 concludes this work with conclusions and future work.

2   mCRL2

In this section we will introduce the necessary preliminaries of a process algebra with multi-actions. The definitions are in line with mCRL2.

We presuppose some abstract data theory. For each sort D we assume the existence of a non-empty semantic domain denoted by $\mathbb{D}$. Additionally, we assume the existence of sorts *Bool* and *Nat* and their associated semantic domains of Booleans ($\mathbb{B}$) and natural numbers ($\mathbb{N}$), respectively. We use $e : D$ to indicate that e is an expression of sort D. Expressions are, as usual, built from variables and function symbols. The set of free variables of an expression e is denoted $\mathsf{FV}(e)$. An expression e is *closed* iff $\mathsf{FV}(e) = \emptyset$. A substitution σ is a total function from variables to closed data expressions of their corresponding sort. We write $\sigma[x \leftarrow e]$ to denote the substitution σ' such that $\sigma'(x) = e$ and for all $y \neq x$, we have $\sigma'(y) = \sigma(y)$. We use $\sigma(e)$ to denote the result of applying substitution σ to expression e: for each variable x each occurrence of x in expression e is replaced by $\sigma(x)$.

We presuppose a fixed *interpretation* function, denoted by $[\![\ldots]\!]$, which maps syntactic objects to values within their corresponding semantic domain. Semantic objects are typeset in *boldface*, e.g., the semantics of expression $1 + 1$ is **2**. We denote *data equivalence* by $e \approx f$, which is true iff $[\![e]\!] = [\![f]\!]$.

We denote a *vector* of length $n + 1$ by $\vec{d} = \langle d_0, \ldots, d_n \rangle$. Two vectors are equivalent, denoted by $\langle d_0, \ldots, d_n \rangle \approx \langle e_0, \ldots, e_n \rangle$, iff their elements are *pairwise equivalent*, i.e., $d_i \approx e_i$ for all $0 \leq i \leq n$. For a vector of data expressions with their corresponding sorts $\langle d_0 : D_0, \ldots, d_n : D_n \rangle$, we write $\vec{d} : \vec{D}$. Let $\vec{d}.i$ denote the ith element of $\vec{d}$. Finally, we define $\mathsf{Vars}(\vec{d}) = \{d_0, \ldots, d_n\}$.

A *multi-set* over a set A is a set with *multiplicity* for each element in A. Formally, a multi-set is a total function $m : A \to \mathbb{N}$, where $m(a)$ is the multiplicity of a. As notation we use $\{\!\!\{ \ldots \}\!\!\}$ for a multi-set where the multiplicity of each element is either written next to it or omitted when it is one. Elements with multiplicity 0 are omitted. For instance, $\{\!\!\{ a : 2, b \}\!\!\}$ over set $\{a, b, c\}$ has elements a and b with multiplicity two and one respectively. For multi-sets m and m' over set A, we write $m \subseteq m'$ iff $m(a) \le m'(a)$ for all $a \in A$. Multi-sets $m + m'$ and $m - m'$ are defined pointwise: $(m + m')(a) = m(a) + m'(a)$ and $(m - m')(a) = \max(m(a) - m'(a), 0)$ for all $a \in A$.

Let Λ be the set of *action labels*. We use D_a to indicate the sort of action label $a \in \Lambda$, $\mathbb{D}_a$ denotes the semantic domain of D_a. The set of all multi-sets over $\{a(\mathbf{e}) \mid a \in \Lambda, \mathbf{e} \in \mathbb{D}_a\}$ is denoted by Ω.

Definition 1. *A labelled transition system, abbreviated LTS, is a tuple $\mathcal{L} = (S, Act, \to)$ where S is a set of states, $Act \subseteq \Omega$ and $\to \subseteq S \times Act \times S$ is a labelled transition relation. Let $s \xrightarrow{\alpha} t$ denote that $(s, \alpha, t) \in \to$.*

Definition 2 (Ref. [16]). Multi-actions *are defined as follows:*

$$\alpha ::= \tau \mid a(e) \mid \alpha | \alpha$$

Constant τ represents the empty *multi-action, $a \in \Lambda$ is an action label, and e is an expression of sort D_a. The semantics of a multi-action α, given a substitution σ, is denoted by $[\![\alpha]\!]_\sigma$ and is an element of Ω. It is defined inductively as follows: $[\![\tau]\!]_\sigma = \{\!\!\{\}\!\!\}$, $[\![a(e)]\!]_\sigma = \{\!\!\{ a([\![\sigma(e)]\!]) \}\!\!\}$ and $[\![\alpha | \beta]\!]_\sigma = [\![\alpha]\!]_\sigma + [\![\beta]\!]_\sigma$. If α is a closed expression then the substitution is typically omitted.*

We will now define two well known equivalence relations on LTSs: strong bisimulation [18] and branching bisimulation [8].

Definition 3. *Let $(S, Act, \to)$ be an LTS. We call a relation $\mathcal{R} \subseteq S \times S$, a strong bisimulation relation if and only if it is symmetric and for all $(s, t) \in \mathcal{R}$ the following condition holds: if $s \xrightarrow{\alpha} s'$, then there exists a state t' such that $t \xrightarrow{\alpha} t'$ and $(s', t') \in \mathcal{R}$. Two states s and t are strongly bisimilar, denoted by $s \leftrightarrow t$, if and only if there exists a strong bisimulation relation $\mathcal{R}$ such that $(s, t) \in \mathcal{R}$.*

Let $\to_*$ denote the reflexive transitive closure of $\to$, i.e. $s \to_* t$ iff t is *reachable* from s. Similarly, let $\twoheadrightarrow$ denote the reflexive transitive closure of the binary relation $\xrightarrow{\{\!\!\{\}\!\!\}}$. Let $s \xrightarrow{(\omega)} t$ be an abbreviation of $s \xrightarrow{\omega} t \lor (\omega = \{\!\!\{\}\!\!\} \land s = t)$.

Definition 4. *Let $(S, Act, \to)$ be an LTS. We call a relation $\mathcal{R} \subseteq S \times S$, a branching bisimulation relation if and only if it is symmetric and for all $(s, t) \in \mathcal{R}$, the following condition holds: if $s \xrightarrow{\alpha} s'$ then there exist states t' and t'' such that $t \twoheadrightarrow t' \xrightarrow{(\alpha)} t''$ and $(s, t') \in \mathcal{R}$ and $(s', t'') \in \mathcal{R}$.*

Two states s and t are branching bisimilar, denoted by $s \leftrightarrow_b t$, if and only if there exists a branching bisimulation relation $\mathcal{R}$ such that $(s, t) \in \mathcal{R}$.

An mCRL2 specification is linearised before other tools can be applied. Linearisation brings the process expression into a normal form, yielding an LPS, consisting of a data specification, action declarations and a *Linear Process Equation* (LPE). LTSs can be compactly represented by LPSs. Let PN be a set of *process names*.

Definition 5 (Ref. [16]). *An LPE is an equation of the form:*

$$P(\vec{d} : \vec{D}) = \sum_{e_0:E_0} c_0 \to \alpha_0 . P(\vec{g_0}) + \ \ldots \ + \sum_{e_n:E_n} c_n \to \alpha_n . P(\vec{g_n})$$

Here $P \in PN$ is the process name, *$\vec{d}$ is the list of* process parameters, *and each:*

- E_i *is a sort ranged over by* sum variable e_i *(where $e_i \notin \vec{d}$),*
- c_i *is the* enabling condition, *a boolean expression s.t. $\mathsf{FV}(c_i) \subseteq \mathsf{Vars}(\vec{d}) \cup \{e_i\}$,*
- α_i *is a multi-action τ or $a_0(f_0)| \ldots |a_n(f_n)$ such that each $a_k \in \Lambda$ and f_k is an expression of sort D_{a_k} such that $\mathsf{FV}(f_k) \subseteq \mathsf{Vars}(\vec{d}) \cup \{e_i\}$,*
- $\vec{g_i}$ *is an* update expression *of sort $\vec{D}$, satisfying $\mathsf{FV}(\vec{g_i}) \subseteq \mathsf{Vars}(\vec{d}) \cup \{e_i\}$.*

The $+$-operator denotes a non-deterministic choice. We use $+_{i \in I}$ for a finite set of *indices* $I \subseteq \mathbb{N}$ as a shorthand for a number of summands. The $\sum$-operator describes a non-deterministic choice among the possible values of the bounded sum variable. We generalise the action sorts and the sum operator in LPEs without increasing expressiveness; we permit ourselves to write $a(f_0, \ldots, f_k)$ and $\sum_{e_0:E_0, \ldots, e_l:E_l}$, respectively.

Let P be the set of expressions $P(\vec{\iota})$, where $P \in PN$ and $\vec{\iota}$ is a closed expression of sort $\vec{D}$ (the sort of P). The labelled transition system induced by an LPE is then formally defined as follows.

Definition 6 (Ref. [16]). *The operational semantics associated with expressions in P is the LTS $(\mathsf{P}, \Omega, \to)$, where the transition relation $\to$ is defined as the smallest relation obtained as follows: for each LPE $P(\vec{d} : \vec{D}) = +_{i \in I} \sum_{e_i:E_i} c_i \to \alpha_i . P(\vec{g_i})$ and for all indices $i \in I$, closed expressions $\vec{\iota} : \vec{D}$ and substitutions σ such that for all j, $\sigma(\vec{d}.j) = \vec{\iota}.j$ there is a transition $P(\vec{\iota}) \xrightarrow{[\![\alpha_i]\!]\sigma} P(\sigma(\vec{g_i}))$ iff $[\![\sigma(c_i)]\!] = \mathsf{true}$.*

We define a minimal language to express parallelism and interaction of LTSs; the operators are taken from mCRL2 [11].

Definition 7 (Ref. [16]). *Let Comm be the set of all possible communication expressions of the form $a_0| \ldots |a_n \to c$ where $a_0, \ldots, a_n, c \in \Lambda$ are action labels. Let St be a set of constants representing states. We introduce the operators com-munication, allow, hide and parallel composition:*

$$S ::= \Gamma_C(S) \ | \ \nabla_A(S) \ | \ \tau_I(S) \ | \ S \parallel S \ | \ s$$

Here, $C \subseteq$ Comm is a finite set of communication expressions, $A \subseteq 2^{A \to \mathbb{N}}$ is a non-empty finite set of finite multi-sets of action labels, $I \subseteq \Lambda$ is a non-empty finite set of action labels and $s \in St$ is a state. Let S denote the set of expressions that can be constructed by the grammar above, which is parametrised with the set of constants St.

Definition 7 is parametrised with a pre-existing LTS. In particular, it can be parametrised with the LTS associated to an LPE. Later we will also need to apply operators on minimised LTSs of components.

Definitions 8 and 9, respectively, define the communication and hiding *functions* on multi-actions. They will be used to define the semantics of the communication and hiding *operators* in Definition 10.

Definition 8 (Ref. [16]). *We define $\gamma_C \colon \Omega \to \Omega$, where $C \subseteq$ Comm, as follows:*

$$\gamma_\emptyset(\omega) = \omega$$
$$\gamma_C(\omega) = \gamma_{C \setminus C_1}(\gamma_{C_1}(\omega)) \text{ for } C_1 \subset C$$
$$\gamma_{\{a_0|\ldots|a_n \to c\}}(\omega) = \begin{cases} \{c(\mathbf{d})\} + \gamma_{\{a_0|\ldots|a_n \to c\}}(\omega - \{a_0(\mathbf{d}),\ldots,a_n(\mathbf{d})\}) \\ \quad \text{if } \{a_0(\mathbf{d}),\ldots,a_n(\mathbf{d})\} \subseteq \omega \\ \omega \quad \text{otherwise} \end{cases}$$

So, for example, note that $\gamma_{\{a|b \to c\}}(a|d|b) = c|d$. Two restrictions are necessary for γ_C to be well-defined. We require that the left-hand sides of the communications do not share labels. For example, $\gamma_{\{a|b \to c, a|d \to c\}}(a|d|b)$ would be problematic as it could evaluate to both $b|c$ and $d|c$. Furthermore, the action label on the right-hand side must not occur in the left-hand side of any other communication. For example, $\gamma_{\{a \to c, c \to a\}}(a)$ would not be well defined because it could evaluate to both a and c.

Definition 9 (Ref. [16]). *Let $\omega \in \Omega$ and $I \subseteq \Lambda$. We define $\theta_I(\omega)$ as the multi-set ω' defined as:*

$$\omega'(a(\mathbf{d})) = \begin{cases} 0 & \text{if } a \in I \\ \omega(a(\mathbf{d})) & \text{otherwise} \end{cases}$$

Given a multi-action α we write $\underline{\alpha}$ to denote the multi-set of action labels that occur in α, i.e. we remove the data expressions from the multi-action. Formally, $\underline{a(e)} = \{a\}$, $\underline{\tau} = \{\}$ and $\underline{\alpha|\beta} = \underline{\alpha} + \underline{\beta}$. We define $\underline{\omega}$ for $\omega \in \Omega$ in a similar way.

Definition 10 (Ref. [16]). *We presuppose an LTS $(St, Act, \to_1)$. We associate an LTS $(\mathsf{S}, \Omega, \to)$ to expressions of S, where S is parametrised with constants St. The relation $\to$ is the least relation including the transition relation $\to_1$ and the rules below. For any $\omega, \omega' \in \Omega$, expressions P, P', Q, Q' of S and sets $C \subseteq$ Comm, $A \subseteq 2^{A \to \mathbb{N}}$ and $I \subseteq \Lambda$:*

$$\text{COM} \ \frac{P \xrightarrow{\omega} P'}{\Gamma_C(P) \xrightarrow{\gamma_C(\omega)} \Gamma_C(P')} \qquad\qquad \text{ALLOW} \ \frac{P \xrightarrow{\omega} P'}{\nabla_A(P) \xrightarrow{\omega} \nabla_A(P')} \ \underline{\omega} \in A \cup \{\langle\rangle\}$$

$$\text{HIDE} \ \frac{P \xrightarrow{\omega} P'}{\tau_I(P) \xrightarrow{\theta_I(\omega)} \tau_I(P')} \qquad\qquad \text{PAR} \ \frac{P \xrightarrow{\omega} P' \quad Q \xrightarrow{\omega'} Q'}{P \parallel Q \xrightarrow{\omega+\omega'} P' \parallel Q'}$$

$$\text{PARR} \ \frac{Q \xrightarrow{\omega} Q'}{P \parallel Q \xrightarrow{\omega} P \parallel Q'} \qquad\qquad \text{PARL} \ \frac{P \xrightarrow{\omega} P'}{P \parallel Q \xrightarrow{\omega} P' \parallel Q}$$

3   Cleave and Combine

We now have the necessary process algebraic preliminaries to discuss the theory of cleaving and combining. This section will define a way to split an LPE, provide requirements on this split, define a recombination process and provide a theorem stating that the recombined process is strongly bisimilar to the original process.

Example 1. Consider the following LPE, which we will use as a running example. Note that states $P(0, \mathsf{false})$ and $P(1, \mathsf{false})$, and states $P(0, \mathsf{true})$ and $P(1, \mathsf{true})$ are branching bisimilar.

$$P(m : Nat, n : Bool) =$$
$$0: \quad (m \approx 0) \rightarrow \tau . P(1, n)$$
$$1: \quad + (m \approx 1) \rightarrow a . P(2, n)$$
$$2: \quad + (m \approx 1) \rightarrow \tau . P(2, n)$$
$$3: \quad + (\neg n) \rightarrow b . P(m, \mathsf{true})$$
$$4: \quad + (m \approx 1 \wedge \neg n) \rightarrow c . P(2, \mathsf{true})$$

Suppose we want to split it into two LPEs, of which one will control parameter m and the other parameter n. Summands 0 to 3 can be split easily as they only depend on parameters of one of the two components. Summand 4, however, poses a challenge as it depends on both m and n; the two LPEs will need to synchronise for the execution of this summand.

Before we define how to split an LPE we need to define projection on vectors. Let $I \subseteq 0, \ldots, n$. We define the *I-projection* of $\langle d_0, ..., d_n \rangle$, denoted by $\langle d_0, \ldots, d_n \rangle_{|I}$, as the vector $\langle d_{i_0}, \ldots, d_{i_l} \rangle$ for the largest $l \in \mathbb{N}$ such that $i_0 < i_1 < \ldots < i_l \leq n$ and $i_k \in I$ for $0 \leq k \leq l$. For a vector of data expressions and their domains $\vec{d} : \vec{D}$ we denote the projection by $\vec{d}_{|I} : \vec{D}_{|I}$.

Below, we define the notion of a *separation tuple*. Intuitively, a separation tuple defines how to split off part of the process parameters and summands of an LPE. To split an LPE into two parts we need two separation tuples.

Definition 11. (Ref. [16]). *Let $P(\vec{d} : \vec{D}) = +_{i \in I} \sum_{e_i : E_i} c_i \rightarrow \alpha_i . P(\vec{g_i})$ be an LPE. A separation tuple for P is a 6-tuple $(U, K, J, c^U, \alpha^U, \vec{h}^U)$ where $U \subseteq \mathbb{N}$ is a set of parameter indices, $K \subseteq J \subseteq I$ are two sets of summand indices, and c^U, α^U and $\vec{h}^U$ are functions with domain $J \setminus K$ and as codomains the sets of conditions, actions and synchronisation expressions, respectively. We require that for all $i \in (J \setminus K)$ it holds that $FV(c^U(i)) \cup FV(\alpha^U(i)) \cup FV(\vec{h}^U(i)) \subseteq \mathsf{Vars}(\vec{d}) \cup \{e_i\}$, and for all $i \in K$ it holds that $FV(c_i) \cup FV(\alpha_i) \cup FV(\vec{g}_{i|U}) \subseteq \mathsf{Vars}(\vec{d}_{|U}) \cup \{e_i\}$. A separation tuple induces an LPE, where $U^c = \mathbb{N} \setminus U$, as follows:*

$$P_U(\vec{d}_{|U} : \vec{D}_{|U}) = +_{i \in (J \setminus K)} \sum_{e_i : E_i, \vec{d}_{|U^c} : \vec{D}_{|U^c}} c^U(i) \rightarrow \alpha^U(i)|\mathsf{sync}_U^i(\vec{h}^U(i)) . P_U(\vec{g}_{i|U})$$

$$+ \ +_{i \in K} \sum_{e_i : E_i} c_i \rightarrow \alpha_i|\mathsf{tag} . P_U(\vec{g}_{i|U})$$

We assume that action labels sync_U^i, for any $i \in I$, and label tag do not occur in the original LPE.

In Definition 11 the set J contains the summand indices of all summands featured in the separation tuple, of which the subset K contains the local summands and subset $J \setminus K$ the summands needing synchronisation.

Example 2. Consider the LPE P from Example 1 again. Suppose we decompose P using the separation tuples $(V, \{0, 1, 2\}, \{0, 1, 2, 4\}, \{4 \mapsto m \approx 1\}, \{4 \mapsto c\}, \{4 \mapsto \langle\rangle\})$ and $(W, \{3\}, \{3, 4\}, \{4 \mapsto \neg n\}, \{4 \mapsto \tau\}, \{4 \mapsto \langle\rangle\})$, where $V = \{0\}$ and $W = \{1\}$ (the parameter indices). These separation tuples would induce the following LPEs:

$$P_V(m : Nat) = (m \approx 0) \rightarrow \tau|\mathsf{tag} . P_V(1)$$
$$+ (m \approx 1) \rightarrow \mathsf{a}|\mathsf{tag} . P_V(2)$$
$$+ (m \approx 1) \rightarrow \tau|\mathsf{tag} . P_V(2)$$
$$+ (m \approx 1) \rightarrow c|\mathsf{sync}_V^4 . P_V(2)$$
$$P_W(n : Bool) = (\neg n) \rightarrow \mathsf{b}|\mathsf{tag} . P_W(\mathsf{true})$$
$$+ (\neg n) \rightarrow \tau|\mathsf{sync}_W^4 . P_W(\mathsf{true})$$

Summands 0 to 3 of the original LPE P are *local*; they only depend on local process parameters. The tag labels are added to prevent local summands from unintended communications in the parallel composition. We revisit the necessity of the tag action in Example 3. Summand 4 of P, which depends on both m and n is split across P_V and P_W, which will need to synchronise on the sync^4 labels. The LTSs induced by the LPEs are given in Example 4.

Note that Definition 11 does not specify how to split an LPE into two sensible separation tuples. To *correctly* split an LPE, a number of requirements need to be fulfilled. Two separation tuples form a *cleave* iff the requirements listed in [16, Definition 4.6] hold. Separation tuples that are a cleave can be recombined in such a way that the result is strongly bisimilar to the original LPE (Theorem 1 below, the main result of [16]). The definition of the notion of cleave is only needed in the proof of Theorem 1, there is no need to repeat it here. The separation tuples from Example 2 constitute a cleave.

Theorem 1 states both how a cleaved LPE can be recombined and that the recombined process is strongly bisimilar to the original LPE. A proof of this theorem is provided in [15].

Theorem 1 (Ref. [16]). *Let $P(\vec{d} : \vec{D}) = +_{i \in I} \sum_{e_i : E_i} c_i \rightarrow \alpha_i \cdot P(\vec{g}_i)$ be an LPE and let $(V, K^V, J^V, c^V, \alpha^V, h^V)$ and $(W, K^W, J^W, c^W, \alpha^W, h^W)$ be a cleave for this LPE. Let P_V and P_W be the LPEs induced by the separation tuples. For every closed expression $\vec{\iota} : \vec{D}$ we get the following correspondence between the original process $P(\vec{\iota})$ and the cleaved and recombined process:*

$$P(\vec{\iota}) \underleftrightarrow{} \tau_{\{\mathsf{tag}\}}(\nabla_{\{\underline{\alpha_i} \mid i \in I\} \cup \{\alpha_i \mid \mathsf{tag} \mid i \in (K^V \cup K^W)\}}($$

$$\tau_{\{\mathsf{sync}^i \mid i \in I\}}(\Gamma_{\{\mathsf{sync}^i_V \mid \mathsf{sync}^i_W \rightarrow \mathsf{sync}^i \mid i \in I\}}(P_V(\vec{\iota}_{|V})) \parallel P_W(\vec{\iota}_{|W}))))$$

Let us review the purpose of each operator in the recombining process expression. Note that each multi-action in $P_V(\vec{\iota}_{|V})$ and $P_W(\vec{\iota}_{|W})$ includes either a tag or sync label. Also note that due to the multi-action semantics, $P_V(\vec{\iota}_{|V}) \parallel P_W(\vec{\iota}_{|W})$ can make any combination of steps from $P_V(\vec{\iota}_{|V})$ and $P_W(\vec{\iota}_{|W})$. The outer hide operator simply hides all tag labels. The allow operator is parametrised with an *allow set*, which contains each multi-action α_i of the summands of the original LPE; it also contains multi-actions $\alpha_i|$tag if α_i is the action in one of the local transitions. Any other action is blocked; in particular, multi-actions containing two tags or a sync label are blocked. The communication and inner operators ensure that matching sync labels communicate and are subsequently hidden.

In the proofs of Sect. 4 we will need that the multiset consisting of a single tag action is always allowed by the allow operator. The allow operator in the process expression of Theorem 1 only includes a tag action if there exists an $i \in I$ such that $\alpha^V(i) = \tau$ or $\alpha^W(i) = \tau$. Corollary 1 (below) states that we can always add the tag action to the allow set.

Corollary 1. *Let $P(\vec{d} : \vec{D}) = +_{i \in I} \sum_{e_i : E_i} c_i \rightarrow \alpha_i \cdot P(\vec{g}_i)$ be an LPE and let $(V, K^V, J^V, c^V, \alpha^V, h^V)$ and $(W, K^W, J^W, c^W, \alpha^W, h^W)$ be a cleave for this LPE. Let P_V and P_W be the LPEs induced by the separation tuples. For every closed expression $\vec{\iota} : \vec{D}$ we get the following correspondence between the original process $P(\vec{\iota})$ and the cleaved and recombined process:*

$$P(\vec{\iota}) \underleftrightarrow{} \tau_{\{\mathsf{tag}\}}(\nabla_{\{\mathsf{tag}\} \cup \{\underline{\alpha_i} \mid i \in I\} \cup \{\alpha_i \mid \mathsf{tag} \mid i \in (K^V \cup K^W)\}}($$

$$\tau_{\{\mathsf{sync}^i \mid i \in I\}}(\Gamma_{\{\mathsf{sync}^i_V \mid \mathsf{sync}^i_W \rightarrow \mathsf{sync}^i \mid i \in I\}}(P_V(\vec{\iota}_{|V})) \parallel P_W(\vec{\iota}_{|W}))))$$

Proof. Let $P'(\vec{\iota})$ be an LPE which includes all summands of $P(\vec{\iota})$ and adds a dummy summand $(\mathsf{false}) \to \tau.P(g)$, where g is some arbitrary update expression. Clearly $P(\vec{\iota})$ and $P'(\vec{\iota})$ are strongly bisimilar. Let the added summand be the last summand with index n. Then there is a cleave of $P'(\vec{\iota})$ consisting of the separation tuples $(V \cup \{n\}, K^V \cup \{n\}, J^V, c^V, \alpha^V, h^V)$ and $(W, K^W, J^W, c^W, \alpha^W, h^W)$. By Theorem 1 we have the following correspondence:

$$P'(\vec{\iota}) \underline{\leftrightarrow} \tau_{\{\mathsf{tag}\}}(\nabla_{\{\alpha_i \mid i \in (I \cup \{n\})\} \cup \{\alpha_i \mid \mathsf{tag} \mid i \in (K^V \cup K^W \cup \{n\})\}}($$
$$\tau_{\{\mathsf{sync}^i \mid i \in (I \cup \{n\})\}}(\Gamma_{\{\mathsf{sync}^i_V \mid \mathsf{sync}^i_W \to \mathsf{sync}^i \mid i \in (I \cup \{n\})\}}(P_V(\vec{\iota}|_V)) \parallel P_W(\vec{\iota}|_W))))$$

The allow set can be rewritten to $\{\mathsf{tag}\} \cup \{\alpha_i \mid i \in I\} \cup \{\alpha_i \mid \mathsf{tag} \mid i \in (K^V \cup K^W)\}$. The extension from I to $I \cup \{n\}$ in the hiding and communication operators does not change the behaviour as $P_{V \cup \{n\}}(\vec{\iota}|_{V \cup \{n\}}))$ and $P_W(\vec{\iota}|_W)$ cannot produce sync^n actions. By transitivity of strong bisimulation we obtain:

$$P(\vec{\iota}) \underline{\leftrightarrow} \tau_{\{\mathsf{tag}\}}(\nabla_{\{\mathsf{tag}\} \cup \{\alpha_i \mid i \in I\} \cup \{\alpha_i \mid \mathsf{tag} \mid i \in (K^V \cup K^W)\}}($$
$$\tau_{\{\mathsf{sync}^i \mid i \in I\}}(\Gamma_{\{\mathsf{sync}^i_V \mid \mathsf{sync}^i_W \to \mathsf{sync}^i \mid i \in I\}}(P_V(\vec{\iota}|_V)) \parallel P_W(\vec{\iota}|_W))))$$

$\square$

For convenience we will from now on use the following shorthand. Note that this is only a well defined process expression in the context of a process and its cleave, which define α, I, V and W.

$$\mathcal{C}(P) := \tau_{\{\mathsf{tag}\}}(\nabla_{\{\mathsf{tag}\} \cup \{\alpha_i \mid i \in I\} \cup \{\alpha_i \mid \mathsf{tag} \mid i \in (K^V \cup K^W)\}}($$
$$\tau_{\{\mathsf{sync}^i \mid i \in I\}}(\Gamma_{\{\mathsf{sync}^i_V \mid \mathsf{sync}^i_W \to \mathsf{sync}^i \mid i \in I\}}(P))))$$

Since strong bisimulation is a congruence for the operators used to construct $\mathcal{C}$, processes $P_V(\vec{\iota}|_V)$ and $P_W(\vec{\iota}|_W)$ can be replaced by any strongly bisimilar process expression or LTS. The theory is implemented in two tools. Given a set of process parameters the tool `lpscleave` correctly cleaves an LPS into two LPSs. Recall that an LPS consists of both an LPE and a data specification. The tool `lpscombine` applies the operators in the context $\mathcal{C}$ to two LPSs or LTSs.

4   Extension to Branching Bisimilarity

The difficulty of supporting branching bisimulation reduction of component LPEs is in the fact that transitions local to a component are made visible by adding a `tag` label. These `tag` labels are necessary to prevent τ-transitions from arbitrarily communicating with other summands, as explained in the following example.

Example 3. Suppose the component LPEs P_V and P_W from Example 2 did not have added **tag** actions. The process expressions $P(1, \text{false})$ and $\mathcal{C}(P_V(1) \parallel P_W(\text{false}))$ would then not be strongly bisimilar, and therefore also not branching bisimilar. As can be seen below, $\mathcal{C}(P_V(1) \parallel P_W(\text{false}))$ can perform a b-labelled transition to $\mathcal{C}(P_V(2) \parallel P_W(\text{true}))$, which cannot be mimicked by $P(1, \text{false})$ as the only b-labelled transition it can perform leads to a state from which an a-labelled transition is enabled. The b-labelled transition stems from a synchronisation between the summands $(m \approx 1) \rightarrow \tau . P_V(2)$ and $(\neg n) \rightarrow \text{b} . P_W(\text{true})$.

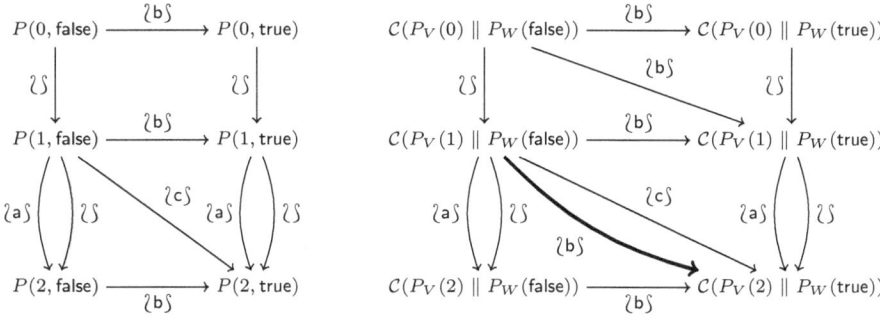

The solution to minimising modulo branching bisimulation without having communicating τs in the parallel composition is presented in Theorem 2. Intuitively, the theorem states that we can replace a component P of a cleaved process by a process P' if P and P' are branching bisimilar *after abstraction from tag-actions*. We will illustrate this with an example.

Example 4. Below we give the LTSs of P_V and P_W from Example 2.

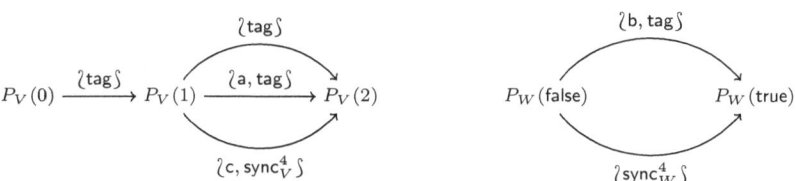

Below we show 3 subsequent transformations on P_V: hiding **tag** actions, minimising modulo branching bisimulation and adding a **tag** action to each transition that does not have a **sync** label.

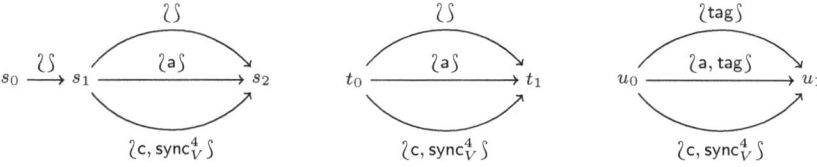

Below we show the states reachable from $\mathcal{C}(u_0 \parallel P_W(\text{false}))$. Note that the LTS is branching bisimilar to the original LPE P (see Example 3), and also smaller.

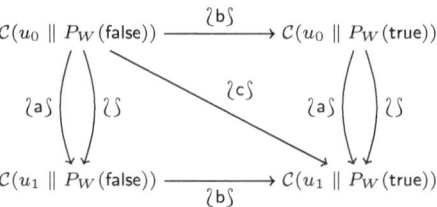

Before we present Theorem 2, which formalises the substitution of $P_W(\text{false})$ with u_0 in Example 4, we will present two lemmas that will shorten the proof of Theorem 2.

For a process expression P, let $A(P)$ denote the *alphabet* of P, i.e., $A(P) = \{\underline{\omega} \in \Lambda \mid P \to_* P' \xrightarrow{\omega} P''\}$.

Definition 12. *Let* $\mathsf{S}_{ts}(V) \subset \mathsf{S}$ *denote the set of* tag/sync *process expressions. A process expression P is in* $\mathsf{S}_{ts}(V)$, $V \subseteq \mathbb{N}$, *if and only if every label in $A(P)$ contains exactly one* tag *or* sync *label:*

$$\forall_{\underline{\omega} \in A(P)}\left(\left(\sum\nolimits_{a \in \{\text{tag}\} \cup \{\text{sync}_V^i \mid i \in \mathbb{N}\}} \underline{\omega}(a)\right) = 1\right).$$

Note that any LPE induced by a cleave is a tag/sync process.

The lemma below relates τ-labelled paths from $\tau_{\{\text{tag}\}}(P)$ and $\mathcal{C}(P \parallel Q)$.

Lemma 1. *Let* $L(\vec{d} : \vec{D}) = +_{i \in I} \sum_{e_i : E_i} c_i \to \alpha_i \, . \, L(\vec{g}_i)$ *be an LPE and let* $(V, K^V, J^V, c^V, a^V, h^V)$ *and* $(W, K^W, J^W, c^W, a^W, h^W)$ *be a cleave. Let $P \in \mathsf{S}_{ts}(V), Q \in \mathsf{S}_{ts}(W)$ be* tag/sync *process expressions. Under these assumptions we have that*

$$\tau_{\{\text{tag}\}}(P) \twoheadrightarrow \tau_{\{\text{tag}\}}(P') \text{ implies } \mathcal{C}(P \parallel Q) \twoheadrightarrow \mathcal{C}(P' \parallel Q)$$

Proof. The proof is by induction on the length of a sequence of τ-transitions from $\tau_{\{\text{tag}\}}(P)$ to $\tau_{\{\text{tag}\}}(P')$. In the base case, when the length of that sequence is 0, there is nothing to prove. For the step case the induction hypothesis is for a sequence of length k:

$$\tau_{\{\text{tag}\}}(P) = \tau_{\{\text{tag}\}}(P_0) \xrightarrow{\wr\wr} \cdots \xrightarrow{\wr\wr} \tau_{\{\text{tag}\}}(P_k)$$

implies $\mathcal{C}(P_0 \parallel Q) \twoheadrightarrow \mathcal{C}(P_k \parallel Q)$. Now consider a sequence $\tau_{\{\text{tag}\}}(P_0) \xrightarrow{\wr\wr} \cdots \xrightarrow{\wr\wr} \tau_{\{\text{tag}\}}(P_{k+1})$ of length $k+1$. By the induction hypothesis $\mathcal{C}(P_0 \parallel Q) \twoheadrightarrow \mathcal{C}(P_k \parallel Q)$. In accordance with the operational rules $P_k \xrightarrow{\{\text{tag}:n\}} P_{k+1}$, with $n \in \mathbb{N}$. Note that by the assumption that P is a tag/sync-process (and P_k as well, since it is reachable from P) we have that $n = 1$. Since $\{\text{tag}\}$ is in the allow set of $\mathcal{C}$, it follows that $\mathcal{C}(P_k \parallel Q) \xrightarrow{\wr\wr} \mathcal{C}(P_{k+1} \parallel Q)$. Hence $\mathcal{C}(P_0 \parallel Q) \twoheadrightarrow \mathcal{C}(P_{k+1} \parallel Q)$. $\square$

The next lemma relates transitions of $\mathcal{C}(P \parallel Q)$ and P and Q.

Lemma 2. *Let* $L(\vec{d} : \vec{D}) = +_{i\in I}\sum_{e_i:E_i} c_i \to \alpha_i \,.\, L(\vec{g}_i)$ *be an LPE and let* $(V, K^V, J^V, c^V, \alpha^V, h^V)$ *and* $(W, K^W, J^W, c^W, \alpha^W, h^W)$ *be a cleave. Let* $P \in \mathsf{S}_{ts}(V)$ *and* $Q \in \mathsf{S}_{ts}(W)$ *be* tag/sync *process expressions. For every transition* $\mathcal{C}(P \parallel Q) \xrightarrow{\omega} \mathcal{C}(P' \parallel Q')$ *we have that either*

- $P \xrightarrow{\omega + \{\mathsf{tag}\}} P'$ *and* $Q = Q'$;
- $Q \xrightarrow{\omega + \{\mathsf{tag}\}} Q'$ *and* $P = P'$ *or*
- $P \xrightarrow{\omega_1 + \{\mathsf{sync}_V^i(\vec{d})\}} P', Q \xrightarrow{\omega_2 + \{\mathsf{sync}_W^i(\vec{d})\}} Q', \omega_1 + \omega_2 = \omega.$

Proof. Suppose $\mathcal{C}(P \parallel Q) \xrightarrow{\omega} \mathcal{C}(P' \parallel Q')$. Unfolding the shorthand $\mathcal{C}$:

$$\tau_{\{\mathsf{tag}\}}\big(\nabla_{\{\mathsf{tag}\}\cup\{\underline{\alpha_i} \mid i\in I\}\cup\{\underline{\alpha_i}|\mathsf{tag} \mid i\in(K^V\cup K^W)\}}\big(\tau_{\{\mathsf{sync}^i \mid i\in I\}}\big($$
$$\Gamma_{\{\mathsf{sync}_V^i|\mathsf{sync}_W^i \to \mathsf{sync}^i \mid i\in I\}}(P \parallel Q)\big)\big)\big) \xrightarrow{\omega} \tau_{\{\mathsf{tag}\}}\big($$
$$\nabla_{\{\mathsf{tag}\}\cup\{\underline{\alpha_i} \mid i\in I\}\cup\{\underline{\alpha_i}|\mathsf{tag} \mid i\in(K^V\cup K^W)\}}\big($$
$$\tau_{\{\mathsf{sync}^i \mid i\in I\}}\big(\Gamma_{\{\mathsf{sync}_V^i|\mathsf{sync}_W^i \to \mathsf{sync}^i \mid i\in I\}}(P' \parallel Q')\big)\big)\big)$$

The last rule applied in the derivation of this transition must be HIDE, with the following premise, for some $n \in \mathbb{N}$:

$$\nabla_{\{\mathsf{tag}\}\cup\{\underline{\alpha_i} \mid i\in I\}\cup\{\underline{\alpha_i}|\mathsf{tag} \mid i\in(K^V\cup K^W)\}}\big(\tau_{\{\mathsf{sync}^i \mid i\in I\}}\big($$
$$\Gamma_{\{\mathsf{sync}_V^i|\mathsf{sync}_W^i \to \mathsf{sync}^i \mid i\in I\}}(P \parallel Q)\big)\big) \xrightarrow{\omega + \{\mathsf{tag}:n\}}$$
$$\nabla_{\{\mathsf{tag}\}\cup\{\underline{\alpha_i} \mid i\in I\}\cup\{\underline{\alpha_i}|\mathsf{tag} \mid i\in(K^V\cup K^W)\}}\big($$
$$\tau_{\{\mathsf{sync}^i \mid i\in I\}}\big(\Gamma_{\{\mathsf{sync}_V^i|\mathsf{sync}_W^i \to \mathsf{sync}^i \mid i\in I\}}(P' \parallel Q')\big)\big)$$

The last rule applied in the derivation of this transition must be ALLOW, with the following premise, and the restriction that $\omega_1 \in \{[\![\alpha]\!] \mid \alpha \in \{\mathsf{tag}\}\cup\{\underline{\alpha_i} \mid i \in I\} \cup \{\underline{\alpha_i}|\mathsf{tag} \mid i \in (K^V \cup K^W)\}\}$:

$$\tau_{\{\mathsf{sync}^i \mid i\in I\}}\big(\Gamma_{\{\mathsf{sync}_V^i|\mathsf{sync}_W^i \to \mathsf{sync}^i \mid i\in I\}}(P \parallel Q)\big) \xrightarrow{\omega_1}$$
$$\tau_{\{\mathsf{sync}^i \mid i\in I\}}\big(\Gamma_{\{\mathsf{sync}_V^i|\mathsf{sync}_W^i \to \mathsf{sync}^i \mid i\in I\}}(P' \parallel Q')\big)$$

The last rule applied in the derivation of this transition must be HIDE, with the following premise.

$$\Gamma_{\{\mathsf{sync}_V^i|\mathsf{sync}_W^i \to \mathsf{sync}^i \mid i\in I\}}(P \parallel Q) \xrightarrow{\omega_1+\omega_2} \Gamma_{\{\mathsf{sync}_V^i|\mathsf{sync}_W^i \to \mathsf{sync}^i \mid i\in I\}}(P' \parallel Q')$$

The same restriction on ω_1 applies. Additionally, there is the restriction that ω_2 consists solely of sync actions, i.e. for all a such that $\underline{\omega_2}(a) > 0$ we have that $a \in \{\mathsf{sync}^i \mid i \in \mathbb{N}\}$.

The last rule applied in the derivation of this transition must be COM, with the following premise.

$$P \parallel Q \xrightarrow{\omega_1 + \omega_3 + \omega_4} P' \parallel Q'$$

The same restriction on ω_1 applies. Additionally, it is required that ω_3 and ω_4 consist of complementing sync actions:

$$\text{For all } a \text{ such that } \underline{\omega_3}(a) > 0 : a \in \{\text{sync}_V^i \mid i \in \mathbb{N}\}$$
$$\text{For all } a \text{ such that } \underline{\omega_4}(a) > 0 : a \in \{\text{sync}_W^i \mid i \in \mathbb{N}\}$$
$$\text{For all } i \in \mathbb{N} \text{ and } \vec{d} \in \mathbb{D}_{\text{sync}_V^i} : \omega_3(\text{sync}_V^i(\vec{d})) = \omega_4(\text{sync}_W^i(\vec{d}))$$

Note that $P \in S_{ts}(V)$ and $Q \in S_{ts}(W)$ are tag/sync process expressions. Hence, any derivation with P or Q at the base contains either a single tag action or a single sync action. We can distinguish three cases for the rule applied in the derivation of $P \parallel Q \xrightarrow{\omega_1 + \omega_3 + \omega_4} P' \parallel Q'$:

- PAR: due to the restrictions on the multiset $\omega_1 + \omega_3 + \omega_4$ and the fact that P and Q are tag/sync process expressions we can exclude a number of possibilities: P and Q cannot both contribute a tag action (since ω_1 contains at most one) and it cannot be the case that either P or Q contributes a tag action and the other a sync action (as syncs must come in matching pairs). The only possibility is that P and Q contribute matching sync actions and ω_1 is split in some way: $P \xrightarrow{\omega_1^1 + \{\text{sync}_V^i(\vec{d})\}} P', Q \xrightarrow{\omega_1^2 + \{\text{sync}_W^i(\vec{d})\}} Q', \omega_1^1 + \omega_1^2 = \omega$.
- PARL: Only P contributes to the derivation. Since P cannot make a step with matching sync actions it must make a step with a single tag action: $P \xrightarrow{\omega + \{\text{tag}\}} P'$ and $Q = Q'$.
- PARR: Only Q contributes to the derivation. Since Q cannot make a step with matching sync actions it must make a step with a single tag action: $Q \xrightarrow{\omega + \{\text{tag}\}} Q'$ and $P = P'$.

$\square$

We are now ready to prove the main contribution of this paper. The theorem states that, in the context $\mathcal{C}(P \parallel Q)$, we can swap out component LPE P with some process expression L if $\tau_{\{\text{tag}\}}(P) \leftrightarrow_b \tau_{\{\text{tag}\}}(L)$.

Theorem 2. *Let $P(\vec{d} : \vec{D}) = +_{i \in I} \sum_{e_i : E_i} c_i \rightarrow \alpha_i . P(\vec{g}_i)$ be an LPE and let $(V, K^V, J^V, c^V, \alpha^V, h^V)$ and $(W, K^W, J^W, c^W, \alpha^W, h^W)$ be a cleave. Let $\vec{\iota} : \vec{D}$ be an arbitrary closed expression. Let $L \in S_{ts}(V)$ and $R \in S_{ts}(W)$ be tag/sync process expressions such that both $\tau_{\{\text{tag}\}}(P_V(\vec{\iota}|_V)) \leftrightarrow_b \tau_{\{\text{tag}\}}(L)$ and $\tau_{\{\text{tag}\}}(P_W(\vec{\iota}|_W)) \leftrightarrow_b \tau_{\{\text{tag}\}}(R)$. We then have the following:*

$$P(\vec{\iota}) \leftrightarrow_b \mathcal{C}(L \parallel R)$$

Proof. By Theorem 1, we have: $P(\bar{\iota}) \leftrightarrow \mathcal{C}(P_V(\bar{\iota}_{|V}) \parallel P_W(\bar{\iota}_{|W}))$ and therefore $P(\bar{\iota}) \leftrightarrow_b \mathcal{C}(P_V(\bar{\iota}_{|V}) \parallel P_W(\bar{\iota}_{|W}))$. It therefore suffices to show that $\mathcal{C}(P_V(\bar{\iota}_{|V}) \parallel P_W(\bar{\iota}_{|W})) \leftrightarrow_b \mathcal{C}(L \parallel R)$, which we prove by showing that

$$\mathcal{R} = \{(\mathcal{C}(P \parallel Q), \mathcal{C}(P' \parallel Q')) \mid P, P' \in \mathsf{S}_{ts}(V) \wedge Q, Q' \in \mathsf{S}_{ts}(W) \text{ tag/sync}$$
$$\text{process expressions s.t. } \tau_{\{\mathsf{tag}\}}(P) \leftrightarrow_b \tau_{\{\mathsf{tag}\}}(P') \wedge \tau_{\{\mathsf{tag}\}}(Q) \leftrightarrow_b \tau_{\{\mathsf{tag}\}}(Q')$$

is a branching bisimulation relation. Note that $P_V(\bar{\iota}_{|V})$ and $P_W(\bar{\iota}_{|W})$ are tag/sync process expressions. To check the transfer conditions of the pairs in $\mathcal{R}$ we pick arbitrary process expressions P, P', Q and Q' meeting the conditions of the definition of $\mathcal{R}$.

Suppose $\mathcal{C}(P \parallel Q) \xrightarrow{\omega} \mathcal{C}(P'' \parallel Q'')$. By Lemma 2 there are 3 options for the contributions of P and Q.

- $P \xrightarrow{\omega + \langle \mathsf{tag}\rangle} P''$ and $Q = Q'$. Since $\tau_{\{\mathsf{tag}\}}(P) \leftrightarrow_b \tau_{\{\mathsf{tag}\}}(P')$, there exist S and S' such that

$$\tau_{\{\mathsf{tag}\}}(P') \twoheadrightarrow \tau_{\{\mathsf{tag}\}}(S) \xrightarrow{(\omega)} \tau_{\{\mathsf{tag}\}}(S')$$

  with both $\tau_{\{\mathsf{tag}\}}(P) \leftrightarrow_b \tau_{\{\mathsf{tag}\}}(S)$ and $\tau_{\{\mathsf{tag}\}}(P'') \leftrightarrow_b \tau_{\{\mathsf{tag}\}}(S')$. By Lemma 1 we have that $\mathcal{C}(P' \parallel Q') \twoheadrightarrow \mathcal{C}(S \parallel Q')$. Since P' is a tag/sync process, so is S; hence a single **tag** action is hidden by the application of the rule HIDE in the derivation of the transition $\tau_{\{\mathsf{tag}\}}(S) \xrightarrow{(\omega)} \tau_{\{\mathsf{tag}\}}(S')$. Therefore $S \xrightarrow{(\omega + \langle \mathsf{tag}\rangle)} S'$. Hence $\mathcal{C}(S \parallel Q') \xrightarrow{(\omega)} \mathcal{C}(S' \parallel Q')$. Due to our definition of $\mathcal{R}$, $\tau_{\{\mathsf{tag}\}}(P) \leftrightarrow_b \tau_{\{\mathsf{tag}\}}(S)$ and $\tau_{\{\mathsf{tag}\}}(P'') \leftrightarrow_b \tau_{\{\mathsf{tag}\}}(S')$ we have that

$$(\mathcal{C}(P \parallel Q), \mathcal{C}(S \parallel Q')) \in \mathcal{R} \text{ and } (\mathcal{C}(P'' \parallel Q), \mathcal{C}(S' \parallel Q')) \in \mathcal{R}.$$

The transfer conditions of Definition 4 are hereby satisfied.
- $Q \xrightarrow{\omega + \langle \mathsf{tag}\rangle} Q'$ and $P = P'$: completely symmetric to the first case.
- $P \xrightarrow{\omega_1 + \langle \mathsf{sync}_V^i(\vec{d})\rangle} P''$, $Q \xrightarrow{\omega_2 + \langle \mathsf{sync}_W^i(\vec{d})\rangle} Q''$, and $\omega = \omega_1 + \omega_2$. Since we have that $\tau_{\{\mathsf{tag}\}}(P) \leftrightarrow_b \tau_{\{\mathsf{tag}\}}(P')$ there must exist S and S' such that

$$\tau_{\{\mathsf{tag}\}}(P') \twoheadrightarrow \tau_{\{\mathsf{tag}\}}(S) \xrightarrow{\omega_1 + \langle \mathsf{sync}_V^i(\vec{d})\rangle} \tau_{\{\mathsf{tag}\}}(S')$$

  with $\tau_{\{\mathsf{tag}\}}(P) \leftrightarrow_b \tau_{\{\mathsf{tag}\}}(S)$ and $\tau_{\{\mathsf{tag}\}}(P'') \leftrightarrow_b \tau_{\{\mathsf{tag}\}}(S')$. Furthermore, since $\tau_{\{\mathsf{tag}\}}(Q) \leftrightarrow_b \tau_{\{\mathsf{tag}\}}(Q')$ there must exist T and T' such that

$$\tau_{\{\mathsf{tag}\}}(Q') \twoheadrightarrow \tau_{\{\mathsf{tag}\}}(T) \xrightarrow{\omega_2 + \langle \mathsf{sync}_W^i(\vec{d})\rangle} \tau_{\{\mathsf{tag}\}}(T')$$

  with $\tau_{\{\mathsf{tag}\}}(Q) \leftrightarrow_b \tau_{\{\mathsf{tag}\}}(T)$ and $\tau_{\{\mathsf{tag}\}}(Q'') \leftrightarrow_b \tau_{\{\mathsf{tag}\}}(T')$.

By Lemma 1 we have that $\mathcal{C}(P' \parallel Q') \twoheadrightarrow \mathcal{C}(S \parallel T)$. Since both $S \xrightarrow{\omega_1 + \{\mathsf{sync}_V^i(\vec{d})\}}$ S' and $T \xrightarrow{\omega_2 + \{\mathsf{sync}_W^i(\vec{d})\}} T'$ (the restrictions on the alphabet of P' and Q' prohibit extra tags in the multi-actions), we have that $\mathcal{C}(S \parallel T) \xrightarrow{\omega_1 + \omega_2}$ $\mathcal{C}(S' \parallel T')$. Due to the definition of $\mathcal{R}$ we also have that

$$(\mathcal{C}(P \parallel Q), \mathcal{C}(S \parallel T)) \in \mathcal{R} \text{ and } (\mathcal{C}(P'' \parallel Q''), \mathcal{C}(S' \parallel T')) \in \mathcal{R}.$$

The transfer conditions of Definition 4 are satisfied.

$\square$

5   Minimisation

Theorem 2 establishes sufficient conditions for correct replacement of a component; we are interested in replacing a component by one that is minimal with respect to branching bisimilarity. In this section we will formalise the process shown in Example 4, in which we hide the tag actions, minimise the LTS and reintroduce tag actions for all transitions that do not have a sync label.

Definition 13. *We extend the process algebra of Definition 7 with a tagging operator $T_H(P)$, where $H \subseteq \Lambda$ is a non-empty finite set of action labels. Let the extended grammar be denoted by S_t and let S_t denote the set of all process expressions that can be constructed from the grammar.*

Definition 14. *We presuppose some LTS $(St, \Omega, \to_1)$. We associate an LTS $(\mathsf{S}_t, \Omega, \to)$ to expressions of S_t, where S_t is parametrised with constants St. The relation $\to$ is the least relation including the transition relation $\to_1$, the rules of Definition 10 and the rules below. For any $\omega \in \Omega$, expressions $P, P' \in \mathsf{S}_t$ and $H \subseteq \Lambda$:*

$$\frac{P \xrightarrow{\omega} P'}{T_H(P) \xrightarrow{\omega} T_H(P')} \; \omega \cap H \neq \emptyset \qquad \frac{P \xrightarrow{\omega} P'}{T_H(P) \xrightarrow{\omega + \{\mathsf{tag}\}} T_H(P')} \; \omega \cap H = \emptyset$$

When the tagging operator is enclosed by a hiding operator that hides tag actions, the tagging operator has no effect, see Lemma 3.

Lemma 3. *For all $H \subseteq \Lambda$ and $P \in \mathsf{S}$: $\tau_{\{\mathsf{tag}\}}(P) \leftrightarrow \tau_{\{\mathsf{tag}\}}(T_H(P))$.*

Proof. Trivial. For any step $P \xrightarrow{\omega} P'$, both $\tau_{\{\mathsf{tag}\}}(P) \xrightarrow{\omega - \{\mathsf{tag}\}} \tau_{\{\mathsf{tag}\}}(P')$ and $\tau_{\{\mathsf{tag}\}}(T_H(P)) \xrightarrow{\omega - \{\mathsf{tag}\}} \tau_{\{\mathsf{tag}\}}(T_H(P'))$. $\square$

Theorem 3 (below) states that the procedure of hiding, minimising and tagging yields components that can be composed in context $\mathcal{C}$ whilst preserving branch bisimilarity. Processes L and R can be chosen to be minimal with respect to branching bisimilarity.

Theorem 3. *Let* $P(\vec{d} : \vec{D}) = +_{i \in I} \sum_{e_i : E_i} c_i \rightarrow \alpha_i \cdot P(\vec{g}_i)$ *be an LPE and let* $(V, K^V, J^V, c^V, \alpha^V, h^V)$ *and* $(W, K^W, J^W, c^W, \alpha^W, h^W)$ *be a cleave. Let* $\vec{\iota} : \vec{D}$ *be an arbitrary closed expression. Let* L *and* R *be process expressions such that they are branching bisimilar to* $\tau_{\{\mathsf{tag}\}}(P_V(\vec{\iota}|_V))$ *and* $\tau_{\{\mathsf{tag}\}}(P_W(\vec{\iota}|_W))$, *respectively. Let* $H = \{\mathsf{sync}_V^i \mid i \in \mathbb{N}\} \cup \{\mathsf{sync}_W^i \mid i \in \mathbb{N}\}$.

$$P(\iota) \underline{\leftrightarrow}_b \mathcal{C}(T_H(L) \parallel T_H(R))$$

Proof. Note that $T_H(L)$ and $T_H(R)$ are tag/sync processes: both have either a tag or sync label in every reachable transition. By Lemma 3

$$\tau_{\{\mathsf{tag}\}}(T_H(L)) \underline{\leftrightarrow} \tau_{\{\mathsf{tag}\}}(L) \text{ and } \tau_{\{\mathsf{tag}\}}(T_H(R)) \underline{\leftrightarrow} \tau_{\{\mathsf{tag}\}}(R).$$

As $L \underline{\leftrightarrow}_b \tau_{\{\mathsf{tag}\}}(P_V(\vec{\iota}|_V))$ and $R \underline{\leftrightarrow}_b \tau_{\{\mathsf{tag}\}}(P_W(\vec{\iota}|_W))$ we have that

$$\tau_{\{\mathsf{tag}\}}(T_H(L)) \underline{\leftrightarrow}_b \tau_{\{\mathsf{tag}\}}(P_V(\vec{\iota}|_V)) \text{ and } \tau_{\{\mathsf{tag}\}}(T_H(R)) \underline{\leftrightarrow}_b \tau_{\{\mathsf{tag}\}}(P_W(\vec{\iota}|_W)).$$

The requirements of Theorem 2 are satisfied, finishing the proof. ☐

The tool `lpscombine` has been extended with the option to automatically add a tag label to every transition that does not contain a sync label in the LTSs it is combining.

6   Experimental Results

Our motivation for studying the extension to branching bisimulation stems from our interest in verifying mCRL2 models that result from an automated translation of SysML State Machines (SMs). In those models there is a natural notion of component (a state machine), and, due to the way the semantics of SMs is defined in mCRL2, these components exhibit quite some internal behaviour. Hence, a significant reduction of the size of the state spaces associated with the components can be achieved by branching bisimulation minimisation. In this section we report on the effect of decompositional branching bisimulation minimisation on several such models. First, we briefly describe the context in which the models were obtained, then we present the results of applying compositional minimisation on them. Finally, we compare compositional minimisation to another technique that aims to support model checking of large models: symbolic model checking.

EULYNX[1] is an initiative of European railway infrastructure managers to develop standardised interfaces for the communication between trackside equipment (points, light signals, etc.) and the interlocking (the device responsible for safe route setting). The EULYNX standard defines the behaviour of the various interfaces through SysML internal block diagrams and SysML SMs. To facilitate

---

[1] https://eulynx.eu.

formal verification of the behaviour of the interfaces, in the FormaSig project[2] the SysML models of the interfaces are translated into mCRL2 (see [1]). A EUL-YNX interface consists of a *generic* part, which is the same for all interfaces, and a *specific* part only relevant for the interface at hand. We ran our experiments on the generic part of the interfaces, on the interface-specific parts of the point interface (SCI-P) and the level crossing interface (SCI-LC), on the combined generic and specific parts of the level crossing interface (full SCI-LC) and on an alternative version of the level crossing interface used in some countries (SCI-LX).

A EULYNX interface consists of several SMs. In the mCRL2 model resulting from the translation of the EULYNX interface, the SMs can be recognised as components, and thus there is a natural partitioning of the parameters resulting in a suitable cleave. Branching bisimulation minimisation is effective on these components (see Table 1). The reason is that the formalisation of SysML state machine semantics involves a computation to determine which state machine transitions can be executed next, and this computation can be deemed internal.

Table 1. Metrics on the state spaces of components of the full SCI-LC interface.

Component	#states	#states (modulo $\leftrightarrow$)	#states (modulo $\leftrightarrow_b$)
scpPrim	150	76	31
scpSec	174	73	24
prim/S_SCI_EfeS_Prim_SR	259	104	38
sec/F_SCI_EfeS_Sec_SR	636	137	37
est/F_EST_EfeS_SR	561	212	77
smi/F_SMI_EfeS_SR	1788	166	51
flc/F_SCI_LC_SR	20,127	10,052	2,373
slc/S_SCI_LC_SR	17,378	11,011	2,058
functions/F_LC_Functions_SR	432,954	64,639	19,940

The tools `lpscleave` and `lpscombine` always work on two components. To support multiple components we cleave recursively, in each step splitting off one component; Combining is performed in reverse order. Table 2 shows by how much the state space is reduced using branching bisimulation minimisation, the number of states for the monolithic models are computed using the symbolic exploration tool `lpsreach`, which is part of the mCRL2 toolset. The reduction achieved by intermediate minimisation is 3 to 6 orders of magnitude and scales with the number of components. For the full SCI-LC model, neither `lpsreach` nor the compositional approach is able to explore it within a day. For most models, intermediate strong bisimulation minimisation was not sufficient to make

---

[2] https://fsa.win.tue.nl/formasig; for more information see also [2].

state space generation tractable. We were only able to obtain the state space of the generic model, which consists of 1.09696×10^9 states.

Table 2. Metrics reduction factor by intermediate branching bisimulation minimisation.

Model	#states monolithic	#states compositional	Reduction factor	#components
Generic	4.90232×10^{10}	1.10505×10^6	4.4363×10^4	6
SCI-LC specific	2.71212×10^{12}	2.59611×10^9	1.0556×10^3	3
SCI-LX	1.33539×10^{12}	1.25395×10^6	1.0649×10^6	6
SCI-P specific	1.21299×10^{12}	1.45627×10^7	8.3294×10^4	3
Full SCI-LC	$>3 \times 10^{18}$	Unknown	Unknown	9

Table 3 shows how much time each step of the compositional approach takes. Linearisation and cleaving are quick operations. State space exploration and minimisation of components can be performed in parallel, making it independent of the number of components. The time to combine the LTSs depends on the size of the final LTS. Any LTS small enough for model checking (less than a billion states) can be generated within approximately a day. All benchmarks were performed on a machine equipped with 4 12-core Intel Xeon Gold 6136 CPUs and 3TB of RAM.

Table 3. Metrics state space exploration compositional approach.

Model	Linearisation + cleaving	Generating + minimising LTSs	Combining LTSs	Total
Generic	5 s	166 s	65 s	236 s
SCI-LC specific	9 s	2883 s	93744 s	96636 s
SCI-LX	7 s	181 s	273 s	460 s
SCI-P specific	3 s	557 s	369 s	929 s

Requirement verification. In an earlier case study of the EULYNX point interface [2] we hit the limits of the model checking capabilities of mCRL2. For the SCI-P specific model we could not generate the entire state space. mCRL2 offers both symbolic model checking [14] and explicit state model checking. The symbolic tools do not explore the state space explicitly; they therefore typically provide much better scalability. A benefit of the explicit tools is that they can provide counterexamples [4,20]. For the point interface we used the symbolic model checking tools but they were still not able to verify any of the requirements.

Below we review whether compositional minimisation combined with the explicit state model checking tools offer better performance than the symbolic model checking tools. In mCRL2, a formula is verified by constructing and subsequently solving a Parametrised Boolean Equation System (PBES) from the combination of the formula and the LPS or LTS. Table 4 shows the time it takes to verify the formula $[true*]\langle true\rangle true$, expressing that every reachable state has some outgoing transition; which we can check for all models since it does not depend on specific action labels. The symbolic tool `pbessolvesymbolic` is applied to a PBES derived from the LPS. The explicit tools are applied to a PBES obtained from the LTS, which was obtained from the compositional approach.

Table 4. Metrics on the time needed for verification using various approaches. Statistics of the explicit tool are provided with and without counterexample generation enabled. The tools add PBES equations to be able to extract a counterexample at the end, which slows down the solver significantly, as can be seen.

Model	Symbolic	Explicit	Explicit + Counter
Generic	>7 days	112 s	3169 s
SCI-LC specific	>7 days	>7 days	>7 days
SCI-LX	>7 days	116 s	4210 s
SCI-P specific	>7 days	1854 s	69621 s

For more involved properties our experience is similar; The explicit verification tool paired with compositional minimisation outperforms the symbolic model checking tool, with the extra benefit of providing counterexamples. The explicit tools cannot be used without compositional minimisation due to the size of the state spaces (see Table 2).

Applied on the examples considered in [16] to illustrate the effect of decompositional strong bisimulation minimisation, the extension to branching bisimulation minimisation presented here does not lead to significantly smaller components. The reason is that the components in those examples have hardly any internal behaviour.

7   Conclusion

We have shown that we are able to reduce the state space induced by the formalised EULYNX models by several orders of magnitude. The extension of `lpscombine` should also be effective for other models with similar features. When only communications between components are renamed to τ, component based minimisation does not reduce the state space. In our models three factors come together enabling a large reduction in the state space:

1. All τ transitions are local to a single component;

2. Most τ transitions are inert;
3. It is difficult to adapt the model to prevent inert τ transitions.

It is likely that other behavioural equivalences (such as weak bisimulation and divergence preserving branching bisimulation) can also be used to safely minimise the LTS of components in which **tag** actions have been hidden. Extending the theory further would entail repeating the proof of Theorem 2 for other behavioural equivalences, which is left as future work.

Acknowledgements. The first author received funding for his research from ProRail and DB Netz AG. The vision presented in this article does not necessarily reflect the strategy of DB Netz AG or ProRail, but reflects the personal views of the authors.

References

1. Bouwman, M., Luttik, B., van der Wal, D.: A formalisation of SysML state machines in mCRL2. In: Peters, K., Willemse, T.A.C. (eds.) FORTE 2021. LNCS, vol. 12719, pp. 42–59. Springer, Cham (2021). https://doi.org/10.1007/978-3-030-78089-0_3
2. Bouwman, M., van der Wal, D., Luttik, B., Stoelinga, M., Rensink, A.: A case in point: verification and testing of a EULYNX interface. Formal Asp. Comput. (2022). https://doi.org/10.1145/3528207
3. Bunte, O., et al.: The mCRL2 toolset for analysing concurrent systems. In: Vojnar, T., Zhang, L. (eds.) TACAS 2019. LNCS, vol. 11428, pp. 21–39. Springer, Cham (2019). https://doi.org/10.1007/978-3-030-17465-1_2
4. Cranen, S., Luttik, B., Willemse, T.A.C.: Evidence for fixpoint logic. In: Kreutzer, S. (ed.) 24th EACSL Annual Conference on Computer Science Logic, CSL 2015, Berlin, Germany, 7–10 September 2015. LIPIcs, vol. 41, pp. 78–93. Schloss Dagstuhl - Leibniz-Zentrum für Informatik (2015). https://doi.org/10.4230/LIPIcs.CSL.2015.78
5. Garavel, H., Lang, F., Mateescu, R., Serwe, W.: CADP 2011: a toolbox for the construction and analysis of distributed processes. Int. J. Softw. Tools Technol. Transf. **15**(2), 89–107 (2013). https://doi.org/10.1007/s10009-012-0244-z
6. Garavel, H., Lang, F., Mounier, L.: Compositional verification in action. In: Howar, F., Barnat, J. (eds.) FMICS 2018. LNCS, vol. 11119, pp. 189–210. Springer, Cham (2018). https://doi.org/10.1007/978-3-030-00244-2_13
7. Glabbeek, R.J.: The linear time — branching time spectrum II. In: Best, E. (ed.) CONCUR 1993. LNCS, vol. 715, pp. 66–81. Springer, Heidelberg (1993). https://doi.org/10.1007/3-540-57208-2_6
8. van Glabbeek, R.J., Weijland, W.P.: Branching time and abstraction in bisimulation semantics. J. ACM **43**(3), 555–600 (1996). https://doi.org/10.1145/233551.233556
9. Graf, S., Steffen, B.: Compositional minimization of finite state systems. In: Clarke, E.M., Kurshan, R.P. (eds.) CAV 1990. LNCS, vol. 531, pp. 186–196. Springer, Heidelberg (1991). https://doi.org/10.1007/BFb0023732
10. Graf, S., Steffen, B., Lüttgen, G.: Compositional minimisation of finite state systems using interface specifications. Formal Asp. Comput. **8**(5), 607–616 (1996). https://doi.org/10.1007/BF01211911

11. Groote, J.F., Mousavi, M.R.: Modeling and Analysis of Communicating Systems. MIT Press (2014). https://mitpress.mit.edu/books/modeling-and-analysis-communicating-systems
12. Keiren, J.J.A., Klabbers, M.: Modelling and verifying IEEE Std 11073-20601 session setup using mCRL2. Electron. Commun. Eur. Assoc. Softw. Sci. Technol. **53** (2012). https://doi.org/10.14279/tuj.eceasst.53.793
13. Lang, F.: Refined interfaces for compositional verification. In: Najm, E., Pradat-Peyre, J.-F., Donzeau-Gouge, V.V. (eds.) FORTE 2006. LNCS, vol. 4229, pp. 159–174. Springer, Heidelberg (2006). https://doi.org/10.1007/11888116_13
14. Laveaux, M., Wesselink, W., Willemse, T.A.C.: On-the-fly solving for symbolic parity games. In: TACAS 2022. LNCS, vol. 13244, pp. 137–155. Springer, Cham (2022). https://doi.org/10.1007/978-3-030-99527-0_8
15. Laveaux, M., Willemse, T.A.C.: Decompositional minimisation of monolithic processes. CoRR abs/2012.06468 (2020). https://arxiv.org/abs/2012.06468
16. Laveaux, M., Willemse, T.A.C.: Decomposing monolithic processes in a process algebra with multi-actions. In: Lange, J., Mavridou, A., Safina, L., Scalas, A. (eds.) Proceedings 14th Interaction and Concurrency Experience, ICE 2021, 18 June 2021. EPTCS, vol. 347, pp. 57–76 (2021). https://doi.org/10.4204/EPTCS.347.4
17. Object Managament Group: OMG System Modeling Language, version 1.6 (2019). https://www.omg.org/spec/SysML/1.6/
18. Park, D.: Concurrency and automata on infinite sequences. In: Deussen, P. (ed.) GI-TCS 1981. LNCS, vol. 104, pp. 167–183. Springer, Heidelberg (1981). https://doi.org/10.1007/BFb0017309
19. Remenska, D., Willemse, T.A.C., Verstoep, K., Templon, J., Bal, H.E.: Using model checking to analyze the system behavior of the LHC production grid. Future Gener. Comput. Syst. **29**(8), 2239–2251 (2013). https://doi.org/10.1016/j.future.2013.06.004
20. Wesselink, W., Willemse, T.A.C.: Evidence extraction from parameterised boolean equation systems. In: Benzmüller, C., Otten, J. (eds.) Proceedings of the 3rd International Workshop on Automated Reasoning in Quantified Non-Classical Logics (ARQNL 2018) affiliated with the International Joint Conference on Automated Reasoning (IJCAR 2018), Oxford, UK, 18 July 2018. CEUR Workshop Proceedings, vol. 2095, pp. 86–100. CEUR-WS.org (2018). http://ceur-ws.org/Vol-2095/paper6.pdf

Types and Choreographies

Realisability of Branching Pomsets

Luc Edixhoven[1,2(✉)] 🆔 and Sung-Shik Jongmans[1,2] 🆔

[1] Open University, Heerlen, Netherlands
{led,ssj}@ou.nl
[2] Centrum Wiskunde & Informatica (CWI), Amsterdam, Netherlands

Abstract. A communication protocol is realisable if it can be faithfully implemented in a distributed fashion by communicating agents. Pomsets offer a way to compactly represent concurrency in communication protocols and have been recently used for the purpose of realisability analysis. In this paper we focus on the recently introduced branching pomsets, which also compactly represent choices. We define well-formedness conditions on branching pomsets, inspired by multiparty session types, and we prove that the well-formedness of a branching pomset is a sufficient condition for the realisability of the represented communication protocol.

Keywords: Realisability · Pomsets · Choreographies

1 Introduction

Designing and implementing distributed systems is difficult. They are becoming ever more important, and yet the complexity resulting from combinations of inter-participant concurrency and dependencies makes the process error-prone and debugging non-trivial. As a consequence, much research has been dedicated to analysing communication patterns, or protocols, in distributed systems. Examples of such research goals are to show the presence or absence of certain safety properties in a given system, to automate such analysis or to guarantee the presence of desirable properties by construction. We are interested in particular in the *realisability* property, i.e., whether a global specification of a protocol can be faithfully implemented in a distributed fashion in the first place. This problem has been well-studied in the last two decades in a variety of settings [2,3,16,25,32].

A recent development has been the use of *partially ordered multisets*, or *pomsets*, for the purpose of realisability analysis [18]. Guanciale and Tuosto use pomsets as a syntax-oblivious specification model for communication protocols and define conditions that ensure realisability directly over pomsets. Pomsets offer a way to compactly represent concurrent behaviour. By using a partial order to explicitly capture causal dependencies between pairs of actions, they avoid the exponential blowup from finite state machines. However, a single pomset does not offer any means to represent choices. Instead, a choice is represented as a set of pomsets, one for each possible branch. Multiple choices result in one pomset for each possible combination of branches, which can yield an exponentially large set of pomsets. Recent work by the authors and Proença and Cledou introduces *branching pomsets* [13]. These extend pomsets with a branching structure to compactly represent both concurrency and choices, avoiding both exponential blowups. In this paper we present a first step in the analysis of the realisability of these branching pomsets.

We consider distributed systems using asynchronous message passing and ordered buffers (FIFO queues) between (ordered pairs of) participants. In the aforementioned paper ([18]) Guanciale and Tuosto consider systems with unordered (non-FIFO) buffers. Using the work by Guanciale and Tuosto as a basis for our analysis would require first adapting their work to a FIFO setting and then adapting it further to branching pomsets. Instead, we choose to draw inspiration from multiparty session types (MPST) [20], which already use FIFO buffers, and thus use MPST as a basis for our analysis. Through its syntax and projection operator, MPST defines a number of well-formedness conditions on global types which ensure realisability. We define similar well-formedness conditions on branching pomsets and prove that they ensure realisability as well. These conditions are sufficient but not necessary, i.e., a protocol may be realisable without being well-formed. We discuss some possible relaxations of the conditions at the end of the paper.

Outline. We recall the concept and definition of branching pomsets in Sect. 2. In Sect. 3 we define our notion of realisability, we define our well-formedness conditions in Sect. 4 and we prove in Sect. 5 that, if a branching pomset is well-formed, then the corresponding communication protocol is realisable. In Sect. 6 we briefly discuss two examples. Finally, we discuss related work in Sect. 7 and we end the paper with our conclusions and a discussion in Sect. 8.

We omit a number of technical lemmas and the majority of proofs in favour of more informal proof sketches or highlights. All omitted content can be found in a separate technical report [12].

2   Preliminaries on Branching Pomsets

In this section we recall the concept and definitions of branching pomsets. This section is heavily based on the original introduction of branching pomsets [13]; a more thorough explanation can be found in that paper.

Notation. Let $\mathcal{A} = \{a, b, \ldots\}$ be the set of participants (or agents). Let $\mathcal{X} = \{x, y, \ldots\}$ be the set of message types. Let $\mathcal{L} = \bigcup_{a,b \in \mathcal{A}, x \in \mathcal{X}} \{ab!x, ab?x\}$ be the set of labels (actions), ranged over by ℓ, where $ab!x$ is a send action from a to b of a message of type x, and $ab?x$ is the corresponding receive action by b. The *subject* of an action ℓ, written $subj(\ell)$, is its active agent: $subj(ab!x) = a$ and $subj(ab?x) = b$.

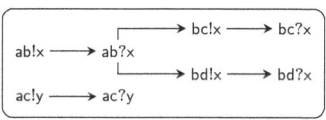 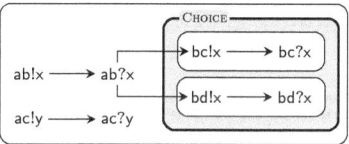

Fig. 1. A pomset (left) and a branching pomset (right) depicting simple communication patterns.

A partially ordered multiset [28], or pomset for short, consists of a set of nodes (events) E, a labelling function λ mapping events to a set of labels (e.g., send and receive actions), and a partial order $\leq$ defining causal dependencies between pairs of events (i.e., an event, or rather its corresponding action, can only fire if all events preceding it in the partial order have already fired). Its behaviour is the set of all sequences of the labels of its events that abide by $\leq$.

An example pomset is shown graphically in Fig. 1 (left), depicting a simple communication pattern between four agents Alice (a), Bob (b), Carol (c) and Dave (d). Alice first sends a message of type x to Bob, who then sends a message of type x to both Carol and Dave. In parallel, Alice sends a message of type y to Carol. The graphical pomset representation shows the labels of the events and the arrows visualising the partial order: an event precedes any other event to which it has an outgoing arrow, either directly or transitively. Formally, $E = \{e_1, \ldots, e_8\}$, λ is such that $e_1, \ldots, e_8$ map to respectively ab!x, ab?x, bc!x, bc?x, bd!x, bd?x, ac!y, ac?y, and $\leq = \{(e_i, e_j) \mid i \leq j \wedge (i, j \in \{1, 2, 3, 4\} \vee i, j \in \{1, 2, 5, 6\} \vee i, j \in \{7, 8\})\}$.

Branching pomsets extend pomsets with a branching structure, which is a tree structure containing events and (binary) choices. Formally, the branching structure is defined below as a tree with root node $\mathcal{B}$, whose children $\mathcal{C}$ are either a single event e or a choice node with children $\mathcal{B}_1, \mathcal{B}_2$. All leaves are events.

$$\mathcal{B} ::= \{\mathcal{C}_1, \ldots, \mathcal{C}_n\}$$
$$\mathcal{C} ::= e \mid \{\mathcal{B}_1, \mathcal{B}_2\}$$

We write $\mathcal{B}_1 \preceq \mathcal{B}_2$ if $\mathcal{B}_1$ is a subtree of $\mathcal{B}_2$, and $\mathcal{B}_1 \prec \mathcal{B}_2$ if $\mathcal{B}_1$ is a strict subtree of $\mathcal{B}_2$, i.e., if $\mathcal{B}_1 \preceq \mathcal{B}_2$ and $\mathcal{B}_1 \neq \mathcal{B}_2$. We use the same notation for $\mathcal{C}$s, es (a special case of $\mathcal{C}$s) and combinations of all the aforementioned.

We formally define branching pomsets in Definition 1.

Definition 1 (Branching pomset [13]). *A branching pomset is a four-tuple* $R = \langle E, \leq, \lambda, \mathcal{B} \rangle$, *where:*

- E *is a set of events;*
- $\leq \subseteq E \times E$ *is a causality relation on events such that* $\leq^*$ *(the transitive closure of* $\leq$*) is a partial order on events;*
- $\lambda : E \mapsto \mathcal{L}$ *is a labelling function assigning an action to every event; and*
- $\mathcal{B}$ *is a branching structure such that the set of leaves of* $\mathcal{B}$ *is* E *and no event in* E *occurs in* $\mathcal{B}$ *more than once.*

We use $R.E$, $R.\leq$, $R.\lambda$ and $R.\mathcal{B}$ to refer to the components of R. We generally omit the prefix if doing so causes no confusion. We also write $e_1 < e_2$ if $e_1 \leq e_2$ and $e_1 \neq e_2$. We say that two events e_1 and e_2 are *causally ordered* if either $e_1 \leq e_2$ or $e_2 \leq e_1$. We say that two events e_1 and e_2 are *mutually exclusive* if there exists some $\mathcal{C} = \{\mathcal{B}_1, \mathcal{B}_2\} \prec R.\mathcal{B}$ such that $e_1 \prec \mathcal{B}_1$ and $e_2 \prec \mathcal{B}_2$.

An example branching pomset is shown graphically in Fig. 1 (right). It depicts the same communication pattern as that in the pomset on the left, except that now Bob sends a message of type x to *either* Carol or Dave instead of to both. This is visualised as a choice box containing two branches. Formally, E, λ and $\leq$ are the same as before. New is $\mathcal{B} = \{e_1, e_2, e_7, e_8, \mathcal{C}\}$, where $\mathcal{C} = \{\{e_3, e_4\}, \{e_5, e_6\}\}$ is Bob's choice between the events corresponding to Carol and Dave. The choice can be resolved by picking one of the branches, e.g., Carol's ($\{e_3, e_4\}$), and merging it with $\mathcal{C}$'s parent, $\mathcal{B}$, resulting in $\mathcal{B}' = \{e_1, e_2, e_3, e_4, e_7, e_8\}$.

Informally, to fire an event whose ancestor in the branching structure is a choice, first the choice must be resolved: it is replaced by one of its children (branches). The other child is discarded and the branching pomset is updated accordingly: the events contained in the discarded subtree are removed, as well as the related entries in the causality relation and the labelling function.

Formally, the semantics of branching pomsets are defined using a *refinement* relation on the branching structure. A structure $\mathcal{B}$ can refine to $\mathcal{B}'$, written $\mathcal{B} \sqsupseteq \mathcal{B}'$, by resolving a number of choices as above. We write $\mathcal{B} \sqsupset \mathcal{B}'$ to specify that $\mathcal{B} \neq \mathcal{B}'$. The refinement rules are formalised in Fig. 2a. The first two rules state that refinement is reflexive and transitive. The third rule, CHOICE, resolves choices. It states that we can replace a choice with one of its branches. Finally, the fourth rule overloads the refinement notation to also apply to branching pomsets themselves: if $R.\mathcal{B}$ can refine to some $\mathcal{B}'$, then R itself can refine to a derived branching pomset with branching structure $\mathcal{B}'$, whose events are restricted to those occurring in $\mathcal{B}'$ and likewise for $\leq$ and λ — as defined in Fig. 2c. We note that we omit one of the rules in [13], since our later well-formedness conditions lead to it never being used.

The reduction and termination rules are defined in Fig. 2b. The first rule states that a branching pomset can terminate if its branching structure can refine to the empty set. The second rule states the conditions for *enabling* an event e, written $R \xrightarrow{\checkmark e} R'$: R can enable e by refining to R' if e is both *minimal* and *active* in R' and if there is no other refinement between R and R' for which this is the case. An event e is minimal if there exists no other event e' such that $e' < e$. It is active if it is not inside a choice, i.e., if $e \in R.\mathcal{B}$. In other words, R may only refine as far as strictly necessary to enable e. Finally, the last two rules state that, if $R \xrightarrow{\checkmark e} R'$, then R can fire e by reducing to $R' - e$, which is the branching pomset obtained by removing e from R' — as defined in Fig. 2c. This reduction is defined both on e's label and on the event itself, the latter for internal use in proofs since $\lambda(e)$ is not necessarily unique but e always is.

For example, let R be the branching pomset (right) in Fig. 1. Its initial active events are those labelled with ab!x, ab!y, ab?x and ab?y, of which the first two are also minimal. After firing either one, the corresponding receive action becomes minimal. In these cases $R \xrightarrow{\checkmark_e} R$ for the relevant event e, i.e., it suffices to refine R to itself. After firing ab?x the two events labelled with bc!x and bd!x become minimal but not yet active. Only now are we allowed to resolve the choice by applying CHOICE to pick one of the branches. After this either the events labelled with bc!x and bc?x or those with bd!x and bd?x will become active, and the first event of the chosen pair can be fired.

$$\frac{}{\mathcal{B} \sqsupseteq \mathcal{B}}[\text{REFL}] \quad \frac{\mathcal{B} \sqsupseteq \mathcal{B}' \sqsupseteq \mathcal{B}''}{\mathcal{B} \sqsupseteq \mathcal{B}''}[\text{TRANS}] \quad \frac{i \in \{1,2\}}{\{\{\mathcal{B}_1, \mathcal{B}_2\}\} \cup \mathcal{B} \sqsupseteq \mathcal{B}_i \cup \mathcal{B}}[\text{CHOICE}] \quad \frac{R.\mathcal{B} \sqsupseteq \mathcal{B}'}{R \sqsupseteq R[\mathcal{B}']}$$

(a) Refinement rules, where we assume for CHOICE that $\{\mathcal{B}_1, \mathcal{B}_2\} \notin \mathcal{B}$.

$$\frac{R \sqsupseteq R' \quad e \in \text{a-min}(R')}{R.\mathcal{B} \sqsupseteq \emptyset} \quad \frac{\forall R'' : R \sqsupseteq R'' \sqsupseteq R' \Rightarrow e \notin \text{a-min}(R'')}{R \xrightarrow{\checkmark_e} R'} \quad \frac{R \xrightarrow{\checkmark_e} R'}{R \xrightarrow{e} R' - e} \quad \frac{R \xrightarrow{e} R'}{R \xrightarrow{\lambda(e)} R'}$$

(b) Reduction and termination rules.

$$\langle E, \leq, \lambda, \mathcal{B} \rangle [\mathcal{B}'] = \langle E|_{\mathcal{B}'}, \leq|_{\mathcal{B}'}, \lambda|_{\mathcal{B}'}, \mathcal{B}' \rangle$$
$$X|_{\mathcal{B}} = \text{restricts } X \text{ only to the events in } \mathcal{B}$$
$$\text{a-min}(R) = \{e \in R.E \mid \nexists e' \in R.E : e' < e\} \wedge e \in R.\mathcal{B}$$
$$\hat{e} - e = \hat{e}$$
$$\{\mathcal{C}_1, \ldots, \mathcal{C}_n\} - e = \begin{cases} \{\mathcal{C}_1, \ldots, \mathcal{C}_{i-1}, \mathcal{C}_{i+1}, \ldots, \mathcal{C}_n\} & \text{if } \mathcal{C}_i = e \\ \{\mathcal{C}_1 - e, \ldots, \mathcal{C}_n - e\} & \text{otherwise} \end{cases}$$
$$\{\mathcal{B}_1, \mathcal{B}_2\} - e = \{\mathcal{B}_1 - e, \mathcal{B}_2 - e\}$$
$$R - e = R[R.\mathcal{B} - e]$$

(c) Operations on branching pomsets.

Fig. 2. Simplified semantics of branching pomsets [13].

3   Realisability

In this section we define our notion of realisability and illustrate it with examples.

We model distributed implementations as compositions of a collection of local branching pomsets $\vec{R}$ and ordered buffers (FIFO queues) $\vec{b}$ containing the messages in transit (sent but not yet received) between directed pairs of agents (or channels), similar to communicating finite-state machines [5]. The local pomsets only contain actions for a single agent; there should be one local branching pomset for each agent and one buffer for each channel.

Composition is formally defined below. We use three auxiliary functions: $add(\text{ab!x}, \vec{b})$ returns $\vec{b}$ with x added to b_{ab}, $remove(\text{ab!x}, \vec{b})$ returns $\vec{b}$ with x removed from b_{ab} and $has(\text{ab!x}, \vec{b})$ returns whether x is pending in b_{ab}. Since we consider ordered buffers, *add* appends message types to the end of the corresponding queue, *remove* removes message types from the front, and *has* only checks whether the first message matches.

We note that our termination condition does not require the buffers to be empty. In practice asynchronous communication channels will typically have some latency, and requiring empty buffers would require processes (the local branching pomsets) to be aware of messages in transit. Instead, in our model the presence or absence of orphan messages (messages unreceived on termination) is a separate property from realisability, to be verified in isolation. It does, however, follow from our well-formedness conditions in Sect. 4 that a well-formed and realisable protocol is also free of orphan messages.

Definition 2 (Composition). *Let $\vec{R}$ be an agent-indexed vector of local branching pomsets. Let $\vec{b}$ be a channel-indexed vector of ordered buffers. Their composition is the tuple $\langle \vec{R}, \vec{b} \rangle$, whose semantics is defined as the labeled transition system defined by the rules below.*

$$
\textit{(Send)} \; \frac{\begin{array}{c} R_a \xrightarrow{\text{ab!x}} R'_a \\ \vec{b}' = add(\text{ab!x}, \vec{b}) \end{array}}{\langle \vec{R}, \vec{b} \rangle \xrightarrow{\text{ab!x}} \langle \vec{R}[R'_a/R_a], \vec{b}' \rangle}
\qquad
\textit{(Receive)} \; \frac{\begin{array}{c} R_b \xrightarrow{\text{ab?x}} R'_b \\ has(\text{ab!x}, \vec{b}) \quad \vec{b}' = remove(\text{ab!x}, \vec{b}) \end{array}}{\langle \vec{R}, \vec{b} \rangle \xrightarrow{\text{ab?x}} \langle \vec{R}[R'_b/R_b], \vec{b}' \rangle}
$$

$$
\textit{(Terminate)} \; \frac{\forall a : R_a {\downarrow}}{\langle \vec{R}, \vec{b} \rangle {\downarrow}}
$$

A protocol is *realisable* if there exists a faithful distributed implementation of it, i.e., one defining the same behaviour. We formally define realisability below. We note that it is typically defined in terms of language (trace) equivalence [18]. However, as the exact branching of choices plays an important part in branching pomsets, we use a more strict notion of equivalence and require the global branching pomset and the composition to be bisimilar [29]. As an example, consider the branching pomsets in Fig. 3. After firing ab!int ab?int, the branching pomset on the left can still fire either bc!yes or bc!no, while the branching pomset on the right has already committed to one of the two upon firing ab!int. We wish to distinguish these two branching pomsets, which cannot be done using language equivalence since they yield the same set of two traces. It is then most natural to compare two branching pomsets using branching equivalence, i.e., bisimilarity. As we wish to be able to make the same distinction in our realisability analysis, we also define realisability in terms of bisimilarity. We note that our well-formedness conditions enforce a deterministic setting, in which bisimilarity agrees with language equivalence. We then choose to prove bisimilarity rather than language equivalence because the proofs are typically more straightforward.

Formally, if two branching pomsets R_1 and R_2 are bisimilar, written $R_1 \sim R_2$, then, for every reduction $R_1 \xrightarrow{\ell} R'_1$ there should exist a reduction $R_2 \xrightarrow{\ell} R'_2$ such

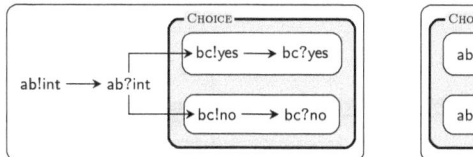 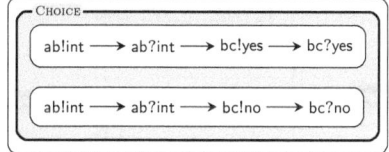

Fig. 3. Two similar yet not *bi*similar branching pomsets.

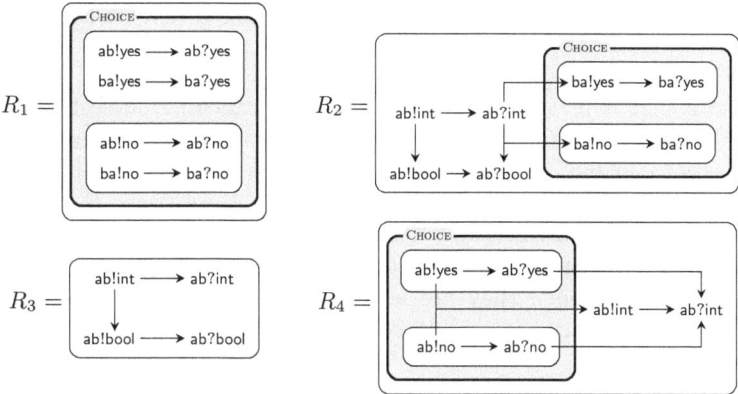

Fig. 4. A collection of realisable and unrealisable branching pomsets.

that R_1' and R_2' are again bisimilar, and vice-versa. Additionally, we require that two bisimilar branching pomsets R_1 and R_2 can terminate iff the other can do so as well.

Definition 3 (Realisability). *Let R be a branching pomset. The protocol it represents is realisable iff there exists a composition $\langle \vec{R}, \vec{b} \rangle$ such that b_{ab} is empty for all* ab *and* $R \sim \langle \vec{R}, \vec{b} \rangle$.

As an example, consider the branching pomsets in Fig. 4:

- R_1 is unrealisable. Alice and Bob both have to send a yes or a no to the other but the two messages must be the same. It is impossible without further synchronisation or communication to prevent a scenario in which one will send a different message than the other.
- R_2 is realisable. Alice first sends an int and then a bool to Bob. After receiving the int, Bob returns either a yes or a no.
- R_3 is unrealisable. Alice sends an int and a bool to Bob, but while they agree that Alice first sends the int and then the bool, the order in which Bob receives the message is unspecified. As we assume ordered buffers, Bob will first receive the int, but the global branching pomset allows an execution in which Bob first receives the bool.
- R_4 is realisable. Alice sends a yes or a no to Bob, followed by an int.

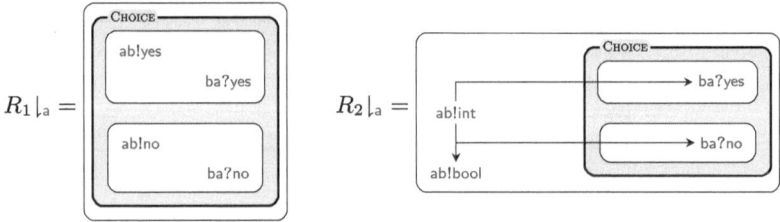

Fig. 5. The projection of the branching pomsets R_1 and R_2 in Fig. 4 on a.

We note that it is easy to go from a global branching pomset R to a local branching pomset for some agent a by *projecting* it on a, written $R\!\restriction_a$. We will use projections in our well-formedness conditions and realisability proof, and we formally define them below. The projection $R\!\restriction_a$ restricts R to the events whose subject is a, and restricts $\leq$ and λ accordingly. The branching structure is pruned by removing all discarded events (leaves), but no inner nodes of the tree are removed, even if they are left without any children. This is done to safeguard the symmetry with the branching structure of R to ease our proofs.

Definition 4 (Projection). $\langle E, \leq, \lambda, \mathcal{B} \rangle\!\restriction_a = \langle E_a, \leq_a, \lambda_a, \mathcal{B}_a \rangle$ *where:*

- $E_a = \{e \in E \mid subj(e) = a\}$
- $\leq_a = \leq \cap (E_a \times E_a)$
- $\lambda_a = \lambda \cap (E_a \times \mathcal{L})$
- $\mathcal{B}_a = \mathcal{B}\!\restriction_a$ *as defined below.*

$$e\!\restriction_a = e \ if \ e \in E_a$$
$$\{\mathcal{C}_1, \ldots, \mathcal{C}_n\}\!\restriction_a = \{\mathcal{C}_i\!\restriction_a \mid 1 \leq i \leq n \wedge \mathcal{C}_i\!\restriction_a \ is \ defined\}$$
$$\{\mathcal{B}_1, \mathcal{B}_2\}\!\restriction_a = \{\mathcal{B}_1\!\restriction_a, \mathcal{B}_2\!\restriction_a\}$$

As an example, Fig. 5 shows the projection of two of the branching pomsets in Fig. 4 on agent a. The events with subject b are removed, as are dependencies involving them. $R_1\!\restriction_a$ is left with no dependencies at all. We note that, as the graphical representation of a branching pomset shows the transitive reduction of the causality relation and not the full relation, it is unclear from just Fig. 4 whether $R_2\!\restriction_a$ should contain dependencies between ab!int and the events ba?yes and ba?no. This is unambiguous in the formal textual definition, which we omitted but which also relates these events.

Finally, we prove that R and its projections can mirror each others refinements. Both proofs are straightforward by induction on the structure of the premise's derivation tree.

Lemma 1. *If* $R \sqsupseteq R'$ *then* $R\!\restriction_a \sqsupseteq R'\!\restriction_a$.

Lemma 2. *If* $R\!\restriction_a \sqsupseteq R'_a$ *then* $R \sqsupseteq R'$ *for some* R' *such that* $R'_a = R'\!\restriction_a$.

4   Well-formedness

In this section we define our well-formedness conditions on branching pomsets.

We define four well-formedness conditions (Definition 10): to be well-formed, a branching pomset must be well-branched, well-channeled, tree-like and choreographic. Well-branchedness, well-channeledness and tree-likeness are inspired by MPST [20] and ensure some safety properties. Choreographicness ensures that the branching pomset represents some sort of meaningful protocol.

- **Well-branched** (Definition 6): every choice is made only on the label of a send-receive pair, i.e., the first events in every branch must be a send and receive between some agents a and b, with the message type being different in every branch. Additionally, the projection on every agent uninvolved in the choice must be the same in every branch. Then a and b are both aware of the chosen branch and all other agents are unaffected by the choice.
  Although the branching pomset model only contains binary choices, an n-ary choice $\mathcal{C}$ can be encoded as a nested binary one, where the n children of $\mathcal{C}$ become the leaves of a binary tree. We call such a leaf $\mathcal{B}$ an *option* of $\mathcal{C}$, written $\mathcal{B} \lhd \mathcal{C}$, which is formally defined below. This allows us to properly interpret $\mathcal{C}$ as an n-ary choice again and reason about it accordingly.
- **Well-channeled** (Definition 7): pairs of sends and pairs of receives on the same channel that can occur in the same execution should be ordered, and the pairs of sends should have the same order as the pairs of their corresponding receives. A branching pomset which is not well-channeled could, for example, yield a trace ab!x ab!y ab?y ab?x, which cannot be reproduced by a composition using ordered buffers.
- **Tree-like** (Definition 8): events inside of choices can only affect future events in the same branch. Graphically speaking, arrows can only enter boxes, not leave them. As a consequence, the causality relation $\leq$ follows the branching structure $\mathcal{B}$ and has a tree-like shape — hence the name.
- **Choreographic** (Definition 9): the branching pomset represents a choreography of some sort, i.e., a communication protocol in which the send and receive events are properly paired and all dependencies can be logically derived. Specifically, all dependencies are between send-receive pairs or between same-subject events, or they can be transitively derived from those. Additionally, there is some correspondence between the send and receive events: every send can be matched to exactly one corresponding receive, and every non-top-level receive has some corresponding send at the same level of the branching structure $\mathcal{B}$. This definition is similar to the definition of well-formedness by Guanciale and Tuosto [18].

Definition 5 (Option). *Let* $\mathcal{C} \prec R.\mathcal{B}$. $\mathcal{B}$ *is an* option *of* $\mathcal{C}$*, written* $\mathcal{B} \lhd \mathcal{C}$*, if* $\mathcal{B} \in \{\mathcal{B}^\dagger \mid \mathcal{B}^\dagger \lhd^\dagger \mathcal{C} \wedge \nexists \mathcal{B}^\ddagger : (\mathcal{B}^\ddagger \lhd^\dagger \mathcal{C} \wedge \mathcal{B}^\ddagger \prec \mathcal{B}^\dagger)\}$*, where* $\lhd^\dagger$ *is defined as follows:*

$$\frac{\mathcal{B} \in \mathcal{C}}{\mathcal{B} \lhd^\dagger \mathcal{C}} \qquad \frac{\mathcal{B} \in \mathcal{C}' \quad \{\mathcal{C}'\} \lhd^\dagger \mathcal{C}}{\mathcal{B} \lhd^\dagger \mathcal{C}}$$

Definition 6 (Well-branched). *A branching pomset R is* well-branched *iff, for every $\mathcal{C} \prec R.\mathcal{B}$ there exist participants a, b such that for every $\mathcal{B}_i \neq \mathcal{B}_j \lhd \mathcal{C}$ there exist events $e_{i1}, e_{i2} \in \mathcal{B}_i$ and $e_{j1}, e_{j2} \in \mathcal{B}_j$ such that:*

- $\lambda(e_{i1}) = ab!x$, $\lambda(e_{i2}) = ab?x$, $\lambda(e_{j1}) = ab!y$ *and* $\lambda(e_{j2}) = ab?y$ *for some* $x \neq y$;
- $e_{i1} \leq e_i$ *for all* $e_i \preceq \mathcal{B}_i$ *and* $e_{j1} \leq e_j$ *for all* $e_j \preceq \mathcal{B}_j$;
- $e_{i2} \leq e_i$ *for all* $e_i \preceq \mathcal{B}_i$ *for which* $subj(e_i) = b$ *and* $e_{j2} \leq e_j$ *for all* $e_j \preceq \mathcal{B}_j$ *for which* $subj(e_j) = b$; *and*
- $R[\mathcal{B}_i]|_c = R[\mathcal{B}_j]|_c$ *for all* $c \neq a, b$[1].

Definition 7 (Well-channeled). *A branching pomset R is* well-channeled *iff, for all events $e_1, e_2, e_3, e_4 \in R.E$:*

- *If e_1 and e_2 are either both sends or both receive events, and if they share the same channel, then they are either causally ordered or mutually exclusive.*
- *If e_1, e_3 and e_2, e_4 are two pairs of matching send-receive events sharing the same channel, and if there exists no $e_5 \in R.E$ such that $e_1 < e_5 < e_3$ or $e_2 < e_5 < e_4$, then $e_1 \leq e_2 \implies e_3 \leq e_4$.*

Definition 8 (Tree-like). *A branching pomset R is* tree-like *iff:*
$$\forall \mathcal{C} = \{\mathcal{B}_1, \mathcal{B}_2\} \prec R.\mathcal{B} : (e_1 \leq e_2 \wedge e_1 \preceq \mathcal{B}_i) \implies e_2 \preceq \mathcal{B}_i, \text{ where } i \in \{1, 2\}.$$

Definition 9 (Choreographic). *A branching pomset R is* choreographic *iff, for every $e \in R.E$:*

- *If there exists $e' \in R.E$ such that $e' < e$ then there exists some event e'' (not necessarily distinct from e') such that $e' \leq e'' < e$ and either $subj(\lambda(e'')) = subj(\lambda(e))$ or $[\lambda(e'') = ab!x$ and $\lambda(e) = ab?x$ for some $a, b, x]$.*
- *If $\lambda(e) = ab?x$ and $e \in \mathcal{B}$ for some $\mathcal{B} \prec R.\mathcal{B}$ then there exists some e' such that $e' \in \mathcal{B}$ and $\lambda(e') = ab!x$ and $e' < e$.*
- *If $\lambda(e) = ab!x$ then there exists exactly one e' such that $e \leq e'$ and that $\lambda(e') = ab?x$ and that $(\lambda(e^\dagger) = ab!x \wedge e^\dagger \leq e') \Rightarrow e^\dagger \leq e$.*

Definition 10 (Well-formed). *A branching pomset R is* well-formed *iff it is well-branched, well-channeled, tree-like and choreographic.*

As an example, recall the branching pomsets in Fig. 4:

- R_1 is not well-formed since it is not well-branched: for example, the branches of the choice have multiple minimal events. It is indeed unrealisable.
- R_2 is both well-formed and realisable.
- R_3 is not well-formed since it is not well-channeled: the two receive events are on the same channel but are unordered. It is indeed unrealisable.

---

[1] Technically $R[\mathcal{B}_i]|_c$ and $R[\mathcal{B}_j]|_c$ have different events and should thus be isomorphic rather than precisely equal. We choose to write it as an equality to not unnecessarily complicate the definition and proofs.

– R_4 is not well-formed since it is not tree-like: there are arrows from events inside branches of a choice to ab!int and ab?int, even though the latter are not part of the same branch. It is, however, realisable, which illustrates that, while we prove in Sect. 5 that our well-formedness conditions are sufficient, they are not necessary.

Finally, we show that well-formedness is retained after a reduction.

Lemma 3. *Let R be a branching pomset and let $R \xrightarrow{\ell} R'$. If R is well-formed then so is R'.*

Proof (sketch). We use that the components of R' are subsets of or (in the case of the branching structure) derived from the components of R. It then follows that a violation of one of the properties in R' would also invariably lead to a violation of one of the properties in R.

5   Bisimulation Proof

In this section we prove that, if a branching pomset R is well-formed, then the corresponding protocol is realisable.

To prove that R's protocol is realisable, we have to show the existence of a bisimilar composition of local branching pomsets and buffers. To do this, we define a canonical decomposition of R by combining our previously defined projections with a buffer construction, and we prove that this canonical decomposition is bisimilar to R. The (re)construction of the buffer contents of channel ab based on R, written $buff_{ab}(R)$, and the canonical decomposition of R, written $cd(R)$, are defined below.

The buffer construction $buff_{ab}(R)$ gathers the receive events in R that have no preceding matching send event. We infer that, since the send has already been fired and the receive has not, the message must be in transit.

Definition 11 (Buffer construction). *Let R be a branching pomset. Let* a *and* b *be agents in R. Let ε be the empty word.*

$$Then\; buff_{ab}(R) = \begin{cases} \mathsf{x} \cdot buff_{ab}(R') & if\; R' = R - e\; and\; \lambda(e) = \mathsf{ab?x} \\ & and\; \forall e'\colon if\; e' < e\; then\; \lambda(e') \neq \mathsf{ab!x} \\ & and\; \forall e', \mathsf{y}\colon if\; e' < e\; then\; \lambda(e') \neq \mathsf{ab?y} \\ \varepsilon & otherwise \end{cases}$$

The corresponding message types are nondeterministically put in some order that respects the order of the gathered receive events — if $e_1 < e_2$ then e_1's message type must precede that of e_2 in the constructed buffer. We note that all unmatched receive events are top-level if R is choreographic, and that the same-channel top-level receive events are totally ordered if R is well-channeled. It follows that, although it may add duplicate messages and is nondeterministic in the general case, $buff_{ab}(R)$ does not add duplicate messages and is deterministic if R is well-formed.

Definition 12 (Canonical decomposition). *Let R be a branching pomset. Let $\vec{R}$ be such that $R_a = R|_a$ for all a. Let $\vec{b}$ be such that $b_{ab} = buff_{ab}(R)$ for all ab. Then $cd(R) = \langle \vec{R}, \vec{b} \rangle$ is the canonical decomposition of R.*

To prove that a well-formed R is bisimilar to $cd(R)$, we define the relation $\mathcal{R} = \{\langle R, \langle \vec{R}^\dagger, \vec{b} \rangle \rangle \mid \langle \vec{R}^\dagger, \vec{b} \rangle \sim \langle \vec{R}, \vec{b} \rangle = cd(R)\}$ and we prove that $\mathcal{R}$ is a bisimulation relation (Theorem 1). Note that the vector of buffers $\vec{b}$ is the same across the definition; we only allow leeway in the vector of local branching pomsets. The proof consists of the two parts mentioned in Sect. 3. Given that $\langle R, \langle \vec{R}^\dagger, \vec{b} \rangle \rangle \in \mathcal{R}$, if one can make some reduction then the other must be able to make the same reduction such that the resulting configurations are again related by $\mathcal{R}$ (Lemma 7, Lemma 8). Additionally, if one can terminate then so should the other (Lemma 9, Lemma 10).

The reason that $\mathcal{R}$ is not simply the set of all $\langle R, cd(R) \rangle$ is that a reduction from $cd(R)$ may not always result in $cd(R')$ for some R'. This is because choices are only resolved in the branching pomset of the agent causing the reduction. For example, consider branching pomset R_4 in Fig. 4. Upon Alice sending yes the global branching pomset would resolve the choice for both agents simultaneously. However, upon Alice sending yes in the canonical decomposition the projection on Bob remains unchanged and still contains receive events for both yes and no. Since yes has been added to the buffer from Alice to Bob, we know that Bob will eventually have to pick the branch containing yes — after all, there is no no to receive. In other words: this configuration is bisimilar to the canonical decomposition of the resulting global branching pomset, in which the choice has also been resolved for Bob. If there were also some additional no being sent from Alice to Bob, e.g., if we replace the messages int in R_4 with no, then R_4 being well-channeled and the buffers being ordered would still ensure that we can safely resolve Bob's choice. This crucial insight is formally proven in Lemma 4.

Lemma 4. *Let R be a well-formed branching pomset. Let $\langle \vec{R}, \vec{b} \rangle = cd(R)$. Let ℓ be some label and let $a = subj(\ell)$. If $R \xrightarrow{\ell} R'$ and if $\langle \vec{R}, \vec{b} \rangle \xrightarrow{\ell} \langle \vec{R}[R'_a/R|_a], \vec{b}^\dagger \rangle$ and if $R'_a = R'|_a$, then $\langle \vec{R}[R'_a/R|_a], \vec{b}^\dagger \rangle \sim \langle \vec{R'}, \vec{b'} \rangle = cd(R')$.*

Proof (sketch). If $\ell = $ ba?x for some b, x then it follows from the well-formedness of R that $R' = R - e$ and the remainder is straightforward. The same is true if $\ell = $ ab!x and e is top-level, i.e., $e \in R.\mathcal{B}$.

Otherwise it follows from the well-formedness of R that e is a minimal send event in one of the options of a top-level choice, i.e., $e \in \mathcal{B} \lhd \mathcal{C} \in R.\mathcal{B}$ for some $\mathcal{B}, \mathcal{C}$, and $R' = R[(R.\mathcal{B} \setminus \mathcal{C}) \cup (\mathcal{B} - e)]$. We proceed to show that the set of unmatched receive events in R' is exactly that of R with the addition of the one corresponding to e, and then $\vec{b'} = add(\text{ab!x}, \vec{b}) = \vec{b}^\dagger$. It follows that $\langle \vec{R}[R'_a/R|_a], \vec{b}^\dagger \rangle = \langle \vec{R}[R'_a/R|_a], \vec{b'} \rangle$. For the projections, we proceed in two steps:

– First we observe that, since R is well-branched, $\mathcal{B}'|_c = \mathcal{B}|_c$ for all $\mathcal{B}' \lhd \mathcal{C}$ and for all $c \neq a, b$. It follows that $R|_c \sim R'|_c$, and then $\langle \vec{R}[R'_a/R|_a], \vec{b'} \rangle \sim \langle \vec{R'}[R|_b/R'|_b], \vec{b'} \rangle$. Note that the projection on a is $R'|_a$ and the projection on b is $R|_b$ on both sides, and the projection on every other c is bisimilar.

- Next we show that, with the new message added to the buffer, no event can ever fire in $R|_b$ in any other option of $\mathcal{C}$ than $\mathcal{B}$. It follows that we can discard all other options, and then $\langle \vec{R}'[R|_b/R'|_b], \vec{b}' \rangle \sim \langle \vec{R}', \vec{b}' \rangle = cd(R')$. $\qquad\square$

To satisfy the preconditions of Lemma 4, we additionally prove that R's reductions can be mirrored by its projection on the reduction label's subject (Lemma 5) and dually that the reductions of R's canonical decomposition can be mirrored by R (Lemma 6). Their proofs rely on the observations that the corresponding event e must be minimal both in R and $R|_a$, and that the branching structures of the two are the same (modulo discarded leaves). It then follows that the same refinement enables e in both R and $R|_a$.

Lemma 5. *Let R be a tree-like branching pomset. If $R \xrightarrow{\ell} R'$ and $a = subj(\ell)$ then $R|_a \xrightarrow{\ell} R'|_a$.*

Lemma 6. *Let R be a well-channeled, tree-like and choreographic branching pomset. Let $\langle \vec{R}, \vec{b} \rangle = cd(R)$. If $\langle \vec{R}, \vec{b} \rangle \xrightarrow{\ell} \langle \vec{R}[R'_a/R|_a], \vec{b}' \rangle$ then $R \xrightarrow{\ell} R'$ for some R' such that $R'_a = R'|_a$.*

Finally, we bring everything together and prove the four necessary steps for bisimulation in the lemmas below, culminating in Theorem 1. The proof for Lemma 7 uses Lemma 5 to show the preconditions of Lemma 4 and then applies the latter. This gives us $cd(R) \xrightarrow{\ell} cd(R)' \sim cd(R')$, and since $\langle \vec{R}^\dagger, \vec{b}^\dagger \rangle \sim cd(R)$ the result is then straightforward. The proof for Lemma 8 is analogous but uses Lemma 6. The proofs for Lemma 9 and Lemma 10 respectively use Lemma 1 and Lemma 2 to show that, if one can terminate by refining to the empty set, then so must the other.

Lemma 7. *Let $\langle R, \langle \vec{R}^\dagger, \vec{b} \rangle \rangle \in \mathcal{R}$. If R is well-formed and $R \xrightarrow{\ell} R'$ then there exist $\vec{R}^\ddagger$ and $\vec{b}^\ddagger$ such that $\langle \vec{R}^\dagger, \vec{b} \rangle \xrightarrow{\ell} \langle \vec{R}^\ddagger, \vec{b}^\ddagger \rangle$ and $\langle R', \langle \vec{R}^\ddagger, \vec{b}^\ddagger \rangle \rangle \in \mathcal{R}$.*

Lemma 8. *Let $\langle R, \langle \vec{R}^\dagger, \vec{b} \rangle \rangle \in \mathcal{R}$. If R is well-formed and $\langle \vec{R}^\dagger, \vec{b} \rangle \xrightarrow{\ell} \langle \vec{R}^\ddagger, \vec{b}^\ddagger \rangle$ then there exists R' such that $R \xrightarrow{\ell} R'$ and $\langle R', \langle \vec{R}^\ddagger, \vec{b}^\ddagger \rangle \rangle \in \mathcal{R}$.*

Lemma 9. *Let $\langle R, \langle \vec{R}^\dagger, \vec{b} \rangle \rangle \in \mathcal{R}$. If R is well-formed and $R\downarrow$ then $\langle \vec{R}^\dagger, \vec{b} \rangle \downarrow$.*

Lemma 10. *Let $\langle R, \langle \vec{R}^\dagger, \vec{b} \rangle \rangle \in \mathcal{R}$. If R is well-formed and $\langle \vec{R}^\dagger, \vec{b} \rangle \downarrow$ then $R\downarrow$.*

Theorem 1. *Let R be a branching pomset. If R is well-formed and $buff_{ab}(R) = \varepsilon$ for all ab then the protocol represented by R is realisable.*

Proof. It follows from Lemma 7, Lemma 8, Lemma 9 and Lemma 10 that $\mathcal{R}$ is a bisimulation relation [29]. Specifically, it then follows that $R \sim cd(R)$. Then, since $buff_{ab}(R) = \varepsilon$ for all ab, by Definition 3 the protocol represented by R is realisable. $\qquad\square$

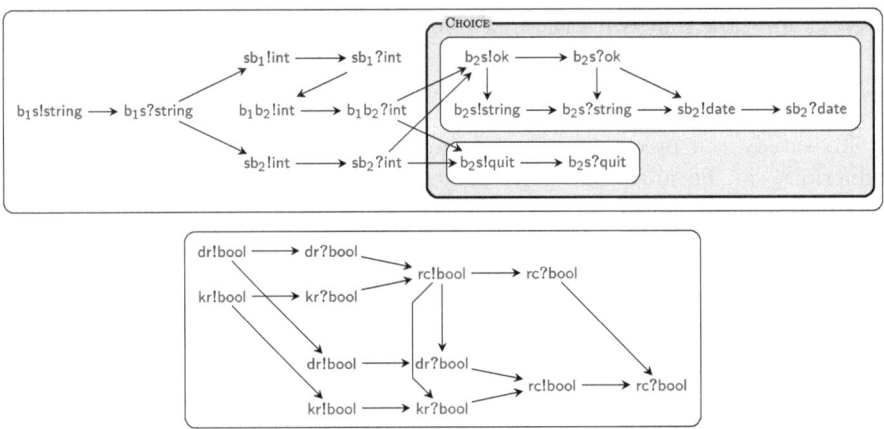

Fig. 6. Branching pomsets representing the two-buyers-protocol (top) and two iterations of the simple streaming protocol (bottom) [20].

6   Examples

In this section we briefly look at two example protocols used by Honda et al. [20]. Both are depicted as branching pomsets in Fig. 6.

The **two-buyers-protocol** (top) features Buyer 1 and Buyer 2 (b_1, b_2) who wish to jointly buy a book from Seller (s). Buyer 1 first sends the title of the book (string) to Seller, Seller sends its quote (int) to both Buyer 1 and Buyer 2, and Buyer 1 sends Buyer 2 the amount they can contribute (int). Buyer 2 then notifies Seller whether they accept (ok) or reject (quit) the quote. If they accept, they also send their address (string), and Seller returns a delivery date (date).

The **simple streaming protocol** (bottom) features Data Producer (d) and Key Producer (k), who continuously respectively send data and keys (both bool) to Kernel (r). Kernel performs some computation and sends the result (bool) to Consumer (c). The protocol in Fig. 6 shows two iterations of this process.

Both branching pomsets are well-formed, and hence the protocols are realisable. We note that, as in the paper by Honda et al., further communication between Buyers 1 and 2 has been omitted in the two-buyers-protocol. Since this is bound to be different in the case of acceptance and rejection, the resulting branching pomset would not be well-branched and thus not well-formed — though still realisable. We discuss relaxed well-branchedness conditions in Sect. 7. Also note that the ok and address (string) messages are sent sequentially; sending these in parallel would violate well-channeledness and make the protocol unrealisable with ordered buffers. The same is true for the streaming protocol: the two iterations are composed sequentially and doing so concurrently would violate well-channeledness and result in unrealisability. The size of the branching pomset for the streaming protocol scales linearly with the number of depicted iterations; showing all (infinitely many) iterations would require an infinitely large branching pomset. We briefly touch upon infinity in Sect. 8.

7   Related Work

Realisability has been well-studied in the last two decades in a variety of settings. For example, Alur et al. study the realisability of message sequence charts [1]. They define the notions of weak and safe realisability of languages, the latter also ensuring deadlock-freedom, and they define closure conditions on languages which they show to precisely capture weakly and safely realisable languages. Lohmann and Wolf define the notion of distributed realisability, where a protocol is distributedly realisable if there exists a set of compositions such that every composition covers a subset of the protocol and the entire protocol is covered by their union [25]. Fu et al. [16], Basu et al. [3], Finkel and Lozes [15] and Schewe et al. [32] all study the realisability of protocols on different network configurations when considering only the sending behaviour — receive events are omitted — showing necessary and/or sufficient conditions and decidability results.

One major source of inspiration for our work has been previous work on pomsets. Pomsets were initially introduced by Pratt [28] for concurrency models and have been widely used, e.g., in the context of message sequence charts by Katoen and Lambert [24]. Recently Guanciale and Tuosto presented a realisability analysis based on sets of pomsets [18], in which they show how to capture the language closure conditions of Alur et al. [1] directly on pomsets, without having to explicitly compute their language. Typically choreography languages are limited in their expressiveness and any analysis on their realisability is then language-dependent. Both Alur et al. and Guanciale and Tuosto perform a syntax-oblivious analysis, which has the benefit of being applicable to any specification which can be encoded as a set of pomsets, regardless of the specification language. The analysis by Guanciale and Tuosto is at a higher level of abstraction than sets of traces. This allows both for a more efficient analysis and for easier identification of design errors, as these can be identified in a more abstract model.

Our approach is similarly syntax-oblivious, though the analysis itself is based on MPST (on which we will elaborate later). A major difference is that Guanciale and Tuosto use unordered buffers (e.g., non-FIFO queues) while ours are ordered. For example, the parallel composition of a→b:x and a→b:y is realisable in the unordered setting and not in the ordered one. The two settings agree on realisability when the two message types are the same (e.g., two concurrent copies of a→b:x); while Guanciale and Tuosto explicitly note that they support concurrently repeated actions, however, our current well-channeledness condition does not make an exception for these. This is an obvious target for relaxation of our conditions. In their paper, Guanciale and Tuosto also separately define termination soundness, i.e., whether participants are not kept waiting indefinitely after a composition terminates. For example, the branching pomset (right) in Fig. 1 is realisable but not termination-sound, as either Carol or Dave will have to wait indefinitely since they do not know that the other received Bob's message. Making this protocol termination-sound would require additional communication between Bob, Carol and/or Dave. Further inspiration for relaxation of well-formedness conditions can be found in an earlier paper by Guanciale and

Tuosto [17]. In particular their definition of well-branchedness, using 'active' and 'passive' agents, could serve as a basis for a relaxed version of our own.

The other major source of inspiration for our work is multiparty session types (MPST), introduced by Honda et al. [20]. Specifically:

- Our well-branchedness condition corresponds to the branching syntax of global types in MPST and its definition of projection. Branching in MPST is done on the label of a single message, and the projection on agents uninvolved in the choice is only defined if it is the same in every branch.
- Our well-channeledness condition corresponds to the principle that same-channeled actions should be ordered. We note that our condition is more lenient: we prohibit concurrent sends or receives on the same channel, while in MPST the projection of a parallel composition on an agent is undefined if the agent occurs in both threads (even if the threads' channels are disjoint).
- Our tree-likeness condition follows from the syntax of global types in MPST, which uses a prefix operator rather than a more general sequential composition. As a consequence all global types are tree-like. The same is true for other languages that use a prefix operator and not sequential composition, such as CCS [26] and π-calculus [30].

Since its introduction, various papers have addressed the strictness of the well-branchedness condition in MPST. One line of research relaxes the condition by using a merge operation to allow all agents to have different behaviour in different branches, as long as they are timely informed of the choice [6,11,31]. Another line relaxes the condition by allowing different branches to start with different receivers, rather than enforcing the same receiver in every branch [4, 7,9,10,22]. These results may also serve as inspiration for relaxations of the well-branchedness condition on branching pomsets.

While our current conditions broadly correspond with well-formedness in MPST, we believe that our approach offers three advantages. First, as discussed before, it is syntax-oblivious, meaning it is not only applicable to MPST but to any specification which can be encoded as a branching pomset. Second, we believe that branching pomsets have the potential to be more expressive than global types in MPST. As mentioned above, our well-channeledness condition is already more lenient than the one in MPST. We have described various sources of possible relaxations for our well-branchedness condition, both in the MPST and in the pomset literature. Lastly, we conjecture that our tree-likeness condition is needed to simplify our proofs, and that it is possible — though more complex — to prove realisability without it or at least with a relaxed version of it.

A proper comparison between the pomset-based approach by Guanciale and Tuosto [18] and ours, both in terms of expressiveness and efficiency, would first require further development of our conditions. In the meantime, one takeaway from their paper is the performance gain they obtain by lifting the analysis from languages (sets of traces) to a higher level of abstraction, i.e., sets of pomsets. Our hope is that a further performance gain can be obtained by lifting the analysis from sets of pomsets to an even higher level of abstraction (e.g., branching pomsets).

Event Structures. Finally, the concept of branching pomsets is reminiscent of event structures [27] and their recent usage in the context of MPST [8]. Both consist of a set of events with some causality relation and a choice mechanism. In event structures the choice mechanism consists of a conflict relation, where two conflicting events may not occur together in the same execution; in branching pomsets choices are modelled as a branching structure. A technical difference seems to be the resolving of choices. In event structures a choice (conflict) may only be resolved by firing an event in one of its branches. In contrast, in branching pomsets a choice may be resolved without doing so; instead, one branch may be discarded to fire an event outside of the choice, which is causally dependent on the discarded branch but not on the other. A thorough formal comparison between the two models, including both technical and conceptual aspects, is ongoing work.

8   Conclusion

We have defined well-formedness conditions on branching pomsets (Definition 10) and have proven that a well-formed branching pomset represents a realisable protocol (Theorem 1). These conditions are sufficient but not necessary, i.e., a protocol may be realisable if its branching pomset is not well-formed. Examples of this are given in Fig. 1 (the branching pomset on the right is realisable but not well-branched) and Fig. 4 (branching pomset R_4 is realisable but not tree-like). Several routes for relaxing our well-channeledness and especially our well-branchedness conditions have been discussed in Sect. 7, in the aim of increasing the number of branching pomsets that are well-formed while retaining well-formedness as a sufficient condition for realisability. In the remainder of this section we share some additional thoughts on well-channeledness and tree-likeness, and then briefly discuss the applicability of our work to branching pomsets of infinite size.

Well-Channeledness. The branching pomset R_3 in Fig. 4 is not well-channeled since the events labelled with ab?int and ab?bool are unordered. It is unrealisable because a local system will force the int to be received before the bool while the global branching pomset allows a different order. However, in this case one may take the view that the problem is not the local system being too strict, but rather the global branching pomset being too lenient in an environment with ordered buffers: it should then in some way allow a user to specify just the two acceptable orderings without having to resort to adding a choice between the two and duplicating all events in the pomset. Therefore, instead of changing the well-channeledness condition, another avenue would be to change the reduction rules in for branching pomsets themselves (Fig. 2) and specifically adapt them to ordered buffers. This could be done in such a way that reducing R_3 by firing ab!int then automatically adds a dependency from ab?int to ab?bool. This might allow the well-channeledness condition to be significantly relaxed or to possibly be removed altogether.

Tree-Likeness. Having the assumption of tree-likeness simplifies our proofs. It is our aim to eventually relax or even remove it and still prove realisability, but this will require a significantly more complex proof. We have noted in Sect. 7 that global types in MPST and expressions in, e.g., CCS and the π-calculus, are tree-like by default. Conceptually a non-tree-like branching pomset could potentially be turned into an equivalent (i.e., bisimilar) tree-like one by distributing the offending events over the branches of the involved choice. For example, consider the branching pomset R_4 in Fig. 4. By duplicating ab!int and ab?int and adding a copy of each with the relevant dependencies to each of the two branches of the choice, we obtain a bisimilar but now tree-like (and well-formed) branching pomset. A more general scheme may be developed based on versions of the CCS expansion theorem [14,19]. However, regaining expressiveness at the cost of duplicating events effectively negates the benefits of using branching pomsets in the first place.

Infinity. The paper introducing branching pomsets [13] supports branching pomsets of infinite size. We note that our theoretical results in this paper also hold for infinite branching pomsets. However, determining the well-formedness of an infinite branching pomset is undecidable due to its size. A solution in the case of infinity through repetition, e.g., loops in choreographies, would be to use a symbolic representation. Alternatively, a solution might be found in the extension from message sequence charts (MSCs) to MSC-graphs [21]. A similar extension could be developed for branching pomsets, where they are sequentially composed in a (possibly cyclic) graph. Finally, it may be possible to leverage the recently introduced pomset automata [23].

References

1. Alur, R., Etessami, K., Yannakakis, M.: Inference of message sequence charts. IEEE Trans. Software Eng. **29**(7), 623–633 (2003). https://doi.org/10.1109/TSE.2003.1214326
2. Alur, R., Etessami, K., Yannakakis, M.: Realizability and verification of MSC graphs. Theor. Comput. Sci. **331**(1), 97–114 (2005). https://doi.org/10.1016/j.tcs.2004.09.034
3. Basu, S., Bultan, T., Ouederni, M.: Deciding choreography realizability. In: Field, J., Hicks, M. (eds.) Proceedings of the 39th ACM SIGPLAN-SIGACT Symposium on Principles of Programming Languages, POPL 2012, Philadelphia, Pennsylvania, USA, 22–28 January 2012, pp. 191–202. ACM (2012). https://doi.org/10.1145/2103656.2103680
4. Bocchi, L., Melgratti, H., Tuosto, E.: Resolving non-determinism in choreographies. In: Shao, Z. (ed.) ESOP 2014. LNCS, vol. 8410, pp. 493–512. Springer, Heidelberg (2014). https://doi.org/10.1007/978-3-642-54833-8_26
5. Brand, D., Zafiropulo, P.: On communicating finite-state machines. J. ACM **30**(2), 323–342 (1983). https://doi.org/10.1145/322374.322380
6. Carbone, M., Yoshida, N., Honda, K.: Asynchronous session types: exceptions and multiparty interactions. In: Bernardo, M., Padovani, L., Zavattaro, G. (eds.) SFM 2009. LNCS, vol. 5569, pp. 187–212. Springer, Heidelberg (2009). https://doi.org/10.1007/978-3-642-01918-0_5

7. Castagna, G., Dezani-Ciancaglini, M., Padovani, L.: On global types and multi-party session. Log. Methods Comput. Sci. **8**(1) (2012). https://doi.org/10.2168/LMCS-8(1:24)2012

8. Castellani, I., Dezani-Ciancaglini, M., Giannini, P.: Event structure semantics for multiparty sessions. In: Boreale, M., Corradini, F., Loreti, M., Pugliese, R. (eds.) Models, Languages, and Tools for Concurrent and Distributed Programming. LNCS, vol. 11665, pp. 340–363. Springer, Cham (2019). https://doi.org/10.1007/978-3-030-21485-2_19

9. Castellani, I., Dezani-Ciancaglini, M., Giannini, P.: Reversible sessions with flexible choices. Acta Informatica **56**(7), 553–583 (2019). https://doi.org/10.1007/s00236-019-00332-y

10. Deniélou, P.-M., Yoshida, N.: Multiparty session types meet communicating automata. In: Seidl, H. (ed.) ESOP 2012. LNCS, vol. 7211, pp. 194–213. Springer, Heidelberg (2012). https://doi.org/10.1007/978-3-642-28869-2_10

11. Deniélou, P., Yoshida, N., Bejleri, A., Hu, R.: Parameterised multiparty session types. Log. Methods Comput. Sci. **8**(4) (2012). https://doi.org/10.2168/LMCS-8(4:6)2012

12. Edixhoven, L., Jongmans, S.S.: Realisability of branching pomsets (technical report), Tech. Rep. OUNL-CS-2022-05, Open University of the Netherlands (2022)

13. Edixhoven, L., Jongmans, S.S., Proença, J., Cledou, G.: Branching pomsets for choreographies. In: Proceedings 15th Interaction and Concurrency Experience, ICE 2022, Lucca, Italy, 17 June 2022, EPTCS (2022). https://doi.org/10.48550/arXiv.2208.04632

14. Ferrari, G.L., Gorrieri, R., Montanari, U.: An extended expansion theorem. In: Abramsky, S., Maibaum, T.S.E. (eds.) TAPSOFT 1991. LNCS, vol. 494, pp. 29–48. Springer, Heidelberg (1991). https://doi.org/10.1007/3540539816_56

15. Finkel, A., Lozes, É.: Synchronizability of communicating finite state machines is not decidable. In: Chatzigiannakis, I., Indyk, P., Kuhn, F., Muscholl, A. (eds.) 44th International Colloquium on Automata, Languages, and Programming, ICALP 2017, 10–14 July 2017, Warsaw, Poland. LIPIcs, vol. 80, pp. 1–14. Schloss Dagstuhl - Leibniz-Zentrum für Informatik (2017). https://doi.org/10.4230/LIPIcs.ICALP.2017.122

16. Fu, X., Bultan, T., Su, J.: Conversation protocols: a formalism for specification and verification of reactive electronic services. Theor. Comput. Sci. **328**(1–2), 19–37 (2004). https://doi.org/10.1016/j.tcs.2004.07.004

17. Guanciale, R., Tuosto, E.: An abstract semantics of the global view of choreographies. In: Bartoletti, M., Henrio, L., Knight, S., Vieira, H.T. (eds.) Proceedings 9th Interaction and Concurrency Experience, ICE 2016, Heraklion, Greece, 8–9 June 2016. EPTCS, vol. 223, pp. 67–82 (2016). https://doi.org/10.4204/EPTCS.223.5

18. Guanciale, R., Tuosto, E.: Realisability of pomsets. J. Log. Algebraic Methods Program. **108**, 69–89 (2019). https://doi.org/10.1016/j.jlamp.2019.06.003

19. Hennessy, M., Milner, R.: Algebraic laws for nondeterminism and concurrency. J. ACM **32**(1), 137–161 (1985). https://doi.org/10.1145/2455.2460

20. Honda, K., Yoshida, N., Carbone, M.: Multiparty asynchronous session types. In: Necula, G.C., Wadler, P. (eds.) Proceedings of the 35th ACM SIGPLAN-SIGACT Symposium on Principles of Programming Languages, POPL 2008, San Francisco, California, USA, 7–12 January 2008. pp. 273–284. ACM (2008). https://doi.org/10.1145/1328438.1328472

21. ITU-TS: ITU-TS Recommendation Z.120: Message Sequence Chart 2011 (MSC11). Tech. rep., ITU-TS, Geneva (2011)

22. Jongmans, S.-S., Yoshida, N.: Exploring type-level bisimilarity towards more expressive multiparty session types. In: ESOP 2020. LNCS, vol. 12075, pp. 251–279. Springer, Cham (2020). https://doi.org/10.1007/978-3-030-44914-8_10

23. Kappé, T., Brunet, P., Luttik, B., Silva, A., Zanasi, F.: On series-parallel pomset languages: rationality, context-freeness and automata. J. Log. Algebraic Methods Program. **103**, 130–153 (2019). https://doi.org/10.1016/j.jlamp.2018.12.001

24. Katoen, J., Lambert, L.: Pomsets for MSC. In: König, H., Langendörfer, P. (eds.) Formale Beschreibungstechniken für verteilte Systeme, 8. GI/ITG-Fachgespräch, Cottbus, 4. und 5. Juni 1998. pp. 197–207. Verlag Shaker (1998)

25. Lohmann, N., Wolf, K.: Realizability is controllability. In: Laneve, C., Su, J. (eds.) WS-FM 2009. LNCS, vol. 6194, pp. 110–127. Springer, Heidelberg (2010). https://doi.org/10.1007/978-3-642-14458-5_7

26. Milner, R. (ed.): A Calculus of Communicating Systems. LNCS, vol. 92. Springer, Heidelberg (1980). https://doi.org/10.1007/3-540-10235-3

27. Nielsen, M., Plotkin, G.D., Winskel, G.: Petri nets, event structures and domains, part I. Theor. Comput. Sci. **13**, 85–108 (1981). https://doi.org/10.1016/0304-3975(81)90112-2

28. Pratt, V.R.: Modeling concurrency with partial orders. Int. J. Parallel Program. **15**(1), 33–71 (1986). https://doi.org/10.1007/BF01379149

29. Sangiorgi, D.: Introduction to bisimulation and coinduction. Cambridge University Press (2011). https://doi.org/10.1017/CBO9780511777110

30. Sangiorgi, D., Walker, D.: The Pi-Calculus - a theory of mobile processes. Cambridge University Press (2001)

31. Scalas, A., Yoshida, N.: Less is more: multiparty session types revisited. Proc. ACM Program. Lang. 3(POPL), 1–29 (2019). https://doi.org/10.1145/3290343

32. Schewe, K.-D., Aït-Ameur, Y., Benyagoub, S.: Realisability of choreographies. In: Herzig, A., Kontinen, J. (eds.) FoIKS 2020. LNCS, vol. 12012, pp. 263–280. Springer, Cham (2020). https://doi.org/10.1007/978-3-030-39951-1_16

Liquidity Analysis in Resource-Aware Programming

Silvia Crafa[1] and Cosimo Laneve[2(✉)]

[1] Department of Mathematics, University of Padova, Padua, Italy
`silvia.crafa@unipd.it`
[2] Department of Computer Science and Engineering, University of Bologna,
Bologna, Italy
`cosimo.laneve@unibo.it`

Abstract. Liquidity is a liveness property of programs managing resources that pinpoints those programs not freezing any resource forever. We consider a simple stateful language whose resources are assets (digital currencies, non fungible tokens, etc.). Then we define a type system that tracks in a symbolic way the input-output behaviour of functions with respect to assets. These types and their composition, which define types of computations, allow us to design an algorithm for liquidity whose cost is exponential with respect to the number of functions. We also demonstrate its correctness.

Keywords: Resource-aware programming · Assets · Liquidity · Type systems · Symbolic analysis

1 Introduction

The proliferation of programming languages that explicitly feature resources has become more and more significant in the last decades. Cloud computing, with the need of providing an elastic amount of resources, such as memories, processors, bandwidth and applications, has pushed the definition of a number of formal languages with explicit primitives for acquiring and releasing them (see [1] and the references therein). More recently, a number of smart contracts languages have been proposed for managing and transferring resources that are assets (usually, in the form of digital currencies, like Bitcoin), such as the Bitcoin Scripting [5], Solidity [8], Vyper [10] and Scilla [13]. Even new programming languages are defined with (linear) types for resources, such as Rust [11].

In all these contexts, the efficient analysis of properties about the usage of resources is central to avoid flaws and bugs of programs that may also have relevant costs at runtime. In this paper, we focus on the *liquidity property*: a program is liquid when no resource remains frozen forever inside it, i.e. it is not redeemable by any party interacting with the program. For example, a program is not liquid if the body of a function does not use the resources transferred during the invocation by the caller. A program is also not liquid if, when it terminates, there is a resource that has not been emptied.

S. L. Tapia Tarifa and J. Proença (Eds.): FACS 2022, LNCS 13712, pp. 205–221, 2022.
https://doi.org/10.1007/978-3-031-20872-0_12

Table 1. Syntax of *Stipula* (X are assets, fields and parameters names)

Functions	F ::=	@Q A : f($\bar{y}$)[k̄] { S } ⟿ @Q$'$
Prefixes	P ::=	$E \rightarrow$ x $\mid$ $E \rightarrow$ A $\mid$ c × h ⊸ h$'$ $\mid$ c × h ⊸ A //$0 \leqslant c \leqslant 1$
Statements	S ::=	_ $\mid$ $P\ S$ $\mid$ if (E) { S } else { S } S
Expressions	E ::=	v $\mid$ X $\mid$ E op E $\mid$ uop E
Values	v ::=	c $\mid$ false $\mid$ true

We analyze liquidity for a simple programming language, a lightweight version of *Stipula*, which is a domain-specific language that has been designed for programming *legal contracts* [7]. In *Stipula*, programs are *contracts* that transit from state to state and a control logic specifies what functionality can be invoked by which caller; the set of callers is defined when the contract is instantiated. Resources are assets (digital currencies, smart keys, non-fungible tokens, etc.) that may be moved with ad-hoc operators from one to another.

Our analyzer is built upon a type system that records the effects of functions on assets by using symbolic names. Then a *correctness* property, whereby the (liquidity) type of the final state of a computation is always an over-approximation of the actual state, allows us to safely reduce our analysis arguments to verifying if liquidity types of computations have assets that are empty.

We identify the liquidity property: *if an asset becomes not-empty in a state then there is a continuation where all the assets are empty in one of its states.* The algorithm for assessing liquidity looks for computations whose liquidity types $\varXi \rightarrow \varXi'$ are such that the contract's assets in $\varXi'$ are empty. The crucial issue of the analysis is termination, given that computations may be *infinitely many* because contracts may have cycles. For this reason we restrict to computations whose length is bound by a value. Actually we found more reasonable computations where every function can be invoked a bounded number of times. It turns out that the computational cost of the algorithm is exponential with respect to the number of functions.

The structure of the paper is as follows. The lightweight *Stipula* language is introduced in Sect. 2 and the semantics is defined in Sect. 3. Section 4 reports the theory underlying our liquidity analyzer and Sect. 5 illustrates the algorithm for verifying liquidity. We end our contribution by discussing the related work in Sect. 6 and delivering our final remarks in Sect. 7. Due to page limits, the technical material has been omitted and can be found in the full paper.

2   The **Stipula** Language

We use disjoint sets of names: *contract names*, ranged over by C, C$'$, $\cdots$; names referring to digital identities, called *parties*, ranged over by A, A$'$, $\cdots$; *function names* ranged over by f, g, $\cdots$; *asset names*, ranged over by h, k, $\cdots$, to be used both as contract's assets and function's asset parameters; *non asset names*, ranged over by x, y, $\cdots$, to be used both as contract's fields and function's non asset parameters. Assets and generic contract's fields are syntactically set apart since they have different semantics; similarly for functions' parameters. Names of assets, fields and parameters are generically ranged over by X. Names Q, Q', $\cdots$ will range over contract states. To simplify the syntax, we often use the vector notation $\overline{x}$ to denote possibly empty *sequences* of elements. With an abuse of notation, in the following sections, $\overline{x}$ will also represent the *set* containing the elements in the sequence.

The code of a *Stipula* contract is[1]

stipula C { parties $\overline{A}$ fields $\overline{x}$ assets $\overline{h}$ init Q $\overline{F}$ }

where C identifies the *contract name*; $\overline{A}$ are the *parties* that can invoke contract's functions, $\overline{x}$ and $\overline{h}$ are the *fields* and the *assets*, respectively, and Q is the *initial state*. The contract body also includes the sequence $\overline{F}$ of functions, whose syntax is defined in Table 1. It is assumed that the names of parties, fields, assets and functions do not contain duplicates and functions' parameters do not clash with the names of contract's fields and assets.

The declaration of a function highlights the state Q when the invocation is admitted, who is the caller party A, and the list of parameters. Function's parameters are split in two lists: the *formal parameters* $\overline{y}$ in brackets and the *asset parameters* $\overline{k}$ in square brackets. The *body* { S } ⇒ @Q′ specifies the *statement part* S and the state Q′ of the contract when the function execution terminates. We write @Q A : f($\overline{y}$) [$\overline{k}$] { S } ⇒ @Q′ ∈ C when $\overline{F}$ contains @Q A : f($\overline{y}$) [$\overline{k}$] { S } ⇒ @Q′ and we will often shorten the above predicate by writing Q A.f Q′ ∈ C. *Stipula* does not admit *internal nondeterminism*: for every Q, A and f, there is at most a Q′ such that Q A.f Q′ ∈ C.

Statements S include the empty statement _ and different prefixes followed by a continuation. Prefixes P use the two symbols → (*update*) and ⊸ (*move*) to differentiate operations on fields and assets, respectively. The prefix E → x updates the field or the parameter x with the value of E – the old value of x is lost – it is *destroyed*; E → A sends the value of E to the party A. The move operations $c \times h$ ⊸ h′ and $c \times h$ ⊸ A define actions that *never destroy resources*. In particular, $c \times h$ ⊸ h′ subtracts the value of $c \times h$ to the asset h and adds this value to h′, where c is a constant between 0 and 1. Notice that, because of this constraint, $c \times h$ is always smaller or equal to h – therefore assets never have negative values. It is also worth to notice that, according to the syntax, the right-hand side of → in E → x is always a field or a non-asset function parameter, while the right-hand side of ⊸ in $c \times h$ ⊸ h′ is always an asset (the left-hand side is an expression that indicates part of an asset). The operation $c \times h$ ⊸ A subtracts the value of $c \times h$ to the asset h and transfers it to A.

Statements also include *conditionals* if (E) { S } else { S' } with the standard semantics. In the rest of the paper we will always abbreviate $1 \times h$ ⊸ h′ and $1 \times h$ ⊸ A (which are very usual, indeed) into h ⊸ h′ and h ⊸ A, respectively.

Expressions E include constant values v, names X of either assets, fields or parameters, and both binary and unary operations. Constant values are

---

[1] Actually this is a lightweight version of the language in [7].

- real numbers n, that are written as nonempty sequences of digits, possibly followed by "." and by a sequence of digits (*e.g.* 13 stands for 13.0). The number may be prefixed by the sign + or -. Reals come with the standard set of binary arithmetic operations (+, -, ×) and the unary division operation E/c where $c \neq 0$, in order to avoid 0-division errors.
- boolean values false and true. The operations on booleans are conjunction &&, disjunction ||, and negation !.
- asset values that represent *divisible* resources (*e.g.* digital currencies). Divisible asset constants are assumed to be identical to positive real numbers (assets cannot have negative values).

Relational operations (<, >, <=, >=, ==) are available between any expression.

The standard definition of *free names* of expressions, statements and functions is assumed and will be denoted $fn(E)$, $fn(S)$ and $fn(F)$, respectively. A contract stipula C { parties $\overline{A}$ fields $\overline{x}$ assets $\overline{h}$ init Q $\overline{F}$ } is *closed* if, for every $F \in \overline{F}$, $fn(F) \subseteq \overline{A} \cup \overline{x} \cup \overline{h}$.

We illustrate relevant features of *Stipula* by means of few examples. Consider the Fill_Move contract

```
stipula Fill_Move { parties Alice,Bob   assets h1,h2   init Q0
    @Q0 Alice: fill()[k]{   k —o h2   } ⇒ @Q1
    @Q1 Bob: move()[]{   h2 —o h1   } ⇒ @Q0
    @Q0 Bob: end()[]{   h1 —o Bob   } ⇒ @Q2
}
```

that regulates interactions between Alice and Bob. It has two assets and three states Q0, Q1 and Q2, with initial state Q0. In Q0, Alice may move part of her asset by invoking fill; the asset is stored in the formal parameter k. That is, the party Alice is assumed to own some asset and the invocation, *e.g.* Alice.fill()[5.0], is removing 5 units from Alice's asset and storing them in k (and then in h2). Said otherwise, the total assets of the system is invariant during the invocation; similarly during the operations h —o h'. The execution of fill moves the assets in k to h2) and makes the contract transit to the state Q1. In this state, the unique admitted function is move by which Bob accumulates in h1 the assets sent by Alice. The contract's state becomes Q0 again and the behaviour may *cycle*. Fill_Move terminates when, in Q0, Bob decides to grab the whole content of h1. Notice the *nondeterministic* behaviour when Fill_Move is in Q0: according to a fill or an end function is invoked, the contract may transit in Q1 or Q2 (this is called *external* nondeterminism in the literature). Notice also that *Stipula* overlooks the details of the interactions with the parties (usually an asset transfer between the parties and the contract is mediated by a bank).

Fill_Move has the property that assets are eventually emptied (whatever it is the state of the contract – *the contract is liquid*). *Stipula* contracts do not always retain this property. For example, the following Ping_Pong contract

```
stipula Ping_Pong { parties Amy,Mary   assets h,k   init Q0
    @Q0 Mary: ping()[u]{   h —∘ Mary   u —∘ k   } ⇒ @Q1
    @Q1 Amy: pong()[v]{   k —∘ Amy   v —∘ h   } ⇒ @Q0
}
```

has a cyclic behaviour where Mary and Amy exchange asset values. By invoking ping, Mary moves her asset into u (and then into the asset field k) and grabs the value stored in h; conversely, by invoking pong, Amy moves her asset into v (and then into the asset field h) and grabs the value stored in k. (Since asset fields are initially empty, the first invocation of ping does not deliver anything to Mary.) Apart the initial state, Ping_Pong never reaches a state where h and k are both empty at the same time (if Mary and Amy invocations carry nonempty assets) – in the terminology of the next section, *the contract is not liquid*. States have either k empty – Q0 – or h empty – Q1.

3   Semantics

Let a *configuration*, ranged over by $\mathbb{C}$, $\mathbb{C}'$, $\cdots$, be a tuple $\mathbb{C}(\mathbb{Q}, \ell, \Sigma)$ where

- $\mathbb{C}$ is the contract name and $\mathbb{Q}$ is one of its states;
- ℓ, called *memory*, is a mapping from names (parties, fields, assets and function's parameters) to values. The values of parties are noted $A, A', \cdots$. These values cannot be passed as function's parameters and cannot be hard-coded into the source contracts, since they do not belong to expressions. We write $\ell[h \mapsto u]$ to specify the memory that binds h to u and is equal to ℓ otherwise;
- Σ is the (possibly empty) residual of a function body, *i.e.* Σ is either _ or a term $S \Rightarrow @\mathbb{Q}$.

Configurations such as $\mathbb{C}(\mathbb{Q}, \ell, \_)$, *i.e.* there is no statement to execute, are called *idle*.

We will use the auxiliary *evaluation function* $[\![E]\!]_\ell$ that returns the value of E in the memory ℓ such that:

- $[\![v]\!]_\ell = v$ for real numbers and asset values, $[\![\mathtt{true}]\!]_\ell = 1.0$ and $[\![\mathtt{false}]\!]_\ell = 0.0$ (booleans are converted to reals); $[\![X]\!]_\ell = \ell(X)$ for names of assets, fields and parameters. i
- let *uop* and *op* be the semantic operations corresponding to uop and op, then $[\![\mathtt{uop}\,E]\!]_\ell = uop\,v$, $[\![E\,\mathtt{op}\,E']\!]_\ell = v\,op\,v'$ with $[\![E]\!]_\ell = v, [\![E']\!]_\ell = v'$. In case of boolean operations, every non-null real corresponds to true and 0.0 corresponds to false; the operations return the reals for true and false. Because of the restrictions on the language, uop and op are always defined.

The semantics of *Stipula* is defined by a *transition relation*, noted $\mathbb{C} \xrightarrow{\mu} \mathbb{C}'$, that is given in Table 2, where μ is either empty or $A : \mathtt{f}(\overline{u})[\overline{v}]$ or $v \rightarrow A$ or $v \multimap A$. Rule [FUNCTION] defines invocations: the label specifies the party A performing the invocation and the function name f with the actual parameters. The transition may occur provided (*i*) the contract is in the state $\mathbb{Q}$ that admits

Table 2. The transition relation of *Stipula*

[FUNCTION]

$$\frac{@Q\ A : f(\overline{y})\,[\overline{k}]\ \{\ S\ \} \circledast @Q' \in C}{\ell(A) = A \qquad \ell' = \ell[\overline{y} \mapsto \overline{u}, \overline{k} \mapsto \overline{v}]}$$
$$\overline{C(Q, \ell, \_) \xrightarrow{A:f(\overline{u})[\overline{v}]} C(Q, \ell', S \circledast @Q')}$$

[STATE-CHANGE]

$$C(Q, \ell, \_ \circledast @Q') \longrightarrow C(Q', \ell, \_)$$

[VALUE-SEND]

$$\frac{[\![E]\!]_\ell = v \quad \ell(A) = A}{C(Q, \ell, E \to A\ \Sigma) \xrightarrow{v \to A} C(Q, \ell, \Sigma)}$$

[ASSET-SEND]

$$\frac{[\![c \times h]\!]_\ell = v \quad \ell(A) = A \quad [\![h - v]\!]_\ell = v'}{C(Q, \ell, c \times h \multimap A\ \Sigma) \xrightarrow{v \multimap A} C(Q, \ell[h \mapsto v'], \Sigma)}$$

[FIELD-UPDATE]

$$\frac{[\![E]\!]_\ell = v}{C(Q, \ell, E \to x\ \Sigma) \longrightarrow C(Q, \ell[x \mapsto v], \Sigma)}$$

[ASSET-UPDATE]

$$\frac{[\![c \times h]\!]_\ell = v \quad [\![h - v]\!]_\ell = v' \quad [\![h' + v]\!]_\ell = v''}{\ell' = \ell[h \mapsto v', h' \mapsto v'']}$$
$$C(Q, \ell, c \times h \multimap h'\ \Sigma) \longrightarrow C(Q, \ell', \Sigma)$$

[COND-TRUE]

$$\frac{[\![E]\!]_\ell = \mathtt{true}}{C(Q, \ell, \mathtt{if}\,(E)\,\{\,S\,\}\,\mathtt{else}\,\{\,S'\,\}\ \Sigma)}$$
$$\longrightarrow C(Q, \ell, S\ \Sigma)$$

[COND-FALSE]

$$\frac{[\![E]\!]_\ell = \mathtt{false}}{C(Q, \ell, \mathtt{if}\,(E)\,\{\,S\,\}\,\mathtt{else}\,\{\,S'\,\}\ \Sigma)}$$
$$\longrightarrow C(Q, \ell, S'\ \Sigma)$$

invocations of f from A such that $\ell(A) = A$ and (*ii*) the configuration is *idle*. Rule [STATE-CHANGE] says that a contract changes state when the execution of the statement in the function's body terminates. To keep simple the operational semantics of *Stipula*, we do not remove garbage names in the memories (the formal parameters of functions once the functions have terminated). Therefore memories retain such names and the formal parameters keep the value they have at the end of the function execution. These values are lost when the function is called again (*c.f.* rule [FUNCTION]: in ℓ', the assets $\overline{k}$ are updated with $\overline{v}$). A function that does not empty asset formal parameters is clearly incorrect and the following analysis will catch such errors.

Regarding statements, we only discuss [ASSET-SEND] and [ASSET-UPDATE] because the other rules are standard. Rule [ASSET-SEND] delivers part of an asset h to A. This part, named v, is removed from the asset, *c.f.* the memory of the right-hand side state in the conclusion. In a similar way, [ASSET-UPDATE] moves a part v of an asset h to an asset h'. For this reason, the final memory becomes $\ell[h \mapsto v', h' \mapsto v'']$, where $v' = \ell(h) - v$ and $v'' = \ell(h') + v$.

A contract $\mathtt{stipula\ C\ \{\ parties\ \overline{A}\ fields\ \overline{x}\ assets\ \overline{h}\ init\ Q\ \overline{F}\ \}}$ is invoked by $C(\overline{A}, \overline{u})$ that corresponds to the initial configuration

$$C(Q, [\overline{A} \mapsto \overline{A}, \overline{x} \mapsto \overline{u}, \overline{h} \mapsto \overline{0}], \_)\,.$$

We remark that no field and asset is left uninitialized, which means that no *undefined-value* error can occur during the execution by accessing to field and assets. Notice that the initial value of assets is 0. In order to keep the notation light we always assume that parties $\overline{A}$ are always instantiated by the corresponding names $\overline{A}$ written with italic fonts.

Below we use the following notation and terminology:

- We write $\mathbb{C} \xRightarrow{A.f(\overline{u})[\overline{v}]} \mathbb{C}'$ if $\mathbb{C} \xrightarrow{A.f(\overline{u})[\overline{v}]} \xrightarrow{\mu_1} \cdots \xrightarrow{\mu_n} \mathbb{C}'$ and μ_i are either empty or $v \multimap A$ or $v \to A$ and $\mathbb{C}'$ is idle.

– We write $\mathbb{C} \Longrightarrow \mathbb{C}'$ if $\mathbb{C} \overset{A_1.\mathtt{f}_1(\overline{u_1})[\overline{v_1}]}{\Longrightarrow} \cdots \overset{A_n.\mathtt{f}_n(\overline{u_n})[\overline{v_n}]}{\Longrightarrow} \mathbb{C}'$, for some $A_1.\mathtt{f}_1(\overline{u_1})[\overline{v_1}]$, $\cdots, A_n.\mathtt{f}_n(\overline{u_n})[\overline{v_n}]$. $\mathbb{C} \Longrightarrow \mathbb{C}'$ will be called *computation*.

An important property of closed contracts guarantees that the invocation of a function never fails. This property immediately follows by the fact that, in such contracts, the evaluation of expressions and statements can never rise an error (operations are total, names are always bound to values and type errors cannot occur because values are always converted to reals).

Theorem 1 (Progress). *Let* $\mathbb{C}$ *be a closed* Stipula *contract with fields* $\overline{x}$, *assets* $\overline{h}$, *parties* $\overline{A}$ *and* $@\mathbb{Q}$ $A\mathtt{:f}(\overline{y})[\overline{k}]\{\ S\ \} \Rightarrow @\mathbb{Q}' \in \mathbb{C}$. *For every* ℓ *such that* $\overline{x}, \overline{h}, \overline{A} \subseteq dom(\ell)$, *there is* ℓ' *such that* $\mathtt{C}(\mathbb{Q}, \ell, \_) \overset{A.\mathtt{f}(\overline{u})[\overline{v}]}{\Longrightarrow} \mathtt{C}(\mathbb{Q}', \ell', \_)$.

We conclude with the definition of liquidity. We use the following notation:

– we write $\ell(\overline{h}) > \overline{0}$ if and only if there is $\mathtt{k} \in \overline{h}$ such that $\ell(\mathtt{k}) > 0$; similarly $\ell(\overline{h}) = \overline{0}$ if and only if, for every $\mathtt{k} \in \overline{h}$, $\ell(\mathtt{k}) = 0$.

Definition 1 (Liquidity). *A* Stipula *contract* $\mathtt{C}$ *with assets* $\overline{h}$ *and initial configuration* $\mathbb{C}$ *is* liquid *if, for every computation* $\mathbb{C} \Longrightarrow \mathtt{C}(\mathbb{Q}, \ell, \_)$, *then*

1. $\ell(\overline{h}') = \overline{0}$ *with* $\overline{h}' = dom(\ell) \setminus \overline{h}$;
2. *if* $\ell(\overline{h}) > \overline{0}$ *then there is* $\mathtt{C}(\mathbb{Q}, \ell, \_) \Longrightarrow \mathtt{C}(\mathbb{Q}', \ell', \_)$ *such that* $\ell'(\overline{h}) = \overline{0}$.

We notice that Progress is critical for reducing liquidity to some form of reachability analysis (otherwise we should also deal with function invocations that terminate into a stuck state because of an error). In the following sections, using a symbolic technique, we define an algorithm for assessing liquidity and demonstrate its correctness.

4   The Theory of Liquidity

We begin with the definition of the *liquidity type system* that returns an abstraction of the input-output behaviour of functions with respect to assets. These abstractions record whether an asset is zero – notation $\mathbb{0}$ – or not – notation $\mathbb{1}$. The values $\mathbb{0}$ and $\mathbb{1}$ are called *liquidity values* and we use the following notation:

– *liquidity expressions* e are defined as follows, where $\xi, \xi', \cdots$ range over (symbolic) liquidity names:

$$e ::= \quad \mathbb{0} \ \mid\ \mathbb{1} \ \mid\ \xi \ \mid\ e \sqcup e \ \mid\ e \sqcap e.$$

They are ordered as $\mathbb{0} \leqslant e$ and $e \leqslant \mathbb{1}$; the operations $\sqcup$ and $\sqcap$ respectively return the maximum and the minimum value of the two arguments; they are monotone with respect to $\leqslant$ (that is $e_1 \leqslant e_1'$ and $e_2 \leqslant e_2'$ imply $e_1 \sqcup e_2 \leqslant e_1' \sqcup e_2'$ and $e_1 \sqcap e_2 \leqslant e_1' \sqcap e_2'$). A liquidity expression that does not contain liquidity names is called *ground*. Two tuples are ordered $\leqslant$ if they are element-wise ordered by $\leqslant$.

Table 3. The Liquidity type system of *Stipula*

[L-SEND]
$$\frac{\text{A}, fn(E) \subseteq \text{X} \cup dom(\Xi)}{\Xi \vdash_\text{x} E \to \text{A} : \Xi}$$

[L-UPDATE]
$$\frac{\text{x}, fn(E) \subseteq \text{X} \cup dom(\Xi)}{\Xi \vdash_\text{x} E \to \text{x} : \Xi}$$

[L-ASEND]
$$\frac{\text{h} \in dom(\Xi) \quad \text{A} \in \text{X}}{\Xi \vdash_\text{x} \text{h} \multimap \text{A} : \Xi[\text{h} \mapsto \mathbb{0}]}$$

[L-EXPASEND]
$$\frac{\text{h} \in dom(\Xi) \quad c \neq 1 \quad \text{A} \in \text{X}}{\Xi \vdash_\text{x} c \times \text{h} \multimap \text{A} : \Xi}$$

[L-AUPDATE]
$$\frac{\text{h}, \text{h}' \in dom(\Xi) \quad e = \Xi(\text{h}) \sqcup \Xi(\text{h}')}{\Xi \vdash_\text{x} \text{h} \multimap \text{h}' : \Xi[\text{h} \mapsto \mathbb{0}, \text{h}' \mapsto e]}$$

[L-EXPAUPD]
$$\frac{\text{h}, \text{h}' \in dom(\Xi) \quad c \neq 1 \quad e = \Xi(\text{h}) \sqcup \Xi(\text{h}')}{\Xi \vdash_\text{x} c \times \text{h} \multimap \text{h}' : \Xi[\text{h}' \mapsto e]}$$

[L-ZERO]
$$\Xi \vdash_\text{x} \_ : \Xi$$

[L-SEQ]
$$\frac{\Xi \vdash_\text{x} P : \Xi' \quad \Xi' \vdash_\text{x} S : \Xi''}{\Xi \vdash_\text{x} P \ S : \Xi''}$$

[L-COND]
$$\frac{fn(E) \subseteq \text{X} \cup dom(\Xi) \qquad \Xi \vdash_\text{x} S : \Xi' \quad \Xi \vdash_\text{x} S' : \Xi'' \qquad \Xi' \sqcup \Xi'' \vdash_\text{x} S'' : \Xi'''}{\Xi \vdash_\text{x} \text{if } (E) \{ S \} \text{ else } \{ S' \} \ S'' : \Xi'''}$$

[L-FUNCTION]
$$\frac{\text{A}, fn(S) \subseteq \text{X} \cup \overline{\text{y}} \qquad \overline{\xi'} \text{ fresh} \qquad \Xi[\overline{\text{k}} \mapsto \overline{\xi'}] \vdash_{\text{x} \cup \overline{\text{y}}} S : \Xi'}{\Xi \vdash_\text{x} @\text{Q A} : \text{f}(\overline{\text{y}})[\overline{\text{k}}]\{ S \} \Rightarrow @\text{Q}' : \text{Q A.f Q}' : \Xi[\overline{\text{k}} \mapsto \overline{\mathbb{1}}] \to \Xi'\{\overline{\mathbb{1}}/\overline{\xi'}\}}$$

[L-CONTRACT]
$$\frac{\overline{\xi} \text{ fresh} \quad \left([\overline{\text{h}} \mapsto \overline{\xi}] \vdash_{\overline{\text{A}} \cup \overline{\text{x}}} F : \mathcal{L}_F \right)^{F \in \overline{F}}}{\vdash \text{stipula C } \{\text{parties } \overline{\text{A}} \text{ fields } \overline{\text{x}} \text{ assets } \overline{\text{h}} \text{ init Q } \overline{F} \} : \bigcup_{F \in \overline{F}} \mathcal{L}_F}$$

- *environments* Ξ map contract's assets and asset parameters to liquidity expressions. Environments that map names to ground liquidity expressions are called *ground environments*.
- *liquidity function types* $\text{Q A.f Q}' : \Xi \to \Xi'$ where $\Xi \to \Xi'$ records the liquidity effects of fully executing the body of $\text{Q A.f Q}'$.
- *judgments* $\Xi \vdash_\text{x} S : \Xi'$ for statements and $\Xi \vdash_\text{x} @\text{Q A} : \text{f}(\overline{\text{x}})[\overline{\text{h}'}] \{ S \} \Rightarrow @\text{Q}' : \mathcal{L}$ for function definitions, where $\mathcal{L}$ is a liquidity function type. The set X contains party and field names.

The liquidity type system is defined in Table 3; below we discuss the most relevant rules. Asset movements have four rules – [L-ASEND], [L-EXPASEND] [L-AUPDATE] and [L-EXPAUPD] – according to whether the constant factor is 0 or not and whether the asset is moved to an asset or a party. According to [L-AUPDATE], the final asset environment of $\text{h} \multimap \text{h}'$ (which is an abbreviation for $1 \times \text{h} \multimap \text{h}'$) has h that is emptied and h' that gathers the value of h, henceforth the liquidity expression $\Xi(\text{h}) \sqcup \Xi(\text{h}')$. Notice that, when both h and h' are $\mathbb{0}$, the overall result is $\mathbb{0}$. In the rule [L-EXPAUPD], the asset h is decreased by an amount that is moved to h'. Since $c \neq 1$, the static analysis (which is independent of the runtime value of h) can only safely assume that the asset h is not emptied by this operation (if it was not empty before). Therefore, after the withdraw, the liquidity value of h has not changed. On the other hand, the asset h' is increased of some amount if both c and h have a non zero liquidity value, henceforth the expression $\Xi(\text{h}) \sqcup \Xi(\text{h}')$. In particular, as before, when both $\Xi(\text{h})$ and $\Xi(\text{h}')$ are $\mathbb{0}$, the overall result is $\mathbb{0}$.

The rule for conditionals is [L-COND], where the operation $\sqcup$ on environments is defined pointwise by $(\Xi' \sqcup \Xi'')(\text{h}) = \Xi'(\text{h}) \sqcup \Xi''(\text{h})$. That is, the liquidity analyzer over-approximates the final environments of if $(E)\,\{\,S\,\}$ else $\{\,S'\,\}$ by taking the maximum values between the results of parsing S (that corresponds to a true value of E) and those of S' (that corresponds to a false value of E). Regarding E, the analyzer only verifies that its names are bound in the contract.

The rule for *Stipula* contracts is [L-CONTRACT]; it collects the liquidity labels $\mathcal{L}_i$ that describe the liquidity effects of each contract's function; each function assumes injective environments that respectively associate contract's assets with fresh symbolic names. In turn, the type produced by [L-FUNCTION] says that the complete execution of Q A.f Q' has liquidity effects $\Xi[\overline{\text{h}'} \mapsto \overline{\mathbb{1}}] \to \Xi'\{\overline{\mathbb{1}}/\overline{\xi'}\}$, assuming that the body S of the function is typed as $\Xi[\overline{\text{h}'} \mapsto \overline{\xi'}] \vdash S : \Xi'$. That is, in the conclusion of [L-FUNCTION] we replace the symbolic values of the liquidity names representing formal parameters with $\mathbb{1}$, because they may be any value when the function will be called.

Example 1. The set $\mathcal{L}$ of the Fill_Move contract contains the following liquidity types:

$$\text{Q0 Alice.fill Q1} : \big[\text{h1} \mapsto \xi_1, \text{h2} \mapsto \xi_2, \text{k} \mapsto \mathbb{1}\big] \to \big[\text{h1} \mapsto \xi_1, \text{h2} \mapsto \xi_2 \sqcup \mathbb{1}, \text{k} \mapsto \mathbb{0}\big]$$
$$\text{Q1 Bob.move Q0} : \big[\text{h1} \mapsto \xi_1, \text{h2} \mapsto \xi_2\big] \to \big[\text{h1} \mapsto \xi_1 \sqcup \xi_2, \text{h2} \mapsto \mathbb{0}\big]$$
$$\text{Q0 Bob.end Q2} : \big[\text{h1} \mapsto \xi_1, \text{h2} \mapsto \xi_2\big] \to \big[\text{h1} \mapsto \mathbb{0}, \text{h2} \mapsto \xi_2\big]$$

In the following we will always shorten $\vdash$ stipula C {parties $\overline{\text{A}}$ fields $\overline{\text{x}}$ assets $\overline{\text{h}}$ init Q $\overline{F}$ } : $\mathcal{L}$ into $\vdash$ C : $\mathcal{L}$. A first property of the liquidity type system is that typed contracts are closed.

Proposition 1. *If* $\vdash$ C : $\mathcal{L}$ *then* C *is closed.*

Therefore typed contracts own the progress property (Theorem 1). The correctness of the system in Table 3 requires the following notions:

- A *(liquidity) substitution* is a map from liquidity names to liquidity expressions (that may contain names, as well). Substitutions will be noted either $\sigma, \sigma', \cdots$ or $\{\overline{\mathbb{e}}/\overline{\text{x}}\}$. A substitution is *ground* when it maps liquidity names to ground liquidity expressions. For example $\{\mathbb{0},\mathbb{1}/\text{x},\varsigma\}$ and $\{\mathbb{0}\sqcup\mathbb{1},\mathbb{1}\sqcap\mathbb{0}/\text{x},\varsigma\}$ are ground substitutions, $\{\mathbb{0}\sqcup\text{x}'/\text{x}\}$ is not.
  We let $\sigma(\Xi)$ be the environment where $\sigma(\Xi)(\text{x}) = \sigma(\Xi(\text{x}))$.
- Let $\llbracket \mathbb{e} \rrbracket$ be the *partial evaluation* of $\mathbb{e}$ by applying the commutativity axioms of $\sqcup$ and $\sqcap$ and the axioms $\mathbb{0} \sqcup \mathbb{e} = \mathbb{e}$, $\mathbb{0} \sqcap \mathbb{e} = \mathbb{0}$, $\mathbb{1} \sqcup \mathbb{e} = \mathbb{1}$, $\mathbb{1} \sqcap \mathbb{e} = \mathbb{e}$. More precisely

$$[\![e]\!] = \begin{cases} e & \text{if } e = 0 \ \text{ or } \ e = 1 \ \text{ or } \ e = \xi \\ [\![e']\!] & \text{if } (e = e' \sqcup e'' \ \text{ or } \ e = e'' \sqcup e') \ \text{ and } \ [\![e'']\!] = 0 \\ [\![e']\!] & \text{if } (e = e' \sqcap e'' \ \text{ or } \ e = e'' \sqcap e') \ \text{ and } \ [\![e'']\!] = 1 \\ 0 & \text{if } e = e' \sqcap e'' \ \text{ and either } [\![e']\!] = 0 \ \text{ or } \ [\![e'']\!] = 0 \\ 1 & \text{if } e = e' \sqcup e'' \ \text{ and either } [\![e']\!] = 1 \ \text{ or } \ [\![e'']\!] = 1 \\ [\![e']\!] \# [\![e'']\!] & \text{otherwise } (\# \ \text{ is either } \ \sqcup \ \text{ or } \ \sqcap) \end{cases}$$

Notice that if e is ground then $[\![e]\!]$ is either 0 or 1.

- We let $[\![\Xi]\!]$ be the environment where $[\![\Xi]\!](x) = [\![\Xi(x)]\!]$. Therefore, when Ξ is ground, $[\![\Xi]\!]$ is ground as well. The converse is false.
- When Ξ and Ξ' are ground, we write $\Xi \leqslant \Xi'$ if and only if, for every $h \in dom(\Xi)$, $[\![\Xi(h)]\!] \leqslant [\![\Xi'(h)]\!]$. Observe that this implies that $dom(\Xi) \subseteq dom(\Xi')$.
- $\Xi|_{\bar{h}}$ is the environment Ξ restricted to the names $\bar{h}$, defined as follows

$$\Xi|_{\bar{h}}(k) = \begin{cases} \Xi(k) & \text{if } k \in \bar{h} \\ \text{undefined} & \text{otherwise} \end{cases}$$

- let $\ell = [\bar{A} \mapsto \bar{A}, \bar{x'} \mapsto \bar{u}, \bar{h'} \mapsto \bar{v}]$ be a memory, where $\bar{x'}$ are contract's fields and non-asset parameters, while $\bar{h'}$ are contract's assets and the asset parameters. We let $\mathbb{E}(\ell)$ be the ground environment defined as follows:

$$\mathbb{E}(\ell)(k) = \begin{cases} 0 & \text{if } k \in \bar{h'} \ \text{and } \ell(k) = 0 \\ 1 & \text{if } k \in \bar{h'} \ \text{and } \ell(k) \neq 0 \\ \text{undefined} & \text{otherwise} \end{cases}$$

Liquidity types are correct, as stated by the following theorem.

Theorem 2 (Correctness of liquidity labels). *Let* $\vdash C : \mathcal{L}$ *and* $@Q\,A : f\,(\bar{y})\,[\bar{k}]\,\{\,S\,\} \Rightarrow @Q' \in C$ *and* $Q\,A.f\,Q' : \Xi \to \Xi'$ *in* $\mathcal{L}$. *If* $C(Q, \ell, \_) \xrightarrow{A.f(\bar{u})[\bar{v}]} C(Q', \ell', \_)$ *then there are* X *and* Ξ'' *such that:*

1. $\mathbb{E}(\ell[\bar{y} \mapsto \bar{u}, \bar{k} \mapsto \bar{v}]) \vdash_X S : \Xi''$;
2. $\mathbb{E}(\ell[\bar{y} \mapsto \bar{u}, \bar{k} \mapsto \bar{v}])|_{dom(\Xi)} \leqslant [\![\sigma(\Xi)]\!]$ *and* $[\![\Xi''|_{dom(\Xi)}]\!] \leqslant [\![\sigma(\Xi')]\!]$, *for a ground substitution* σ;
3. $\mathbb{E}(\ell')|_{dom(\Xi)} \leqslant [\![\Xi'']\!]$.

For example, consider the transition

$$\texttt{Fill\_Move}(Q1, \ell, \_) \xrightarrow{Bob.\texttt{move}()[]} \texttt{Fill\_Move}(Q0, \ell', \_)$$

of `Fill_Move` where $\ell = [\texttt{h1} \mapsto 25.0, \texttt{h2} \mapsto 5.0]$ and $\ell' = [\texttt{h1} \mapsto 30.0, \texttt{h2} \mapsto 0.0]$. By definition, $\mathbb{E}(\ell) = [\texttt{h1} \mapsto 1, \texttt{h2} \mapsto 1]$. Letting $X = \{\texttt{Alice}, \texttt{Bob}\}$, by the liquidity type system we obtain $\mathbb{E}(\ell) \vdash_X \texttt{h2} \multimap \texttt{h1} : \Xi''$, $\Xi'' = [\texttt{h1} \mapsto 1 \sqcup 1, \texttt{h2} \mapsto 0]$. Since $Q1\ \texttt{Bob.move}\ Q0 : [\texttt{h1} \mapsto \xi_1, \texttt{h2} \mapsto \xi_2] \to [\texttt{h1} \mapsto \xi_1 \sqcup \xi_2, \texttt{h2} \mapsto 0]$ (see Example 1), the ground substitution σ that satisfies Theorem 2.2 is $[\xi_1 \mapsto 1, \xi_2 \mapsto 1]$ (actually, in this case, the "$\leqslant$" are equalities). Regarding the last item, $\mathbb{E}(\ell') = [\texttt{h1} \mapsto 1, \texttt{h2} \mapsto 0]$ and $\mathbb{E}(\ell') \leqslant [\![\Xi'']\!]$ follows by definition.

A basic notion of our theory is the one of abstract computation and its liquidity type.

Definition 2. *An* abstract computation *of a* Stipula *contract, ranged over by* $\varphi, \varphi', \cdots$, *is a finite sequence* $\mathsf{Q}_1 \; \mathsf{A}_1.\mathsf{f}_1 \; \mathsf{Q}_2 \; ; \; \cdots \; ; \; \mathsf{Q}_n \; \mathsf{A}_n.\mathsf{f}_n \; \mathsf{Q}_{n+1}$ *of contract's functions, shortened into* $\{ \mathsf{Q}_i \; \mathsf{A}_i.\mathsf{f}_i \; \mathsf{Q}_{i+1} \}^{i \in 1..n}$. *We use the notation* $\mathsf{Q} \overset{\varphi}{\curvearrowright} \mathsf{Q}'$ *to highlight the initial and final states of* φ *and we let* $\{ \mathsf{Q}_i \; \mathsf{A}_i.\mathsf{f}_i \; \mathsf{Q}_{i+1} \}^{i \in 1..n}$ *be the abstract computation of* $\left(\mathsf{C}(\mathsf{Q}_i \, , \, \ell_i \, , \, \_) \overset{A_i:\mathsf{f}_i(\overline{u_i})[\overline{v_i}]}{\Longrightarrow} \mathsf{C}(\mathsf{Q}_{i+1} \, , \, \ell_{i+1} \, , \, \_) \right)^{i \in 1..n}$.

An abstract computation φ *is* κ-canonical *if functions occur at most* κ-*times in* φ.

We notice that abstract computations do not mind of memories. Regarding canonical computations, every prefix of a κ-canonical computation is κ-canonical as well, including the empty computation.

Definition 3 (Liquidity type of an abstract computation). *Let* $\vdash \mathsf{C} : \mathcal{L}$ *and* $\overline{\mathsf{h}}$ *be the assets of* C. *Let also* $\mathsf{Q}_i \; \mathsf{A}_i.\mathsf{f}_i \; \mathsf{Q}_{i+1} : \Xi_i \to \Xi_i' \in \mathcal{L}$ *for every* $i \in 1..n$. *The* liquidity type *of* $\varphi = \{ \mathsf{Q}_i \; \mathsf{A}_i.\mathsf{f}_i \; \mathsf{Q}_{i+1} \}^{i \in 1..n}$, *noted* L_φ, *is* $\Xi_1^{(b)}|_{\overline{\mathsf{h}}} \to \Xi_n^{(e)}|_{\overline{\mathsf{h}}}$ *where* $\Xi_1^{(b)}$ *and* $\Xi_n^{(e)}$ *("b" stays for* begin, *"e" stays for* end*) are defined as follows*

$$\Xi_1^{(b)} = \Xi_1 \qquad \Xi_{i+1}^{(b)} = \Xi_{i+1}\{\Xi_i^{(e)}(\overline{\mathsf{h}})/\overline{\xi}\} \qquad \Xi_i^{(e)} = \Xi_i'\{\Xi_i^{(b)}(\overline{\mathsf{h}})/\overline{\xi}\} \; .$$

Notice that, by definition, the initial environment of the i-th type is updated so that it maps assets to the values computed at the end of the $(i-1)$-th transition. These values are also propagated to the final environment of the i-th transitions by substituting the occurrence of a liquidity name with the computed value of the corresponding asset. Notice also that the domains of the environments $\Xi_i^{(b)}$, $1 \leqslant i \leqslant n$, are in general different because they are also defined on the asset parameters of the corresponding function. However, formal parameters are not relevant because they are always replaced by $\mathbb{1}$ and are therefore dropped in the liquidity types of computations.

For example, consider the computation of the `Fill_Move` contract

$$\varphi = \mathsf{Q0} \; \mathsf{Alice.fill} \; \mathsf{Q1} \; ; \; \mathsf{Q1} \; \mathsf{Bob.move} \; \mathsf{Q0} \; ; \; \mathsf{Q0} \; \mathsf{Bob.end} \; \mathsf{Q2}$$

(we refer to Example 1 for the types of the contract). Let $H = \{\mathsf{h1}, \mathsf{h2}\}$ and $\Xi = [\mathsf{h1} \mapsto \xi_1, \mathsf{h2} \mapsto \xi_2]$. φ has liquidity type $\Xi_1^{(b)}|_H \to \Xi_3^{(e)}|_H$ where:

$$\Xi_1^{(b)} = \Xi[\mathsf{k} \mapsto \mathbb{1}]$$
$$\Xi_2^{(b)} = \Xi\{\xi_2 \sqcup \mathbb{1}/\xi_2\}$$
$$\qquad = \Xi[\mathsf{h2} \mapsto \xi_2 \sqcup \mathbb{1}]$$
$$\Xi_3^{(b)} = \Xi\{\xi_1 \sqcup \xi_2 \sqcup \mathbb{1}, 0/\xi_1, \xi_2\}$$
$$\qquad = [\mathsf{h1} \mapsto \xi_1 \sqcup \xi_2 \sqcup \mathbb{1}, \mathsf{h2} \mapsto 0]$$

$$\Xi_1^{(e)} = \Xi[\mathsf{h2} \mapsto \xi_2 \sqcup \mathbb{1}, \mathsf{k} \mapsto 0]$$
$$\Xi_2^{(e)} = [\mathsf{h1} \mapsto \xi_1 \sqcup \xi_2, \mathsf{h2} \mapsto 0]\{\xi_2 \sqcup \mathbb{1}/\xi_2\}$$
$$\qquad = [\mathsf{h1} \mapsto \xi_1 \sqcup \xi_2 \sqcup \mathbb{1}, \mathsf{h2} \mapsto 0]$$
$$\Xi_3^{(e)} = [\mathsf{h1} \mapsto 0, \mathsf{h2} \mapsto \xi_2]\{\xi_1 \sqcup \xi_2 \sqcup \mathbb{1}, 0/\xi_1, \xi_2\}$$
$$\qquad = [\mathsf{h1} \mapsto 0, \mathsf{h2} \mapsto 0]$$

Therefore $\mathsf{L}_\varphi = \Xi \to [\mathsf{h1} \mapsto 0, \mathsf{h2} \mapsto 0]$. That is, whatever they are the initial values of $\mathsf{h1}$ and $\mathsf{h2}$, which are represented in Ξ by the liquidity names ξ_1 and

ξ_2, respectively, their liquidity values after the computation φ are 0 (henceforth they are 0 by the following Theorem 3). Notice also the differences between L_φ and $\varXi \to [\mathtt{h1} \mapsto 0, \mathtt{h2} \mapsto \xi_2]$, which is the type of $\mathtt{Bob.end}$: from this last type we may derive that $\mathtt{h1}$ is 0 in the final environment, while the value of $\mathtt{h2}$ is the same of the initial environment (in fact, $\mathtt{Bob.end}$ only empties $\mathtt{h1}$ and does not access to $\mathtt{h2}$).

We recall that the operational semantics of *Stipula* in Table 2 does not remove garbage names in the memories (the formal parameters of functions once the functions have terminated, see Sect. 3). However, these names do not exist in environments of the liquidity types of abstract computations. For this reason, in the following statement, we restrict the inequalities to the names of the contract's assets.

Theorem 3 (Correctness of an abstract computation). *Let* $\vdash \mathtt{C} : \mathcal{L}$ *and*
$$\left(\mathtt{C}(\mathtt{Q}_i\,,\,\ell_i\,,\,\_)\xrightarrow{\;A_i : \mathtt{f}_i\,(\overline{u_i})\,[\overline{v_i}]\;}\mathtt{C}(\mathtt{Q}_{i+1}\,,\,\ell_{i+1}\,,\,\_))\right)^{i \in 1..n} \quad and\ the\ abstract\ computation$$
$\varphi = \{\ \mathtt{Q}_i\ A_i.\mathtt{f}_i\ \mathtt{Q}_{i+1}\ \}^{i \in 1..n}$ *have liquidity type* $L_\varphi = \varXi \to \varXi'$.

Then there is a substitution σ *such that* $\mathbb{E}(\ell_1)|_{\overline{\mathtt{h}}} \leqslant [\![\sigma(\varXi)]\!]$ *and* $\mathbb{E}(\ell_{n+1})|_{\overline{\mathtt{h}}} \leqslant [\![\sigma(\varXi')]\!]$.

5   The Algorithm for Liquidity

Analyzing the liquidity of a *Stipula* contract amounts to verifying the two constraints of Definition 1. Checking constraint 1 is not difficult: for every transition $\mathtt{Q\ A.f\ Q'}$ of the contract with assets $\overline{\mathtt{h}}$, we consider its liquidity type $\varXi \to \varXi'$ and verify whether, for every parameter $\mathtt{k} \notin \overline{\mathtt{h}}$, $[\![\varXi'(\mathtt{k})]\!] = 0$. Since there are finitely many transitions, this analysis is exhaustive. The correctness is the following: if $\mathtt{k} \notin \overline{\mathtt{h}}$ implies $[\![\varXi'(\mathtt{k})]\!] = 0$ then, for every substitution σ, $[\![\sigma(\varXi')]\!](\mathtt{k}) = 0$. Specifically for the substitution σ' such that $\mathbb{E}(\ell') \leqslant [\![\sigma'(\varXi')]\!]$, which is guaranteed by Theorem 2.

On the contrary, verifying the constraint 2 of Definition 1 is harder because the transition system of a *Stipula* contract may be complex (cycles, absence of final states, nondeterminism).

We first define the notions of reachable function and reachable state of a *Stipula* contract $\mathtt{C}$. Let $\vdash \mathtt{C} : \mathcal{L}$; $\mathbb{T}_\mathtt{Q}^\kappa$ is *the set of κ-canonical liquidity types* $\mathtt{Q} \overset{\varphi}{\curvearrowright} \mathtt{Q}' : L_\varphi$ where φ is a κ-canonical abstract computation starting at $\mathtt{Q}$ in the contract (the contract is left implicit). Notice that, by definition, the empty computation $\mathtt{Q} \overset{\varepsilon}{\curvearrowright} \mathtt{Q} : \varXi \to \varXi$ belongs to $\mathbb{T}_\mathtt{Q}^\kappa$. We say that $\mathtt{Q}'$ *is reachable* from $\mathtt{Q}$ if there is φ such that $\mathtt{Q} \overset{\varphi}{\curvearrowright} \mathtt{Q}' : L_\varphi$ is in $\mathbb{T}_\mathtt{Q}^\kappa$. For example, in the $\mathtt{Fill\_Move}$ contract, the set $\mathbb{T}_{\mathtt{Q0}}^\kappa$ is the following

$$\mathbb{T}_{\mathtt{Q0}}^\kappa = \bigcup_{i \leqslant \kappa, j \leqslant 1}\left(\mathtt{Q0\ Alice.fill\ Q1}\ ;\ \mathtt{Q1\ Bob.move\ Q0}\right)^i ; \left(\mathtt{Q0\ Bob.move\ Q2}\right)^{j \leqslant 1} : L_{i,j}$$

$$\cup \bigcup_{i \leqslant \kappa-1}\left(\mathtt{Q0\ Alice.fill\ Q1}\ ;\ \mathtt{Q1\ Bob.move\ Q0}\right)^i ; \mathtt{Q0\ Alice.fill\ Q1} : L_i$$

(with suitable $L_{i,j}$ and L_i).

Table 4. The algorithm for liquidity – $\mathcal{Z}$ contains pairs $(\mathbb{Q}, \overline{k})$

Let $\mathbb{Q}$ be the initial state of C whose assets are $\overline{h}$.

step 1. Compute $\mathbb{T}_{\mathbb{Q}'}^{\kappa}$ for every $\mathbb{Q}'$ reachable from $\mathbb{Q}$; let $\mathcal{Z} = \varnothing$.

step 2. For every $\mathbb{Q}'$ and $\mathbb{Q}' \overset{\varphi}{\curvearrowright} \mathbb{Q}'' : \Xi \to \Xi' \in \mathbb{T}_{\mathbb{Q}'}^{\kappa}$ and $\varnothing \subsetneq \overline{k} \subseteq \overline{h}$ such that

    (*a*) for every $k' \in \overline{k}$, $[\![\Xi'(k')]\!] \neq 0$ and $[\![\Xi'(k')]\!] \neq [\![\Xi(k')]\!]$

    (*b*) for every $k'' \in \overline{h} \setminus \overline{k}$, $[\![\Xi'(k'')]\!] = [\![\Xi(k'')]\!]$

    (*c*) $(\mathbb{Q}'', \overline{k}) \notin \mathcal{Z}$:

  2.1 If there is no $\mathbb{Q}'$, $\mathbb{Q}' \overset{\varphi}{\curvearrowright} \mathbb{Q}'' : \Xi \to \Xi'$ and $\overline{k}$ then exit: *the contract is liquid.*

  2.2 otherwise verify whether there is $\mathbb{Q}'' \overset{\varphi'}{\curvearrowright} \mathbb{Q}''' : \Xi'' \to \Xi''' \in \mathbb{T}_{\mathbb{Q}'}^{\kappa}$, such that $[\![\Xi'''(\overline{k})]\!] = \overline{0}$ and, for every $k' \in \overline{h} \setminus \overline{k}$, either $[\![\Xi'''(k')]\!] = 0$ or $[\![\Xi'''(k')]\!] = [\![\Xi''(k')]\!]$. If this is the case, add $(\mathbb{Q}'', \overline{k})$ to $\mathcal{Z}$ and reiterate **step 2**, otherwise exit: *the contract is not liquid.*

Below, without loss of generality, we assume that every state in the contract is reachable from the initial state. A straightforward optimization allows us to reduce to this case. We also assume that our contracts satisfy item 1 of liquidity. Therefore we focus on item 2.

Verifying liquidity is complex because a single asset or a tuple of assets may become 0 during *a computation*, rather than just one transition. Let us discuss the case with an example. Consider the Ugly contract with assets w1 and w2 and functions:

```
@Q0 Mark: get()[u]{    u ⊸ w2    } ⇒ @Q1
@Q1 Sam: shift()[]{    w1 ⊸ Sam    w2 ⊸ w1    } ⇒ @Q1
@Q1 Sam: end()[]{ } ⇒ @Q2
```

In this case there is no single transition that empties all the assets. However there is a liquid computation (a computation that empties all the assets), which is the one invoking shift two times: Q1 Sam.shift Q1 ; Q1 Sam.shift Q1. In particular, we have

Q1 Sam.shift Q1 : $[\text{w1} \mapsto \xi_1, \text{w2} \mapsto \xi_2] \to [\text{w1} \mapsto 0 \sqcup \xi_2, \text{w2} \mapsto 0]$

Q1 Sam.shift Q1 ; Q1 Sam.shift Q1 : $[\text{w1} \mapsto \xi_1, \text{w2} \mapsto \xi_2] \to [\text{w1} \mapsto 0, \text{w2} \mapsto 0]$

(we have simplified the final environment). That is, in this case, liquidity requires the analysis of 2-canonical computations to be assessed. (When the contract has no cycle, 1-canonical computations are sufficient to verify liquidity.) Since we have to consider cycles, in order to force termination, we restrict our analysis to κ-canonical abstract computations (with a finite value of κ).

The algorithm uses the set $\mathbb{T}_{\mathbb{Q}'}^{\kappa}$, for every state $\mathbb{Q}'$ of the contract that is reachable from $\mathbb{Q}$ – see step 1 of Table 4. Step 2 identifies the "critical pairs" $(\mathbb{Q}'', \overline{k})$ such that there is a computation updating the assets $\overline{k}$ and terminating in the state $\mathbb{Q}''$. Assume that $(\mathbb{Q}'', \overline{k}) \notin \mathcal{Z}$. Then we must find $\mathbb{Q}'' \overset{\varphi'}{\curvearrowright} \mathbb{Q}''' : \Xi'' \to \Xi'''$ in $\mathbb{T}_{\mathbb{Q}'}^{\kappa}$, such that $\Xi'''(\overline{k}) = \overline{0}$ and the other assets in $\overline{h} \setminus \overline{k}$ are either 0 or equal to the corresponding value in Ξ''. That is, as for the efficient algorithms, assets $\overline{h} \setminus \overline{k}$ have not been modified by φ' and may be overlooked. Notice that these checks are exactly those defined in step 2.2. If no liquidity type $Q'' \overset{\varphi'}{\curvearrowright} Q''' : \Xi'' \to \Xi'''$ is found in $\mathbb{T}_{Q'}^{\kappa}$, such that $\Xi'''(\overline{k}) = 0$, the liquidity cannot be guaranteed and the algorithm exits stating that the contract is not liquid (which might be a false negative because the liquidity type might exist in $\mathbb{T}_{Q''}^{\kappa+1}$).

For example, in case of the Fill_Move contract, the liquidity algorithm spots $Q0 \overset{\text{Alice.fill}}{\curvearrowright} Q1 : \Xi \to \Xi'$ because $[\![\Xi'(\text{h1})]\!] \neq [\![\Xi(\text{h1})]\!]$. Therefore it parses the liquidity types in $\mathbb{T}_{Q1}^{1}$ and finds the type $Q1 \overset{\varphi}{\curvearrowright} Q2 : [\text{h1} \mapsto \xi_1, \text{h2} \mapsto \xi_2] \to [\text{h1} \mapsto 0, \text{h2} \mapsto 0]$, where $\varphi' = Q1$ Bob.move $Q0$; $Q0$ Bob.end $Q2$. Henceforth $(Q1, \text{h1})$ is added to $\mathcal{Z}$. There is also another problematic type: $Q1 \overset{\text{Bob.move}}{\curvearrowright} Q0 : \Xi'' \to \Xi'''$, because $[\![\Xi'''(\text{h1})]\!] \neq [\![\Xi''(\text{h1})]\!]$. In this case, the liquidity type of the abstract computation $Q0$ Bob.end $Q2$ (still in $\mathbb{T}_{Q0}^{1}$) satisfies the liquidity constraint. We leave this check to the reader.

The correctness of the algorithm follows from Theorem 3. For example, assume that step 2.2 returns $Q \overset{\varphi}{\curvearrowright} Q' : \Xi \to \Xi'$ where $\Xi'(\overline{k}) = \overline{0}$ and $\Xi'(\overline{h} \setminus \overline{k}) = \Xi(\overline{h} \setminus \overline{k})$. Then, for every initial memory ℓ, the concrete computation corresponding to φ ends in a memory ℓ' such that $\mathbb{E}(\ell')|_{\overline{k}} \leqslant [\![\sigma(\Xi'|_{\overline{k}})]\!] = [\![\sigma([\overline{k} \mapsto \overline{0}])]\!] = [\overline{k} \mapsto \overline{0}]$ (for every σ) and $\mathbb{E}(\ell')|_{\overline{h} \setminus \overline{k}} \leqslant [\![\sigma(\Xi'|_{\overline{h} \setminus \overline{k}})]\!] = [\![\sigma([\overline{h} \setminus \overline{k} \mapsto \overline{\xi}])]\!] = [\overline{k} \mapsto \overline{1}]$, by taking a $\sigma = [\overline{\xi} \mapsto \overline{1}]$. As regards termination, the set $\mathcal{Z}$ increases at every iteration. When no other pair can be added to $\mathcal{Z}$ (and we have not already exited) the algorithm terminates by declaring the contract as liquid.

Proposition 2. *Let* $\vdash C : \mathcal{L}$. *If the algorithm of Table 4 returns that* C *is liquid then it is liquid. Additionally, the algorithm always terminates.*

The computational cost of liquidity is the following. Let n be the size of the *Stipula* contract (the number of functions, prefixes and conditionals in the code), h be the number of assets, m be the number of states and m' be the number of functions. Then

- the cost of the inference of liquidity types is linear with respect to the size of the contract, *i.e.* $O(n)$;
- the length of κ-canonical traces starting in a state is less than $\kappa \times m'$; therefore the cardinality of $\mathbb{T}_Q^{\kappa}$ is bounded by $\sum_{0 \leqslant i \leqslant \kappa \times m'} i!$. The cost of computing $\mathbb{T}_Q^{\kappa}$, for every Q, and the liquidity types of the elements therein is proportional to the number of κ-canonical traces, which are $N = m \times (\sum_{0 \leqslant i \leqslant \kappa \times m'} i!)$;
- the cost for verifying steps 2 and 3 of the algorithm is $O(N \times N \times 2^h)$ because, for every κ-canonical trace and every subset of $\overline{h}$, we must look for a κ-canonical trace satisfying step 3.

Therefore the overall cost of the algorithm is $O(n + N + N^2 \times 2^h)$, which means it is $O(N^2)$, *i.e.* exponential with respect to m', assuming the other values are in linear relation with m'.

6   Related Works

Liquidity properties have been put forward by Tsankov et al. in [14] as the property of a smart contract to always admit a trace where its balance is decreased (so, the funds stored within the contract do not remain frozen). Later, Bartoletti and Zunino in [3] discussed and extended this notion to a general setting – the Bitcoin language – that takes into account the strategy that a participant (which is possibly an adversary) follows to perform contract actions. More precisely, they observe that there are many possible flavours of liquidity, depending on which participants are assumed to be honest and on what are the strategies. In the taxonomy of [2,3], the notion of liquidity that we study in this work is the so-called *multiparty strategyless liquidity*, which assumes that all the contract's parties cooperate by actually calling the functions provided by the contract. We notice that, without cooperation, there is no guarantee that a party that has the permission to call a function will actually call it.

Both [14] and [2,3] adopt a model checking technique to verify properties of contracts. However, while [14] uses finite state models and the Uppaal model checker to verify the properties, [2,3] targets infinite state system and reduces them to finite state models that are consistent and complete with respect to liquidity. This mean that the technique of [2,3] is close to ours (we also target infinite state models and reduce to finite sets of abstract computations that over-approximate the real ones), even if we stick to a symbolic approach. Last, the above contributions and the ones we are aware of in the literature always address programs with one asset only (the contract balance). In this work we have understood that analyzing liquidity in programs with several different assets is way more complex than the case with a single asset.

A number of research projects are currently investigating the subject of resource-aware programming, as the prototype languages Obsidian [6] Nomos [4, 9], Marlowe [12] and Scilla [13]. As discussed in the empirical study [6], programming with linear types, ownership and assets is difficult and the presence of strong type systems can be an effective advantage. In fact, the above languages provide type systems that guarantee that assets are not accidentally lost, even if none of them address liquidity. More precisely, Obsidian uses types to ensure that owning references to assets cannot be lost unless they are explicitly disowned by the programmer. Nomos uses a linear type system to prevent the duplication or deletion of assets and amortized resource analysis to statically infer the resource cost of transactions. Marlowe [12], being a language for financial contracts, does not admit that money be locked forever in a contract. In particular, Marlowe's contracts have a finite lifetime and, at the end of the lifetime, any remaining money is returned to the participants. In other terms, all contracts are liquid by construction. In the extension of *Stipula* with events, the finite lifetime constraint can be explicitly programmed: a contract issues an event at the beginning so that at the timeout all the contract's assets are sent to the parties. Finally, Scilla is an intermediate-level language for safe smart contracts that is based on System F and targets a blockchain. It is unquestionable that a blockchain implementation of *Stipula* would bring in the advantages of a public and decentralised platform, such as traceability and the enforcement of contractual conditions. Being Scilla a minimalistic language with a formal semantics and a powerful type system, it seems an excellent candidate for implementing *Stipula*.

7 Conclusions

We have studied liquidity, a property of programs managing resources that pin-points those programs not freezing any resource forever. In particular we have designed an algorithm for liquidity and demonstrated its correctness.

We are currently prototyping the algorithm. Our prototype takes in input an integer value κ and verifies liquidity by sticking to types in $\mathbb{T}_0^\kappa$. This allows us to tune the precision of the analysis according to the contract to verify. We are also considering optimisations that improve both the precision of the algorithms and the performance. For example, the precision of the checks $[\![\Xi'(\mathsf{k})]\!] \neq \mathbb{0}$ and $[\![\Xi'(\mathsf{k})]\!] \neq [\![\Xi(\mathsf{k})]\!]$ may be improved by noticing that the algebra of liquidity expressions is a distributive lattice with min ($\mathbb{0}$) and max ($\mathbb{1}$). This algebra has a complete axiomatization that we may implement (for simplicity sake, in this paper we have only used min-max rules – see definition of $[\![e]\!]$). Other optimizations we are studying allow us to reduce the number of canonical computations to verify (such as avoiding repetition of cycles that modify only one asset).

Another research objective addresses the liquidity analysis in languages featuring *conditional transitions* and *events*, such as the full *Stipula* [7]. These primitives introduce *internal nondeterminism*, which may undermine state reachability and, for this reason, they have been dropped in this paper. In particular, our analysis might synthesize a computation containing a function whose execution depends on values of fields that never hold. Therefore the computation will never be executed (it is a false positive) and must be discarded (and the contract might be not liquid). To overcome these problems, we will try to complement our analysis with an (off-the-shelf) constraint solver technique that guarantees the reachability of states of the computations synthesized by our algorithms.

Acknowledgments. We thank the FACS 2022 referees for their careful reading and many constructive suggestions on the submitted paper.

References

1. Albert, E., de Boer, F.S., Hähnle, R., Johnsen, E.B., Laneve,C.: Engineering virtualized services. In: NordiCloud 2013, vol. 826 of ACM International Conference Proceeding Series, pp. 59–63. ACM (2013)
2. Bartoletti, M., Lande, S., Murgia, M., Zunino, R.: Verifying liquidity of recursive bitcoin contracts. Log. Methods Comput. Sci. **18**(1) (2022)
3. Bartoletti, M., Zunino, R.: Verifying liquidity of bitcoin contracts. In: Nielson, F., Sands, D. (eds.) POST 2019. LNCS, vol. 11426, pp. 222–247. Springer, Cham (2019). https://doi.org/10.1007/978-3-030-17138-4_10
4. Blackshear, S., et al.: Resources: a safe language abstraction for money. CoRR, abs/2004.05106 (2020). https://arxiv.org/abs/2004.05106
5. Brakmić, H.: Bitcoin Script, pp. 201–224. Apress, Berkeley, CA (2019)
6. Coblenz, M.J., Aldrich, J., Myers, B.A., Sunshine, J.: Can advanced type systems be usable? an empirical study of ownership, assets, and typestate in Obsidian. Proceed. ACM Program. Lang. **4**(OOPSLA), 1–28 (2020)

7. Crafa, S., Laneve, C., Sartor, G.: Pacta sunt servanda: legal contracts in Stipula. Technical report. arXiv:2110.11069 (2021)
8. Dannen, C.: Introducing ethereum and solidity: foundations of cryptocurrency and blockchain programming for beginners. Apress, Berkely, USA (2017)
9. Das, A., Balzer, S., Hoffmann, J., Pfenning, F., Santurkar, I.: Resource-aware session types for digital contracts. In: IEEE 34th CSF, pp. 111–126. IEEE Computer Society (2021)
10. Kaleem, M., Mavridou, A., Laszka, A.: Vyper: a security comparison with Solidity based on common vulnerabilities. In: BRAINS 2020, pp. 107–111. IEEE (2020)
11. Klabnik, S., Nichols, C.: The RUST programming language. No Starch Press (2019)
12. Lamela Seijas, P., Nemish, A., Smith, D., Thompson, S.: Marlowe: implementing and analysing financial contracts on blockchain. In: Bernhard, M., et al. (eds.) FC 2020. LNCS, vol. 12063, pp. 496–511. Springer, Cham (2020). https://doi.org/10. 1007/978-3-030-54455-3_35
13. Sergey, I., Nagaraj, V., Johannsen, J., Kumar, A., Trunov, A., Hao, K.C.G.: Safer smart contract programming with scilla. Proc. ACM Program. Lang. **3**(OOPSLA), 1–30 (2019)
14. Tsankov, P., Dan, A.M., Drachsler-Cohen, D., Gervais, A., Bünzli, F., Vechev, M.T.: Securify: practical security analysis of smart contracts. In: Proceedings ACM SIGSAC Conference on Computer and Communications Security, pp. 67–82. ACM (2018)

Open Compliance in Multiparty Sessions

Franco Barbanera[1]([✉])[iD], Mariangiola Dezani-Ciancaglini[2][iD],
and Ugo de'Liguoro[2][iD]

[1] Dipartimento di Matematica e Informatica, University of Catania,
95125 Catania, Italy
barba@dmi.unict.it
[2] Dipartimento di Informatica, University of Torino, 10149 Torino, Italy
{dezani,deligu}@di.unito.it

Abstract. Multiparty sessions are a foundational model for distributed entities interacting through message passing. Communication is disciplined by global types, which ensures lock-freedom for participants following the described protocols. A key issue is the composition of well-typed sessions, that we face via the *participants-as-interfaces* approach. We study session composition when a client system is connected with compliant server systems, where compliance is naturally biased towards the client. We prove that, if the sessions are well-typed and the compliance relation can be proved, then a unique session can be constructed by transforming the interface participants of the client and the servers into gateways (forwarders). Such a session has a global type that can be derived from the global types of the composing sessions and the proof of compliance among the client and the servers. We consider the present study as a further step toward a theory of *Open* MultiParty Session Types.

Keywords: Communication-centric systems · System composition · Process calculi · Multiparty session types

1 Introduction

The key issue in multiparty distributed systems is the composition of independent entities such that a sensible behaviour of the whole emerges from those ones of the components, while avoiding type errors of exchanged messages and ensuring good properties like lock-freedom.

The shortcoming of many formalisms for the description and analysis of such systems, for instance MultiParty Session Types (MPST), introduced in [20, 21], or Communicating Finite State Machines (CFSM) introduced in [10], is that they consider *closed* systems only, with a fixed set of participants and roles, where one needs full knowledge of the whole to discipline the composing parts' behaviour. In scenarios such as the Internet, the cloud and various forms of decentralised computing, such assumption is less and less realistic.

S. L. Tapia Tarifa and J. Proença (Eds.): FACS 2022, LNCS 13712, pp. 222–243, 2022.
https://doi.org/10.1007/978-3-031-20872-0_13

In contrast, it is desirable to specify systems as modules that are ready to interact with the environment in a disciplined way, hence treating them as *open* systems. An approach to the composition of systems has been proposed in [3] enabling composition even if the systems have been designed and developed as closed ones. The main idea of the approach is to take two systems, select two of their components (participants) – one per system – and transform them, in case they exhibit "compatible behaviours", into coupled gateways connecting the two systems. Such gateways work simply as "forwarders": if a message of one system is sent to the system's gateway it gets immediately forwarded to the coupled gateway in the other system, which, in turn, sends it to the other components of its own system. In order to clarify the underlying idea, let us

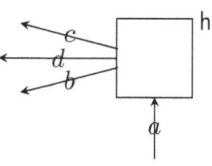

consider a system M_1 made of several communicating participants, among which a participant h (graphically represented on the left) is present. Participant h can receive messages of type a from other participants of M_1 and send messages of type b, c or d. We abstract here from the way communications are performed and from the logical order of the exchanged messages.

Now let us consider a second system M_2 where a participant k is present, and assume that k (graphically represented below on the right) can send to other participants messages of type a, and receive messages of type b, c or d.

Participants h and k are *compatible* in the sense that,
if some participant of the system M_1 needs a message
"produced" by h, this message can actually be the one
received in system M_2 by k. Symmetrically, the messages
that h receives from other participants of M_1 could be forwarded to M_2 as it were messages sent by k to participants

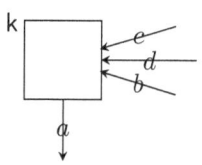

of M_2. From this point of view, the behaviours of h and k can be looked at as "interfaces" which describe what one system expects from the other in case they interact. Notice that *any* participant of a system can be looked at as an interface as far as it is possible to find a compatible participant in another system.

The composition mechanism consists then in replacing the two participants h and k by, respectively, two forwarders: $\mathbf{gw}(h, k)$ and $\mathbf{gw}(k, h)$, and connecting them as follows:

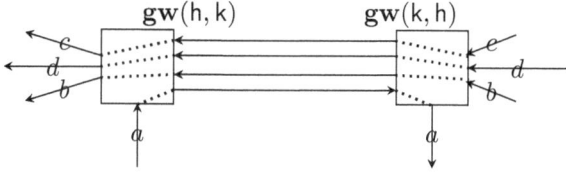

Using gateways, open systems can be seen as a special kind of closed ones, satisfying suitable compatibility conditions when composing them into a whole,

so that known techniques and formalisms from the theory of multiparty distributed systems can be put to use, encompassing modular development. This has been exploited both in the context of MPST [5] and of CFSM [3,6], where the behaviour of a participant can be described by non-trivial communication protocols, taking into account the order in which messages can be exchanged, as well as the possibility of branching. Notably, relevant communication properties like lock-freedom have been shown to be preserved by composition.

The compatibility relation considered in the above-mentioned references is some kind of bisimulation, since there must be a correspondence between the messages of h and k. In the present paper, we push forward the idea of composition via gateways, exploring the setting of client/server interactions where the natural bias towards the client leads from the symmetric notion of compatibility to the asymmetric relation of *compliance*. In fact, a client connected to a server is expected to have all its "requests" satisfied by the server, whereas it is not so for the server with respect to the client. This notion has been widely investigated in the literature for binary interactions [7–9,12,26].

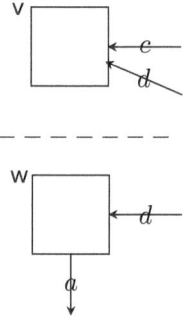

Actually, we propose a more general notion of compliance, so that: (i) more than one server can be used to satisfy the "client's requests"; (ii) part of these "requests" can be internally managed, i.e. without recurring to any outer server. In the informal graphical formalism used before, these two points can be described by the following example. Let us assume that, in system $\mathbb{M}_1$, we consider h as an interface, i.e., its behaviour describes what $\mathbb{M}_1$ expects from some outer server. Let us now assume to have at our disposal two systems: $\mathbb{M}_3$ and $\mathbb{M}_4$ containing, respectively, participants v and w (graphically represented on the right). It is possible to connect $\mathbb{M}_1$ to both $\mathbb{M}_3$ and $\mathbb{M}_4$ via gateways as follows:

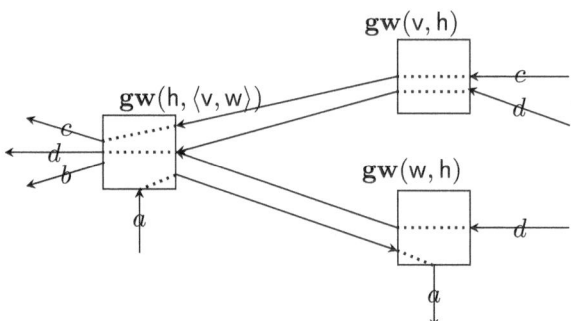

In this way, part of the communications represented by the behaviour of h is provided collectively by the systems $\mathbb{M}_3$ and $\mathbb{M}_4$. Notice how in this example h has not been completely replaced by a forwarder, since messages of type b do come from neither of the server systems. In general we allow the client to deal directly with messages also when they could be forwarded/received to/from one of the servers. Moreover, messages of type d sent to participants of system $\mathbb{M}_1$ can be forwarded by either $\mathbb{M}_3$ or $\mathbb{M}_4$. As a consequence, the compliance of h with v and w is not uniquely determined.

We shall exploit this idea in the setting of MPST, where the behaviours of the participants of a system are described by a simplified form of processes – communicating synchronously – in a suitable calculus. A type system is devised to guarantee a system's overall behaviour to adhere to a given communication protocol. Typability also ensures lock-freedom. Throughout the paper, we shall use a simple running example to give a better intuition of the notions and properties we shall introduce.

Contributions and Structure of the Paper. In Sect. 2 we shall introduce our calculus of multiparty sessions. Section 3 will be devoted to the presentation of our type system – assigning global types to multiparty sessions – and to the statements of relevant properties of typable sessions: subject reduction, session fidelity and lock-freedom. In Sect. 4 we shall define the relation of *open compliance*, formalising the possibility for a client system to be (partially or completely) satisfied by two server systems (our technique can actually be extended to any number of these). The definition of gateways in our approach will then be given in terms of derivations for the open compliance relation. Systems connection will hence be defined and lock-freedom preservation will be ensured by a theorem showing that the connection of typable systems is still typable. A section summing up our results, discussing related works and possible directions for future work will conclude the paper.

2   Calculus

We use the following base sets and notation: *messages*, ranged over by $\lambda, \lambda', \ldots$; *session participants*, ranged over by h, p, q, r, v, w, t, ...; *processes*, ranged over by $H, P, Q, R, V, W, \ldots$; *multiparty sessions*, ranged over by $\mathbb{M}, \mathbb{M}', \ldots$; *integers*, ranged over by $i, j, l, n, \ldots$; sets of indexes, ranged over by $I, J, \ldots$.

Definition 1 (Processes). Processes *are defined by:*

$$P ::=^{coind} \mathbf{0} \ \mid \ \mathsf{p}!\{\lambda_i.P_i\}_{i \in I} \ \mid \ \mathsf{p}?\{\lambda_i.P_i\}_{i \in I}$$

where $I \neq \emptyset$ *and* $\lambda_j \neq \lambda_l$ *for* $j, l \in I$ *and* $j \neq l$.

The symbol $::=^{coind}$ in Definition 1 and in later definitions indicates that the productions are interpreted *coinductively*, so that the set of processes is the greatest fixed point of the (monotonic) functor over sets defined by the grammar above. That is, there are possibly infinite processes. However, we assume such processes to be *regular*, i.e., with finitely many distinct sub-processes. In this way, we only obtain processes which are solutions of finite sets of equations, see [14]. We choose this formulation as we will use coinduction in some definitions and proofs and, moreover, it allows us to avoid explicitly handling variables, thus simplifying a lot the technical development.

Processes implement the behaviour of participants. The output process $p!\{\lambda_i.P_i\}_{i \in I}$ non-deterministically chooses one message λ_i for some $i \in I$, and sends it to the participant p, thereafter continuing as P_i. Symmetrically, the input process $p?\{\lambda_i.P_i\}_{i \in I}$ waits for one of the messages λ_i from the participant p, then continues as P_i after receiving it. When there is only one output we write $p!\lambda.P$ and similarly for one input. We use $\mathbf{0}$ to denote the terminated process.

In a full-fledged calculus, messages would carry values, that we avoid for the sake of simplicity; hence no selection operation over values is included in the syntax.

Definition 2 (Multiparty sessions). Multiparty sessions *are defined by:*

$$\mathbb{M} = p_1[P_1] \parallel \cdots \parallel p_n[P_n]$$

where $p_j \neq p_l$ for $1 \leq j, l \leq n$ and $j \neq l$.

Multiparty sessions (sessions, for short) are parallel compositions of located processes of the form $p[P]$, each enclosed within a different participant p. We assume the standard structural congruence $\equiv$ on multiparty sessions, stating that parallel composition is associative and commutative and has neutral elements $p[\mathbf{0}]$ for any p. If $P \neq \mathbf{0}$ we write $p[P] \in \mathbb{M}$ as short for $\mathbb{M} \equiv p[P] \parallel \mathbb{M}'$ for some $\mathbb{M}'$. This abbreviation is justified by the associativity and commutativity of parallel composition.

The *set of participants* of a session $\mathbb{M}$, notation $\mathsf{prt}(\mathbb{M})$, is as expected:

$$\mathsf{prt}(\mathbb{M}) = \{p \mid p[P] \in \mathbb{M}\}$$

To define the *synchronous operational semantics* of sessions we use an LTS, whose transitions are decorated by communications.

Definition 3 (LTS for Multiparty Sessions). *The* labelled transition system (LTS) for multiparty sessions *is the closure under structural congruence of the reduction specified by the unique rule:*

$$[\text{COMM-T}] \quad \frac{l \in I \subseteq J}{p[q!\{\lambda_i.P_i\}_{i \in I}] \parallel q[p?\{\lambda_j.Q_j\}_{j \in J}] \parallel \mathbb{M} \xrightarrow{p\lambda_l q} p[P_l] \parallel q[Q_l] \parallel \mathbb{M}}$$

Rule [COMM-T] makes the communication possible: participant p sends message λ_l to participant q. This rule is non-deterministic in the choice of messages. The condition $I \subseteq J$ (borrowed from [4,5]) ensures that the sender can freely choose the message, since the receiver must offer all sender messages and possibly more. This allows us to distinguish in the operational semantics between internal and external choices. Note that this condition will always be true in well-typed sessions.

Let Λ range over communications of the form $p\lambda q$. We define *traces* as (possibly infinite) sequences of communications, by:

$$\sigma ::=^{coind} \epsilon \mid \Lambda \cdot \sigma$$

where ϵ is the empty sequence. We use $|\sigma|$ to denote the length of the trace σ, where $|\sigma| = \infty$ when σ is an infinite trace. We define the participants of communications and traces:

$$\mathsf{prt}(\mathsf{p}\lambda\mathsf{q}) = \{\mathsf{p},\mathsf{q}\} \qquad \mathsf{prt}(\epsilon) = \emptyset \qquad \mathsf{prt}(\Lambda \cdot \sigma) = \mathsf{prt}(\Lambda) \cup \mathsf{prt}(\sigma)$$

When $\sigma = \Lambda_1 \cdot \ldots \cdot \Lambda_n$ $(n \geq 0)$ we write $\mathbb{M} \xrightarrow{\sigma} \mathbb{M}'$ as short for

$$\mathbb{M} \xrightarrow{\Lambda_1} \mathbb{M}_1 \cdots \xrightarrow{\Lambda_n} \mathbb{M}_n = \mathbb{M}'$$

We give now a very simple example that shall be used throughout the paper in order to clarify the notions we introduce.

Example 1 (A simple running example). Let us consider a two-participants system represented by the session:

$$\mathbb{M}_1 = \mathsf{p}[P] \parallel \mathsf{h}[H]$$

Process H controls the entrance of customers in a mall (via some sensor). As soon as a customer enters, H sends a message START to the process P which controls a display for advertisements. After the start message, H sends to P a general advertising image (through message IMG). Process P does control also a sensor detecting emotional reactions and sends to H the reaction (REACTN) of the customer to the general advertisement. Using that information H sends to P a customised image (again through message IMG). These interactions go on until H sends to P an HALT message to shutdown the system. Process P is also able to receive a RESET message even if H cannot ever send it. The processes P and H can then be defined as follows

$$P = \mathsf{h}?\text{START}.\mathsf{h}? \begin{cases} \text{IMG}.\,\mathsf{h}!\text{REACTN}.\,\mathsf{h}?\text{IMG}.\,P \\ \text{HALT}.\mathbf{0} \\ \text{RESET}.P \end{cases}$$

$$H = \mathsf{p}!\text{START}.\mathsf{p}! \begin{cases} \text{IMG}.\,\mathsf{p}?\text{REACTN}.\,\mathsf{p}!\text{IMG}.\,H \\ \text{HALT}.\mathbf{0} \end{cases}$$

where sets of alternatives are denoted by branchings.   ◇

We end this section by defining the property of lock-freedom for multiparty sessions as in [22,27]. In words, each participant ready to communicate will eventually find her partner exposing a dual communication action in a possible evolution of the system. Lock-freedom guarantees progress for each participant, and hence deadlock-freedom.

Definition 4 (Lock-freedom). *A multiparty session $\mathbb{M}$ is a lock-free session if $\mathbb{M} \xrightarrow{\sigma} \mathbb{M}'$ and $\mathsf{p}[P] \in \mathbb{M}'$ imply $\mathbb{M}' \xrightarrow{\sigma' \cdot \Lambda} \mathbb{M}''$ for some σ' and Λ such that $\mathsf{p} \in \mathsf{prt}(\Lambda)$.*

3   Type System

The behaviour of multiparty sessions can be disciplined by means of types. Global types describe the whole conversation scenarios of multiparty sessions. As in [13,15], we shorten the technical treatment by directly assigning global types to multiparty sessions without using projections, session types and subtyping [20,21]. We resort to subject reduction as the fundamental tool to establish the correctness of the type system.

Definition 5 (Global types). Global types *are defined by:*

$$\mathsf{G} ::=^{coind} \mathbf{End} \mid \mathsf{p} \to \mathsf{q} : \{\lambda_i.\mathsf{G}_i\}_{i \in I}$$

where $I \neq \emptyset$ *and* $\lambda_j \neq \lambda_l$ *for* $j, l \in I$ *and* $j \neq l$.

As for processes, we allow only *regular* global types. The type $\mathsf{p} \to \mathsf{q} : \{\lambda_i.\mathsf{G}_i\}_{i \in I}$ formalises a protocol where participant p must send to q a message λ_j for some $j \in I$, (and q must receive it) and then, depending on which λ_j was chosen by p, the protocol continues as G_j. We write $\mathsf{p} \to \mathsf{q} : \lambda.\mathsf{G}$ when there is only one message. We use $\mathbf{End}$ to denote the terminated protocol.

We define the *set of paths of a global type*, notation $paths(\mathsf{G})$, as the greatest set such that:

$$paths(\mathbf{End}) = \{\epsilon\} \qquad paths(\mathsf{p} \to \mathsf{q} : \{\lambda_i.\mathsf{G}_i\}_{i \in I}) = \bigcup_{i \in I} \{\mathsf{p}\lambda_i\mathsf{q} \cdot \sigma \mid \sigma \in paths(\mathsf{G}_i)\}$$

Clearly, paths of global types are traces as defined after Definition 3. The *set of participants of a global type* is the set of participants of its paths:

$$\mathsf{prt}(\mathsf{G}) = \bigcup_{\sigma \in paths(\mathsf{G})} \mathsf{prt}(\sigma)$$

The regularity of global types ensures that the sets of participants are finite.

To guarantee lock-freedom by typing, we require that each participant occurs either in all the paths or in no path from any node. However, this is different if the path is finite or infinite; in the latter case, we must forbid the existence of an infinite path that definitely defers any action involving some participant in the type. Notably this condition is a form of fairness. Technically, we use the notions of *depth* and of *bounded* types below. We denote by $\sigma[n]$ with $n \in \mathbb{N}$ the n-th communication in the path σ, where $1 \leq n \leq |\sigma|$.

Definition 6 (Depth). *Let* G *be a global type. For* $\sigma \in paths(\mathsf{G})$ *we define*

$$depth(\sigma, \mathsf{p}) = \text{infimum}\{n \mid \mathsf{p} \in \mathsf{prt}(\sigma[n])\}$$

and define $depth(\mathsf{G}, \mathsf{p})$, *the* depth *of* p *in* G, *as follows:*

$$depth(\mathsf{G}, \mathsf{p}) = \begin{cases} \text{supremum}\{depth(\sigma, \mathsf{p}) \mid \sigma \in paths(\mathsf{G})\} & \text{if } \mathsf{p} \in \mathsf{prt}(\mathsf{G}) \\ 0 & \text{otherwise} \end{cases}$$

Note that $depth(\mathsf{G}, \mathsf{p}) = 0$ iff $\mathsf{p} \notin \mathsf{prt}(\mathsf{G})$. Moreover, if $\mathsf{p} \in \mathsf{prt}(\mathsf{G})$, but for some $\sigma \in paths(\mathsf{G})$ it is the case that $\mathsf{p} \notin \mathsf{prt}(\sigma[n])$ for all $n \leq |\sigma|$, then $depth(\sigma, \mathsf{p}) = $ infimum $\emptyset = \infty$. Hence, if p is a participant of a global type G and there is some path in G where p does not occur, then $depth(\mathsf{G}, \mathsf{p}) = \infty$.

Definition 7 (Boundedness). *A global type* G *is* bounded *if* $depth(\mathsf{G}', \mathsf{p})$ *is finite for all participants* $\mathsf{p} \in \mathsf{prt}(\mathsf{G}')$ *and all types* G' *which occur in* G.

Intuitively, this means that if $\mathsf{p} \in \mathsf{prt}(\mathsf{G}')$ for a type G' which occurs in G, then the search for an interaction of the shape $\mathsf{p}\lambda\mathsf{q}$ or $\mathsf{q}\lambda\mathsf{p}$ along a path $\sigma \in paths(\mathsf{G}')$ terminates (and recall that G' can be infinite, in which case G is such). Hence the name.

Example 2. The following example shows the necessity of considering all types occurring in a global type when defining boundedness. Let

$$\mathsf{G} = \mathsf{r} \to \mathsf{q} : \lambda.\mathsf{G}' \quad \text{where} \quad \mathsf{G}' = \mathsf{p} \to \mathsf{q} : \{\lambda_1.\mathsf{q} \to \mathsf{r} : \lambda'.\text{End}, \lambda_2.\mathsf{G}'\}$$

Any $\sigma \in paths(\mathsf{G})$ is such that $\sigma = \mathsf{r}\lambda\mathsf{q} \cdot \sigma'$, for some $\sigma' \in paths(\mathsf{G}')$, then we get $depth(\mathsf{G}, \mathsf{r}) = depth(\mathsf{r}\lambda\mathsf{q} \cdot \sigma', \mathsf{r}) = 1$. On the other hand $\mathsf{r} \in \mathsf{prt}(\mathsf{G}')$, but the infinite path $\sigma'' = \mathsf{p}\lambda_2\mathsf{q} \cdot \mathsf{p}\lambda_2\mathsf{q} \cdot \ldots$ belongs to $paths(\mathsf{G}')$ and $depth(\mathsf{G}', \mathsf{r}) = $ supremum $\{depth(\mathsf{p}\lambda_1\mathsf{q} \cdot \mathsf{q}\lambda'\mathsf{r}, \mathsf{r}), depth(\sigma'', \mathsf{r})\} = \infty$, since $depth(\sigma'', \mathsf{r}) = $ infimum $\emptyset = \infty$. We conclude that G is unbounded; in fact, and equivalently, r does not occur in the infinite path σ'' of G'.
Also a finite global type can be unbounded; for example, r is a participant of

$$\mathsf{p} \to \mathsf{q} : \{\lambda_1.\mathsf{q} \to \mathsf{r} : \lambda'.\text{End}, \lambda_2.\text{End}\}$$

but r does not occur in the path $\mathsf{p}\lambda_2\mathsf{q}$.   ◇

Since global types are regular, the boundedness condition is decidable. We shall allow only bounded global types in typing sessions.

Example 3 (A global type for $\mathbb{M}_1$*).* The overall behaviour of the multiparty session $\mathbb{M}_1$ in Example 1 adheres to what the global type G_1 below describes:

$$\mathsf{G}_1 = \mathsf{h} \to \mathsf{p} : \text{START}.\mathsf{h} \to \mathsf{p} : \begin{cases} \text{IMG}.\, \mathsf{p} \to \mathsf{h} : \text{REACTN}.\, \mathsf{h} \to \mathsf{p} : \text{IMG}.\, \mathsf{G}_1 \\ \text{HALT}.\, \textbf{End} \end{cases}$$

The derivability of the global type G_1 for the multiparty session $\mathbb{M}_1$ will be formally guaranteed by means of the type system defined below.   ◇

The simplicity of our calculus allows us to formulate a type system deriving directly global types for multiparty sessions, i.e. judgments of the form $\mathsf{G} \vdash \mathbb{M}$ (where G is bounded). Here and in the following the double line indicates that the rules are interpreted coinductively [28, Chapter 21].

Definition 8 (Type system). *The type system is defined by the following axiom and rule:*

$$(\text{End}) \quad \text{End} \vdash \mathsf{p}[\mathbf{0}]$$

$$[\text{Comm}] \quad \frac{G_i \vdash \mathsf{p}[P_i] \parallel \mathsf{q}[Q_i] \parallel \mathbb{M} \quad \mathsf{prt}(G_i) \setminus \{\mathsf{p}, \mathsf{q}\} = \mathsf{prt}(\mathbb{M}) \quad \forall i \in I}{\mathsf{p} \to \mathsf{q} : \{\lambda_i.G_i\}_{i \in I} \vdash \mathsf{p}[\mathsf{q}!\{\lambda_i.P_i\}_{i \in I}] \parallel \mathsf{q}[\mathsf{p}?\{\lambda_j.Q_j\}_{j \in J}] \parallel \mathbb{M}} \quad I \subseteq J$$

Rule [Comm] just adds simultaneous communications to global types and to corresponding processes inside sessions. Notice that this rule allows more inputs than corresponding outputs, in agreement with the condition in Rule [Comm-T]. It also allows more branches in the input process than in the global type, just mimicking the subtyping for session types [16]. Instead, the number of branches in the output process and the global type must be the same. This does not restrict typability as shown in [5], while it improves Session Fidelity as discussed after Theorem 2. The condition $\mathsf{prt}(G_i) \setminus \{\mathsf{p}, \mathsf{q}\} = \mathsf{prt}(\mathbb{M})$ for all $i \in I$ ensures that the global type and the session have exactly the same set of participants. In this way we forbid for example to derive $\mathsf{p} \to \mathsf{q} : \lambda.\text{End} \vdash \mathsf{p}[\mathsf{q}!\lambda.\mathbf{0}] \parallel \mathsf{q}[\mathsf{p}?\lambda.\mathbf{0}] \parallel \mathsf{r}[R]$ with R arbitrary.

The regularity of processes and global types ensures the decidability of type checking.

Example 4 (Typing $\mathbb{M}_1$ with G_1). It is easy to check that, for the multiparty session $\mathbb{M}_1$ of Example 1 and the global type G_1 of Example 3, we can derive $G_1 \vdash \mathbb{M}_1$.   ◇

To formalise the properties of Subject Reduction and Session Fidelity [20, 21], we use the standard LTS for global types given below.

Definition 9 (LTS for Global Types). *The* labelled transition system (LTS) *for global types is specified by the rules:*

$$[\text{Ecomm}] \quad \frac{}{\mathsf{p} \to \mathsf{q} : \{\lambda_i.G_i\}_{i \in I} \xrightarrow{\mathsf{p}\lambda_j\mathsf{q}} G_j} \quad j \in I$$

$$[\text{Icomm}] \quad \frac{G_i \xrightarrow{\mathsf{p}\lambda\mathsf{q}} G_i' \quad \forall i \in I \quad \{\mathsf{p}, \mathsf{q}\} \cap \{\mathsf{r}, \mathsf{s}\} = \emptyset}{\mathsf{r} \to \mathsf{s} : \{\lambda_i.G_i\}_{i \in I} \xrightarrow{\mathsf{p}\lambda\mathsf{q}} \mathsf{r} \to \mathsf{s} : \{\lambda_i.G_i'\}_{i \in I}}$$

Rule [Icomm] makes sense since, in a global type $\mathsf{r} \to \mathsf{s} : \{\lambda_i.G_i\}_{i \in I}$, behaviours involving participants p and q, ready to interact with each other uniformly in all branches, can do so if neither of them is involved in a previous interaction between r and s. In this case, the interaction between p and q is independent of the choice of r, and may be executed before it.

Subject Reduction guarantees that the transitions of well-typed sessions are mimicked by those of global types.

Theorem 1 (Subject Reduction). *If* $G \vdash M$ *and* $M \xrightarrow{p\lambda q} M'$, *then* $G \xrightarrow{p\lambda q} G'$ *and* $G' \vdash M'$.

Session Fidelity ensures that the communications in a session typed by a global type proceed as prescribed by the global type.

Theorem 2 (Session Fidelity). *If* $G \vdash M$ *and* $G \xrightarrow{p\lambda q} G'$, *then* $M \xrightarrow{p\lambda q} M'$ *and* $G' \vdash M'$.

Combining Subject Reduction and Session Fidelity theorems, we conclude that sessions and global types are bisimilar with respect to their reduction relations.

Notice that, if Rule [COMM] had allowed more branches in the global type than in the output process as the subtyping of [16] does, then Theorem 2 would have failed. An example is

$$p \rightarrow q : \{\lambda.\mathsf{End}, \lambda'.\mathsf{End}\} \vdash p[q!\lambda.\mathbf{0}] \parallel q[p?\{\lambda.\mathbf{0}, \lambda'.\mathbf{0}\}]$$

since $p \rightarrow q : \{\lambda.\mathsf{End}, \lambda'.\mathsf{End}\} \xrightarrow{p\lambda'q} \mathsf{End}$, but there is no transition labelled $p\lambda'q$ from $p[q!\lambda.\mathbf{0}] \parallel q[p?\{\lambda.\mathbf{0}, \lambda'.\mathbf{0}\}]$.

We can show that typability ensures lock-freedom. By Subject Reduction it is enough to prove that in well-typed sessions no active participant waits forever. This follows from Session Fidelity thanks to the boundedness condition.

Theorem 3 (Lock-freedom). *If* M *is typable, then* M *is lock-free.*

4   Open Compliance

In the present paper we exploit the "participants-as-interfaces" approach to the composition of closed systems from a client/(multi)server perspective, as explained in the introduction.

In the general case, a *client system* is such when we look at the behaviour (a process in our formalism) of any participant of it, say h, as what the system would expect *at least* from some outer server (usually another system). Dually, any *server system* is such as far as the behaviour of any of its participants, say k, is looked at as what a client could expect *at most* from the client system. By looking at h as an interface of a client then, whenever h receives a message λ, this has to be interpreted as a message to be sent to the server. Dually k can be considered as the interface of a *compliant* server in case k can send some messages, among which λ is present, since any message sent by k is interpreted as a message sent by a possible client. Similar reasoning applies to messages sent by h and received by k. Participant k can be soundly considered as the interface of a compliant server also in case the client terminates notwithstanding k does not. It is worth noticing that the senders and receivers of messages from and to h and k are irrelevant in defining the above notion of compliance.

In the present paper, we show how to relax the notion of compliance by enabling the "requests" of a client to be (partially) satisfied by more than one server, as intuitively shown in the introduction. For the sake of readability, we shall consider compliance with just two servers, but this notion can be generalised to the case of any finite but non-empty set of servers.

We need to state which communications are interactions with the left or right server and which communications are dealt with by the client without involving the servers. This is realised by the open compliance relation.

Definition 10 (Open Compliance). *The* open compliance relation $H \dashv \langle V, W \rangle$ *on processes (compliance for short), is the largest asymmetric relation coinductively defined by the rules in Fig. 1.*

$$[\text{COMP-0}] \; \overline{\overline{\mathbf{0} \dashv \langle V, W \rangle}}$$

$$[\text{COMP-O/I-L}] \; \frac{H_i \dashv \langle V_i, W \rangle \quad \forall i \in I}{\mathsf{p}!\{\lambda_j.H_j\}_{j \in J} \dashv \langle \mathsf{q}?\{\lambda_i.V_i\}_{i \in I}, W \rangle} \; I \subseteq J$$

$$[\text{COMP-I/O-L}] \; \frac{H_i \dashv \langle V_i, W \rangle \quad \forall i \in I}{\mathsf{p}?\{\lambda_i.H_i\}_{i \in I} \dashv \langle \mathsf{q}!\{\lambda_j.V_j\}_{j \in J}, W \rangle} \; I \subseteq J$$

$$[\text{COMP-O/I-R}] \; \frac{H_i \dashv \langle V, W_i \rangle \quad \forall i \in I}{\mathsf{p}!\{\lambda_j.H_j\}_{j \in J} \dashv \langle V, \mathsf{q}?\{\lambda_i.W_i\}_{i \in I} \rangle} \; I \subseteq J$$

$$[\text{COMP-I/O-R}] \; \frac{H_i \dashv \langle V, W_i \rangle \quad \forall i \in I}{\mathsf{p}?\{\lambda_i.H_i\}_{i \in I} \dashv \langle V, \mathsf{q}!\{\lambda_j.W_j\}_{j \in J} \rangle} \; I \subseteq J$$

$$[\text{COMP-O/I-A}] \; \frac{H_i \dashv \langle V, W \rangle \quad \forall i \in I}{\mathsf{p}!\{\lambda_i.H_i\}_{i \in I} \dashv \langle V, W \rangle} \qquad [\text{COMP-I/O-A}] \; \frac{H_i \dashv \langle V, W \rangle \quad \forall i \in I}{\mathsf{p}?\{\lambda_i.H_i\}_{i \in I} \dashv \langle V, W \rangle}$$

Fig. 1. Compliant processes.

Rule [COMP-0] says that all servers are compliant with a terminated client.

In Rules [COMP-O/I-L] and [COMP-I/O-L] the client interacts with the left server, with the proviso that they offer dual communications and the set of possibly sent labels contains the set of possibly received ones. This last condition (which is the opposite of the condition in the typing Rule [COMM]) is needed since - when the processes will be replaced by gateways - the received labels will be forwarded, see Fig. 2. The following two rules are similar, the only difference is that the client interacts with the right server.

Rules [COMP-O/I-A] and [COMP-I/O-A] deal with the case when some messages are handled directly by the client; namely, she is free to handle the messages by herself, without forwarding/getting them to/from the servers. For this reason the processes of the client and of the servers could be completely unrelated.

Notably in all rules the participants p and q are different and do not communicate, since they occur in different sessions.

Open compliance requires conditions similar to the subtyping of [18], and so it can be checked by an easy adaptation of the algorithm given in that paper.

Example 5 (Two server systems). Continuing Example 1, let us suppose that two further systems are available. The first system M_2 is a simple two-participants system for advertising. In it, the participant s continually produces general advertising images that sends (through message IMG) to an advertising display controlled by the participant v. Participant s can send, instead of an image, a message HALT forcing the reboot of the display controlled by v.

The behaviour of such a system can be formally described by the following global type.

$$G_2 = s \rightarrow v : \begin{cases} \text{IMG}.G_2 \\ \text{HALT}.G_2 \end{cases}$$

A possible implementation of the system is provided by the multiparty session

$$M_2 = v[V] \parallel s[S] \quad \text{where} \quad V = s? \begin{cases} \text{IMG}.V \\ \text{HALT}.V \end{cases} \quad \text{and} \quad S = v! \begin{cases} \text{IMG}.S \\ \text{HALT}.S \end{cases}$$

The session M_2 behaves as described by G_2 since we can derive $G_2 \vdash M_2$.

The second system (two-participant again, for the sake of simplicity) has a behaviour described by the following global type.

$$G_3 = w \rightarrow r : \text{REACTN}.\, r \rightarrow w : \text{IMG}.\, G_3$$

Such a system controls a coupled pair of advertising panels. The first panel displays a standard advertising image. Participant w controls an emotional-reaction detector in the first panel and sends a message about the detected reaction (through message REACTN) to the participant r which, on the base of that, sends to w (through message IMG) a customised version of the standard advertising image to be displayed in the second panel. And so on.

A possible implementation of the system can be provided by the following multiparty session:

$$M_3 \quad = \quad w[W] \parallel r[R]$$

where $W = r!\text{REACTN}.r?\text{IMG}.W$ and $R = w?\text{REACTN}.r!\text{IMG}.R$

The session M_3 behaves as described by G_3 since we can derive $G_3 \vdash M_3$.

Using the rules in Fig. 1 we can also build the following derivation $\mathcal{D}$ of $H \dashv \langle V, W \rangle$:

$$\mathcal{D} = \cfrac{\cfrac{\cfrac{\cfrac{\mathcal{D}}{H_3 \dashv \langle V, W_1 \rangle} \,[\text{COMP-O/I-R}]}{H_2 \dashv \langle V, W \rangle} \,[\text{COMP-I/O-R}]}{H_1 \dashv \langle V, W \rangle} \qquad \mathbf{0} \dashv \langle V, W \rangle}{H \dashv \langle V, W \rangle} \,[\text{COMP-O/I-L}]}{H \dashv \langle V, W \rangle} \,[\text{COMP-O/I-A}]$$

where H is as in Example 1, V and W as in Example 5, and

$$H_1 = \mathsf{p}! \begin{cases} \text{IMG.}\, H_2 \\ \text{HALT.}\mathbf{0} \end{cases} \qquad H_2 = \mathsf{p}?\text{REACTN.}\, H_3 \qquad H_3 = \mathsf{p}!\text{IMG.}\, H \qquad W_1 = \mathsf{r}?\text{IMG.}W$$

In particular, in this derivation the START sent by h to p is of h's own, whereas the choice between IMG and HALT (the latter makes h and p stop) depends on which of the two is actually received by v from s. In case the branch IMG is chosen, the REACTN message sent by w to r is actually the one received by h from p, whereas the following IMG sent by h to p is the one received by w from r.

The above open compliance derivation showing the given processes to be in the open compliance relation is not unique. In particular h could choose a different order for interacting with v and w on message IMG.   ◇

The communication among a client system hosting h[H] and two server systems hosting v[V] and w[W], such that $H \dashv \langle V, W \rangle$, is enabled by replacing the respective processes by *forwarders*. In our setting the client, for each of her communications, can either involve one of the two servers or deal with it autonomously. Of course the servers must be ready to collaborate. This sequence of client decisions in agreement with the servers is provided by the specific way client and servers are shown to be compliant. For this reason the construction of the gateways depends on open compliance derivations. Therefore we get a system of judgements of the form

$$(\widehat{H}, \widehat{V}, \widehat{W}) : H \dashv \langle V, W \rangle$$

relating the construction of the gateways $\widehat{H}$, $\widehat{V}$ and $\widehat{W}$ to the compliance derivation for processes H, V and W.

Definition 11 (Gateway). *Given three processes H, V and W, a compliance derivation $\mathcal{D}$ of $H \dashv \langle V, W \rangle$, and three distinct participants names h, v and w, we define the gateway processes induced by $\mathcal{D}$ as*

$$\mathsf{gw}_{\mathcal{D}}(H, \langle \mathsf{v}, \mathsf{w} \rangle) = \widehat{H} \qquad \mathsf{gw}_{\mathcal{D}}(V, \mathsf{h}) = \widehat{V} \qquad \mathsf{gw}_{\mathcal{D}}(W, \mathsf{h}) = \widehat{W}$$

where $(\widehat{H}, \widehat{V}, \widehat{W}) : H \dashv \langle V, W \rangle$ is obtained coinductively by a derivation that coincides with $\mathcal{D}$ but for the gateway decorations. Figure 2 shows the most relevant rules for gateway building.

If the client process is $\mathbf{0}$, then no communication is added (Rule [GW-0]).

In all other rules we distinguish if the processes of the client in the premises are different or equal to $\mathbf{0}$ by splitting the set I of indexes into I_1 and I_2. This is needed since compliance allows the client to stop also when the servers are ready to go on.

$$[\text{GW-0}] \; \frac{}{(0, V, W) : \mathbf{0} \dashv \langle V, W \rangle}$$

$$[\text{GW-O/I-L}] \; \frac{\begin{array}{c}(\widehat{H_i}, \widehat{V_i}, \widehat{W_0}) : H_i \dashv \langle V_i, W \rangle \; \forall i \in I_1 \\ (\mathbf{0}, V_i, W) : \mathbf{0} \dashv \langle V_i, W \rangle \; \forall i \in I_2 \end{array}}{(\widehat{H}, \widehat{V}, \widehat{W}) : \mathsf{p}!\{\lambda_j.H_j\}_{j \in J} \dashv \langle \mathsf{q}?\{\lambda_i.V_i\}_{i \in I}, W \rangle} \quad \begin{array}{c} I_1 \cup I_2 = I \subseteq J \\ H_i \neq \mathbf{0} \quad \forall i \in I_1 \\ H_i = \mathbf{0} \quad \forall i \in I_2 \end{array}$$

where $\widehat{H} = \begin{cases} \mathsf{v}?\{\lambda_i.\mathsf{p}!\lambda_i.\widehat{H_i}\}_{i \in I_1} & \text{if } I_2 = \emptyset \\ \mathsf{v}?\{\lambda_i.\mathsf{p}!\lambda_i.\mathbf{0}\}_{i \in I_2} & \text{if } I_1 = \emptyset \\ \mathsf{v}?\{\lambda_i.\mathsf{p}!\lambda_i.\mathsf{w}!go.\widehat{H_i}\}_{i \in I_1} \cup \{\lambda_i.\mathsf{p}!\lambda_i.\mathsf{w}!stop.\mathbf{0}\}_{i \in I_2} & \text{otherwise} \end{cases}$

$\widehat{V} = \mathsf{q}?\{\lambda_i.\mathsf{h}!\lambda_i.\widehat{V_i}\}_{i \in I_1} \cup \{\lambda_i.\mathsf{h}!\lambda_i.V_i\}_{i \in I_2}$

$\widehat{W} = \begin{cases} \widehat{W_0} & \text{if } I_2 = \emptyset \\ W & \text{if } I_1 = \emptyset \\ \mathsf{h}?\{go.\widehat{W_0}, \; stop.W\} & \text{otherwise} \end{cases}$

$$[\text{GW-I/O-A}] \; \frac{\begin{array}{c}(\widehat{H_i}, \widehat{V_0}, \widehat{W_0}) : H_i \dashv \langle V, W \rangle \; \forall i \in I_1 \\ (\mathbf{0}, V, W) : \mathbf{0} \dashv \langle V, W \rangle \; \forall i \in I_2 \end{array}}{(\widehat{H}, \widehat{V}, \widehat{W}) : \mathsf{p}?\{\lambda_i.H_i\}_{i \in I} \dashv \langle V, W \rangle} \quad \begin{array}{c} I_1 \cup I_2 = I \\ H_i \neq \mathbf{0} \quad \forall i \in I_1 \\ H_i = \mathbf{0} \quad \forall i \in I_2 \end{array}$$

where $\widehat{H} = \begin{cases} \mathsf{p}?\{\lambda_i.\widehat{H_i}\}_{i \in I_1} & \text{if } I_2 = \emptyset \\ \mathsf{p}?\{\lambda_i.\mathbf{0}\}_{i \in I_2} & \text{if } I_1 = \emptyset \\ \mathsf{p}?\{\lambda_i.\mathsf{v}!go.\mathsf{w}!go.\widehat{H_i}\}_{i \in I_1} \cup \{\lambda_i.\mathsf{v}!stop.\mathsf{w}!stop.\mathbf{0}\}_{i \in I_2} & \text{otherwise} \end{cases}$

$\widehat{V} = \begin{cases} \widehat{V_0} & \text{if } I_2 = \emptyset \\ V & \text{if } I_1 = \emptyset \\ \mathsf{h}?\{go.\widehat{V_0}, \; stop.V\} & \text{otherwise} \end{cases}$

$\widehat{W} = \begin{cases} \widehat{W_0} & \text{if } I_2 = \emptyset \\ W & \text{if } I_1 = \emptyset \\ \mathsf{h}?\{go.\widehat{W_0}, \; stop.W\} & \text{otherwise} \end{cases}$

Fig. 2. Gateway construction rules (excerpt).

Rule [GW-O/I-L] corresponds to Rule [COMP-O/I-L] of Fig. 1. The gateway of the left server sends the message received from participant q to the gateway of the client, which forwards it to participant p. After these communications the processes of the servers must be different according to the index i belongs to I_1 (i.e., the client will stop) or I_2 (i.e., the client will continue interacting). More precisely:
- the process of the left server must be $\widehat{V_i}$ if $i \in I_1$ and V_i if $i \in I_2$;
- the process of the right server must be $\widehat{W_0}$ if $i \in I_1$ and W if $i \in I_2$.

The left server is informed by receiving the message λ_i with $i \in I_1$ or $i \in I_2$. In case both I_1 and I_2 are not empty the client advises the right server by sending the message go if $i \in I_1$ and the message $stop$ if $i \in I_2$.

In Rule [GW-I/O-A] the only communications between different systems are the message *go* if $i \in I_1$ and the message *stop* if $i \in I_2$, which the client sends to both servers to let them know if she will require other interactions.

In all rules the uses of *stop* and *go* messages are needful to prevent server starvation. Indeed, these messages and their specific continuations are essential for getting a bounded global type by the construction given in Theorem 4.

Example 6 (Gateways for the running example). Let H be defined as in Example 1 and V, W as in Example 5. The gateway processes $\widehat{H}, \widehat{V}$ and $\widehat{W}$ induced by the derivation $\mathcal{D}$ of Example 5 are built by means of the derivation $\widehat{\mathcal{D}}$ of $(\widehat{H}, \widehat{V}, \widehat{W}) : H \dashv \langle V, W \rangle$, which is

$$
\widehat{\mathcal{D}} = \cfrac{\cfrac{\cfrac{\widehat{\mathcal{D}}}{(\widehat{H}_3, \widehat{V}, \widehat{W}_1) : H_3 \dashv \langle V, W_1 \rangle} \ [\text{GW-O/I-R}]}{(\widehat{H}_2, \widehat{V}, \widehat{W}) : H_2 \dashv \langle V, W \rangle} \ [\text{GW-I/O-R}] \quad (0, V, W) : 0 \dashv \langle V, W \rangle}{\cfrac{(\widehat{H}_1, \widehat{V}, \widehat{W}) : H_1 \dashv \langle V, W \rangle}{(\widehat{H}, \widehat{V}, \widehat{W}) : H \dashv \langle V, W \rangle} \ [\text{GW-O/I-A}]} \quad [\text{GW-O/I-L}]
$$

where H_1, H_2, H_3, W_1 are as in Example 5, and

$$
\widehat{H} = \mathsf{p}!\text{START}.\widehat{H}_1 \qquad \widehat{H}_1 = \mathsf{v}? \begin{cases} \text{IMG. } \mathsf{p}!\text{IMG}.\mathsf{w}!go.\widehat{H}_2 \\ \text{HALT. } \mathsf{p}!\text{HALT. } \mathsf{w}!stop.\, \mathbf{0} \end{cases}
$$

$$
\widehat{H}_2 = \mathsf{p}?\text{REACTN. } \mathsf{w}!\text{REACTN}.\widehat{H}_3 \qquad \widehat{H}_3 = \mathsf{w}?\text{IMG. } \mathsf{p}!\text{IMG. } \widehat{H}
$$

$$
\widehat{V} = \mathsf{s}? \begin{cases} \text{IMG. } \mathsf{h}!\text{IMG. } \widehat{V} \\ \text{HALT}.\mathsf{h}!\text{HALT. } V \end{cases}
$$

$$
\widehat{W} = \mathsf{h}? \begin{cases} go.\, \mathsf{h}?\text{REACTN. } \mathsf{r}!\text{REACTN. } \widehat{W}_1 \\ stop.\, W \end{cases} \qquad \widehat{W}_1 = \mathsf{r}?\text{IMG. } \mathsf{h}!\text{IMG. } \widehat{W}
$$

As said in Example 5 there are other derivations proving processes H, V and W to be compliant. Clearly different derivations would induce different gateway processes. $\diamond$

We can now formally define the system obtained by composing a client system with two server systems provided that they have compliant processes.

Definition 12 (Composition via gateways). *Let* $\mathbb{M}_1 \equiv \mathsf{h}[H] \parallel \mathbb{M}_1'$ *and* $\mathbb{M}_2 \equiv \mathsf{v}[V] \parallel \mathbb{M}_2'$ *and* $\mathbb{M}_3 \equiv \mathsf{w}[W] \parallel \mathbb{M}_3'$ *and* $\mathcal{D}$ *be a compliance derivation of* $H \dashv \langle V, W \rangle$. *We define the composition of* $\mathbb{M}_1$ *and* $\langle \mathbb{M}_2, \mathbb{M}_3 \rangle$ *via* h *and* $\langle \mathsf{v}, \mathsf{w} \rangle$ *induced by* $\mathcal{D}$ *(dubbed* $\mathbb{M}_1{}^{\mathcal{D}:\mathsf{h} \to \langle \mathsf{v}, \mathsf{w} \rangle} \langle \mathbb{M}_2, \mathbb{M}_3 \rangle$*) by*

$$
\mathbb{M}_1{}^{\mathcal{D}:\mathsf{h} \to \langle \mathsf{v}, \mathsf{w} \rangle} \langle \mathbb{M}_2, \mathbb{M}_3 \rangle \quad = \quad \mathbb{M} \parallel \mathbb{M}_1' \parallel \mathbb{M}_2' \parallel \mathbb{M}_3'
$$

where $\mathbb{M} \equiv \mathsf{h}[\mathsf{gw}_{\mathcal{D}}(H, \langle \mathsf{v}, \mathsf{w} \rangle)] \parallel \mathsf{v}[\mathsf{gw}_{\mathcal{D}}(V, \mathsf{h})] \parallel \mathsf{w}[\mathsf{gw}_{\mathcal{D}}(W, \mathsf{h})]$.

Example 7 (Composition via gateways for the running example). Let the systems $\mathbb{M}_1$, $\mathbb{M}_2$, $\mathbb{M}_3$ and the processes $\widehat{H}$, $\widehat{V}$, $\widehat{W}$ be defined as in Examples 1, 5, and 6. The system composition is then:

$$\mathsf{h}[\widehat{H}] \parallel \mathsf{v}[\widehat{V}] \parallel \mathsf{w}[\widehat{W}] \parallel \mathsf{p}[P] \parallel \mathsf{s}[S] \parallel \mathsf{r}[R]$$

As previously mentioned, the systems $\mathbb{M}_1$, $\mathbb{M}_2$ and $\mathbb{M}_3$ are lock-free and behave as prescribed by G_1, G_2 and G_3 respectively. Indeed the following are derivable: $\mathsf{G}_1 \vdash \mathbb{M}_1$, $\mathsf{G}_2 \vdash \mathbb{M}_2$, and $\mathsf{G}_3 \vdash \mathbb{M}_3$.   ◇

In the following, we will show that lock-freedom is preserved by the composition of compliant and typable systems. We start with the simple case in which the systems are just put in parallel.

Definition 13. *We define the* simple composition *of two global types* G *and* G' *(notation* $\mathsf{G} \circ \mathsf{G}'$*) coinductively by:*

$$\mathsf{End} \circ \mathsf{G} = \mathsf{G} \qquad \mathsf{p} \to \mathsf{q} : \{\lambda_i.\mathsf{G}_i\}_{i \in I} \circ \mathsf{G} = \mathsf{p} \to \mathsf{q} : \{\lambda_i.\mathsf{G} \circ \mathsf{G}_i\}_{i \in I}$$

Lemma 1. *If* $\mathsf{G} \vdash \mathbb{M}$ *and* $\mathsf{G}' \vdash \mathbb{M}'$ *and* $\mathsf{prt}(\mathsf{G}) \cap \mathsf{prt}(\mathsf{G}') = \emptyset$, *then* $\mathsf{G} \circ \mathsf{G}' \vdash \mathbb{M} \parallel \mathbb{M}'$.

We are now in place to state and prove the main result, i.e. that by composing typable systems as prescribed by Definition 12 we get a typable system.

Theorem 4. *Let* $\mathbb{M}_1 = \mathbb{M}_1' \parallel \mathsf{h}[H]$ *and* $\mathbb{M}_2 = \mathbb{M}_2' \parallel \mathsf{v}[V]$ *and* $\mathbb{M}_3 = \mathbb{M}_3' \parallel \mathsf{w}[W]$ *be such that* $H \dashv \langle V, W \rangle$ *can be proved with a derivation* $\mathcal{D}$. *If* $\mathsf{G}_n \vdash \mathbb{M}_n$ *for* $n = 1, 2, 3$, *then* $\mathsf{G} \vdash \mathbb{M}_1 {}^{\mathcal{D}:\mathsf{h} \to \langle \mathsf{v}, \mathsf{w} \rangle} \langle \mathbb{M}_2, \mathbb{M}_3 \rangle$, *where* G *can be built out of the* G_n*'s and* $\mathcal{D}$.

Proof. We use $\mathcal{G}(\mathcal{D}, \mathsf{G}_1, \mathsf{G}_2, \mathsf{G}_3)$ to denote the global type G built out of $\mathcal{D}, \mathsf{G}_1, \mathsf{G}_2$ and G_3. We show that

$$\mathcal{G}(\mathcal{D}, \mathsf{G}_1, \mathsf{G}_2, \mathsf{G}_3) \vdash \mathbb{M}_1 {}^{\mathcal{D}:\mathsf{h} \to \langle \mathsf{v}, \mathsf{w} \rangle} \langle \mathbb{M}_2, \mathbb{M}_3 \rangle$$

by coinduction on $\mathcal{D}$ and induction on $depth(\mathsf{G}_1, \mathsf{h}) + depth(\mathsf{G}_2, \mathsf{v}) + depth(\mathsf{G}_3, \mathsf{w})$. The regularity of the processes ensures that we get a regular global type. We show only the two most interesting cases. The first one is the simplest: the last applied rule is [GW-0]. Then $depth(\mathsf{G}_1, \mathsf{h}) = 0$, the gateways are the processes themselves and

$$\mathsf{G}_1 \circ (\mathsf{G}_2 \circ \mathsf{G}_3) \vdash \mathbb{M}_1 {}^{\mathcal{D}:\mathsf{h} \to \langle \mathsf{v}, \mathsf{w} \rangle} \langle \mathbb{M}_2, \mathbb{M}_3 \rangle$$

The theorem follows from Lemma 1.

The second case we consider is when the last applied rule is [GW-0/I-L], $\mathsf{G}_1 = \mathsf{h} \to \mathsf{p} : \{\lambda_j.\mathsf{G}_j'\}_{j \in J}$, $\mathsf{G}_2 = \mathsf{q} \to \mathsf{v} : \{\lambda_i.\mathsf{G}_i''\}_{i \in I'}$, $I_1 \neq \emptyset$ and $I_2 \neq \emptyset$. It is paradigmatic of all other cases. We have that $\mathcal{D}$ ends with

$$[\text{GW-O/I-L}] \frac{\begin{array}{l} (\widehat{H}_i, \widehat{V}_i, \widehat{W}_0) : H_i \dashv \langle V_i, W \rangle \; \forall i \in I_1 \\ (\mathbf{0}, V_i, W) : \mathbf{0} \dashv \langle V_i, W \rangle \; \forall i \in I_2 \end{array}}{(\widehat{H}, \widehat{V}, \widehat{W}) : \mathsf{p}!\{\lambda_j.H_j\}_{j \in J} \dashv \langle \mathsf{q}?\{\lambda_i.V_i\}_{i \in I}, W \rangle} \quad \begin{array}{l} I_1 \cup I_2 = I \subseteq J \\ H_i \neq \mathbf{0} \quad \forall i \in I_1 \\ H_i = \mathbf{0} \quad \forall i \in I_2 \end{array}$$

where

$$\widehat{H} = \mathsf{v}?\{\lambda_i.\mathsf{p}!\lambda_i.\mathsf{w}!go.\widehat{H}_i\}_{i \in I_1} \cup \{\lambda_i.\mathsf{p}!\lambda_i.\mathsf{w}!stop.\mathbf{0}\}_{i \in I_2}$$
$$\widehat{V} = \mathsf{q}?\{\lambda_i.\mathsf{h}!\lambda_i.\widehat{V}_i\}_{i \in I_1} \cup \{\lambda_i.\mathsf{h}!\lambda_i.V_i\}_{i \in I_2} \qquad \widehat{W} = \mathsf{h}?\{go.\widehat{W}_0, \; stop.W\}$$

Let $\mathcal{D}_i$ be the derivation of $H_i \dashv \langle V_i, W \rangle$ for $i \in I$. Since $\mathsf{G}_n \vdash \mathbb{M}_n$ for $n = 1, 2$, we have that

$$\mathbb{M}_1 \equiv \mathsf{h}[\mathsf{p}!\{\lambda_j.H_j\}_{j \in J}] \parallel \mathsf{p}[\mathsf{h}?\{\lambda_j.P_j\}_{j \in J'}] \parallel \mathbb{M}_1'$$

$$\mathbb{M}_2 \equiv \mathsf{q}[\mathsf{v}!\{\lambda_i.Q_i\}_{i \in I'}] \parallel \mathsf{v}[\mathsf{q}?\{\lambda_i.V_i\}_{i \in I}] \parallel \mathbb{M}_2'$$

with $J \subseteq J'$ and $I' \subseteq I$, and that, for each $i \in I'$

$$\mathsf{G}_i' \vdash \mathsf{h}[H_i] \parallel \mathsf{p}[P_i] \parallel \mathbb{M}_1' \qquad \mathsf{G}_i'' \vdash \mathsf{q}[Q_i] \parallel \mathsf{v}[V_i] \parallel \mathbb{M}_2'$$

and, by coinduction,

$$\mathcal{G}(\mathcal{D}_i, \mathsf{G}_i', \mathsf{G}_i'', \mathsf{G}_3) \vdash (\mathsf{h}[H_i] \parallel \mathsf{p}[P_i] \parallel \mathbb{M}_1')^{\mathcal{D}:\mathsf{h} \to \langle \mathsf{v}, \mathsf{w} \rangle} \langle \mathsf{q}[Q_i] \parallel \mathsf{v}[V_i] \parallel \mathbb{M}_2', \mathbb{M}_3 \rangle$$

Let $I_1' = I_1 \cap I'$ and $I_2' = I_2 \cap I'$. Since, by construction, $\mathbb{M}_3 \equiv \mathsf{w}[W] \parallel \mathbb{M}_3'$ we get for all $i \in I_1'$

$$\mathcal{G}(\mathcal{D}_i, \mathsf{G}_i', \mathsf{G}_i'', \mathsf{G}_3) \vdash \mathsf{h}[\widehat{H}_i] \parallel \mathsf{p}[P_i] \parallel \mathbb{M}_1' \parallel \mathsf{q}[Q_i] \parallel \mathsf{v}[\widehat{V}_i] \parallel \mathbb{M}_2' \parallel \mathsf{w}[\widehat{W}_0] \parallel \mathbb{M}_3' \quad (1)$$

and for all $i \in I_2'$

$$\mathcal{G}(\mathcal{D}_i, \mathsf{G}_i', \mathsf{G}_i'', \mathsf{G}_3) \vdash \mathsf{h}[\mathbf{0}] \parallel \mathsf{p}[P_i] \parallel \mathbb{M}_1' \parallel \mathsf{q}[Q_i] \parallel \mathsf{v}[V_i] \parallel \mathbb{M}_2' \parallel \mathsf{w}[W] \parallel \mathbb{M}_3' \quad (2)$$

We can use the derivations for (1) - that we have by coinduction hypothesis - to derive for all $i \in I_1'$

$$\frac{\dfrac{\dfrac{\mathcal{G}(\mathcal{D}_i, \mathsf{G}_i', \mathsf{G}_i'', \mathsf{G}_3) \vdash \mathsf{h}[\widehat{H}_i] \parallel \mathsf{w}[\widehat{W}_0] \parallel \cdots}{\mathsf{h} \to \mathsf{w} : go.\mathcal{G}(\mathcal{D}_i, \mathsf{G}_i', \mathsf{G}_i'', \mathsf{G}_3) \vdash \mathsf{h}[\mathsf{w}!go.\widehat{H}_i] \parallel \mathsf{p}[P_i] \parallel \mathsf{w}[\mathsf{h}?\{go.\widehat{W}_0, stop.W\}] \parallel \cdots}}{\mathsf{h} \to \mathsf{p} : \lambda_i.\mathsf{h} \to \mathsf{w} : go.\mathcal{G}(\mathcal{D}_i, \mathsf{G}_i', \mathsf{G}_i'', \mathsf{G}_3) \vdash \mathsf{h}[\mathsf{p}!\lambda_i.\mathsf{w}!go.\widehat{H}_i] \parallel \mathsf{p}[\mathsf{h}?\{\lambda_j.P_j\}_{j \in J'}] \parallel \cdots}}{}$$

and the derivations for (2) - that we have by coinduction hypothesis - to derive for all $i \in I_2'$

$$\frac{\dfrac{\dfrac{\mathcal{G}(\mathcal{D}_i, \mathsf{G}_i', \mathsf{G}_i'', \mathsf{G}_3) \vdash \mathsf{h}[\mathbf{0}] \parallel \mathsf{w}[W] \parallel \cdots}{\mathsf{h} \to \mathsf{w} : stop.\mathcal{G}(\mathcal{D}_i, \mathsf{G}_i', \mathsf{G}_i'', \mathsf{G}_3) \vdash \mathsf{h}[\mathsf{w}!stop.\mathbf{0}] \parallel \mathsf{p}[P_i] \parallel \mathsf{w}[\mathsf{h}?\{go.\widehat{W}_0, stop.W\}] \parallel \cdots}}{\mathsf{h} \to \mathsf{p} : \lambda_i.\mathsf{h} \to \mathsf{w} : stop.\mathcal{G}(\mathcal{D}_i, \mathsf{G}_i', \mathsf{G}_i'', \mathsf{G}_3) \vdash \mathsf{h}[\mathsf{p}!\lambda_i.\mathsf{w}!stop.\mathbf{0}] \parallel \mathsf{p}[\mathsf{h}?\{\lambda_j.P_j\}_{j \in J'}] \parallel \cdots}}{}$$

where we only write the processes which change from the premises to the conclusions. Starting from the conclusions of these derivations we get for all $i \in I'_1$:

$$\widehat{G}_i \vdash h[v?\{\lambda_i.p!\lambda_i.w!go.\widehat{H}_i\}_{i\in I_1} \cup \{\lambda_i.p!\lambda_i.w!stop.\mathbf{0}\}_{i\in I_2}] \parallel v[h!\lambda_i.\widehat{V}_i] \parallel \cdots$$

and, for all $i \in I'_2$:

$$\widehat{G}_i \vdash h[v?\{\lambda_i.p!\lambda_i.w!go.\widehat{H}_i\}_{i\in I_1} \cup \{\lambda_i.p!\lambda_i.w!stop.\mathbf{0}\}_{i\in I_2}] \parallel v[h!\lambda_i.V_i] \parallel \cdots$$

where

$$\widehat{G}_i = \begin{cases} v \to h : \lambda_i.h \to p : \lambda_i.h \to w : go.\mathcal{G}(\mathcal{D}_i, G'_i, G''_i, G_3) & \text{if } i \in I'_1 \\ v \to h : \lambda_i.h \to p : \lambda_i.h \to w : stop.\mathcal{G}(\mathcal{D}_i, G'_i, G''_i, G_3) & \text{if } i \in I'_2 \end{cases}$$

Then we can derive

$$q \to v : \{\lambda_i.\widehat{G}_i\}_{i\in I'} \vdash q[v!\{\lambda_i.Q_i\}_{i\in I'}] \parallel v[q?\{\lambda_i.h!\lambda_i.\widehat{V}_i\}_{i\in I_1}\cup\{\lambda_i.h!\lambda_i.V_i\}_{i\in I_2}] \parallel \cdots$$

and defining

$$\mathcal{G}(\mathcal{D}, G_1, G_2, G_3) = q \to v : \{\lambda_i.\widehat{G}_i\}_{i\in I'}$$

we can conclude

$$\mathcal{G}(\mathcal{D}, G_1, G_2, G_3) \vdash \mathbb{M}_1 \xrightarrow{\mathcal{D}:h\to\langle v,w\rangle} \langle \mathbb{M}_2, \mathbb{M}_3\rangle$$

In order to complete the proof we have to show that $\mathcal{G}(\mathcal{D}, G_1, G_2, G_3)$ is bounded. It is enough to prove that there is d such that $depth(\sigma, t) \leq d$ for each path $\sigma \in paths(\mathcal{G}(\mathcal{D}, G_1, G_2, G_3))$ and all $t \in prt(\mathcal{G}(\mathcal{D}, G_1, G_2, G_3))$. The case $t \in \{p, q, h, v, w\}$ is immediate. Otherwise, by coinduction, $depth(\sigma, t) \leq d_i$ for some d_i, for each path $\sigma \in paths(\mathcal{G}(\mathcal{D}_i, G'_i, G''_i, G_3))$, all $t \in prt(\mathcal{G}(\mathcal{D}_i, G'_i, G''_i, G_3))$ and all $i \in I'$. From $\sigma \in paths(\mathcal{G}(\mathcal{D}, G_1, G_2, G_3))$ we get

$$\sigma = q\lambda_i v \cdot v\lambda_i h \cdot h\lambda_i p \cdot h\lambda w \cdot \sigma'$$

where $\lambda \in \{go, stop\}$ and $\sigma' \in paths(\mathcal{G}(\mathcal{D}_i, G'_i, G''_i, G_3))$ for some $i \in I'$. This implies $depth(\sigma, t) \leq 4 + max\{d_i\}_{i\in I'}$. □

Example 8. (A global type for $\mathbb{M}_1 \xrightarrow{\mathcal{D}:h\to\langle v,w\rangle} \langle \mathbb{M}_2, \mathbb{M}_3\rangle$*).* The global type which can be obtained out of G_1, G_2 and G_3, using the derivation $\mathcal{D}$ described in Example 5 is

$$G = h \to p : \text{START}. s \to v : \begin{cases} \text{IMG}. v \to h : \text{IMG}. h \to p : \text{IMG}. h \to w : go. G' \\ \text{HALT}.v \to h : \text{HALT}. h \to p : \text{HALT}. h \to w : stop. G'' \end{cases}$$

where

$$G' = p \rightarrow h : \text{REACTN}. \, h \rightarrow w : \text{REACTN}. \, w \rightarrow r : \text{REACTN}.$$
$$r \rightarrow w : \text{IMG}. \, w \rightarrow h : \text{IMG}. \, h \rightarrow p : \text{IMG}. \, G$$

and $\quad G'' = G_2 \circ G_3.$ $\qquad\qquad\qquad\qquad\qquad\qquad\qquad\qquad\quad \diamond$

Thanks to the regularity of global types we can give an inductive formulation of the rules for the global type construction in the proof of the previous theorem, obtaining in this way an algorithm.

Notice that, in the particular case in which the client does not use the servers, the composition by gateways is just the parallel composition. In this case Lemma 1 implies that the parallel composition is typable by the simple composition of types.

5 Concluding Remarks, Related and Future Works

A minor contribution of this paper is a simple type system for synchronous multiparty sessions, typing networks with global types without recurring to local types and projections. Instead, our main result lies in constructing a unique multiparty session starting from three participants of different typable sessions. One participant plays the role of the client and two participants play the role of servers, proviso that the servers are compliant with the client. The feature of our construction is the typability of the obtained session since this ensures that we get a well-behaved system.

Any formalism for the design and development of concurrent and distributed systems must cope with the fact that, in the real world, systems can rarely be considered standalone entities. They are composed modularly and/or need to be dynamically connected to other systems offering specific services. In a nutshell, they need to be treated as *open* entities, i.e. providing access points for communication with the environment. Some formalisms are developed just to directly deal with the notion of system composition, like in [19]. Other formalisms, previously developed for specifying closed systems, have been extended or modified to cope with inter-systems communication. This is the case, for instance, of automata description of systems and the development of the theory of *interface automata* [1,2].

In [29] compositional issues for the theory of MPST are faced by devising a specific syntax for partial (multiparty) session types, enabling merging sessions with the same name but observed by different participants, proviso their types are compatible. In general, the study of compositional approaches in MPST settings is still in a preliminary stage. Since its introduction in [20], MPST proved to be a theoretically solid choreographic formalism but, as such, mostly suitable to handle closed systems. However, this is not necessarily a shortcoming as far as one manages to treat the notion of closure/openness as a matter of "perspective". This is precisely the essence of the "participants as interfaces" approach introduced in [3] where the behaviour of any component (participant) of a system can be considered - depending on the current needs - as an interface modelling the environment.

Once in two systems two "compatible" interfaces are identified, the former can be connected by replacing the latter with two coupled forwarders (dubbed gateways). Such an approach is quite general and hence naturally applicable to different formalisms. It has been exploited in the formalism of CFSM in [3,6]. For what concerns MPST, the participants-as-interfaces approach has been investigated in [5], which we took as a starting basis for the present paper.

Here we shifted from the notion of compatibility to that of *compliance*. Compliance arose from *contract theory* [12,17] where it denotes the asymmetric relation of a client with a server, such that any action from the client is matched by a dual action of the server: in this way any possible interaction will never prevent the client from completing. See [8] and [25] for a survey and a comparison of several (non-equivalent) definitions of this concept. In the previous references compliance has been only defined as a binary relation; here we have considered a mild generalisation to the case of a client complying with two servers and possibly dealing independently with some messages. We believe that such a notion smoothly extends to the case of arbitrarily many servers as well.

Our type system derives directly global types for multiparty sessions without typing processes and projecting global types as usual [21]. It is inspired by the type system in [15], also used in [13], the differences being both in the syntax of global types and in the communications which are asynchronous instead of synchronous.

As future work we plan to investigate compliance and composition when the communication between processes is asynchronous, possibly using the type system of [15]. An advantage of that type system is the possibility of anticipating outputs over inputs without requiring the asynchronous subtyping of [24], which is known to be undecidable [11,23]. A difficulty will come from the larger freedom in choosing the order of interactions due to the splitting between writing and reading messages on a queue.

Acknowledgements. We are grateful to anonymous referees for their comments and suggestions to improve the readability of the paper. This research has been partially supported by Progetto di Ateneo "Piaceri" UNICT; Progetto DE_U_RILO_19_01 UNITO; INdAM (GNCS).

References

1. de Alfaro, L., Henzinger, T.A.: Interface automata. In: Tjoa, A.M., Gruhn, V. (eds.) ESEC / SIGSOFT FSE, pp. 109–120. ACM Press (2001). https://doi.org/10.1145/503209.503226

2. de Alfaro, L., Henzinger, T.A.: Interface-based design. In: Broy, M., Grünbauer, J., Harel, D., Hoare, T. (eds.) Engineering Theories of Software Intensive Systems. NSS, vol. 195, pp. 83–104. Springer, Dordrecht (2005). https://doi.org/10.1007/1-4020-3532-2_3

3. Barbanera, F., de Liguoro, U., Hennicker, R.: Connecting open systems of communicating finite state machines. J. Logical Algebraic Methods Program. **109**, 100476 (2019). https://doi.org/10.1016/j.jlamp.2019.07.004

4. Barbanera, F., Dezani-Ciancaglini, M., de'Liguoro, U.: Reversible client/server interactions. Formal Aspects Comput. **28**(4), 697–722 (2016). https://doi.org/10.1007/s00165-016-0358-2

5. Barbanera, F., Dezani-Ciancaglini, M., Lanese, I., Tuosto, E.: Composition and decomposition of multiparty sessions. J. Logic Algebraic Methods Program. **119**, 100620 (2021). https://doi.org/10.1016/j.jlamp.2020.100620

6. Barbanera, F., Lanese, I., Tuosto, E.: Composing communicating systems, synchronously. In: Margaria, T., Steffen, B. (eds.) ISoLA 2020. LNCS, vol. 12476, pp. 39–59. Springer, Cham (2020). https://doi.org/10.1007/978-3-030-61362-4_3

7. Bartoletti, M., Cimoli, T., Pinna, G.M.: A note on two notions of compliance. In: Lanese, I., Lluch-Lafuente, A., Sokolova, A., Vieira, H.T. (eds.) ICE. EPTCS, vol. 166, pp. 86–93. Open Publishing Association (2014). https://doi.org/10.4204/eptcs.166.9

8. Bartoletti, M., Cimoli, T., Zunino, R.: Compliance in behavioural contracts: a brief survey. In: Bodei, C., Ferrari, G.-L., Priami, C. (eds.) Programming Languages with Applications to Biology and Security. LNCS, vol. 9465, pp. 103–121. Springer, Cham (2015). https://doi.org/10.1007/978-3-319-25527-9_9

9. Bernardi, G., Hennessy, M.: Compliance and testing preorders differ. In: Counsell, S., Núñez, M. (eds.) SEFM 2013. LNCS, vol. 8368, pp. 69–81. Springer, Cham (2014). https://doi.org/10.1007/978-3-319-05032-4_6

10. Brand, D., Zafiropulo, P.: On communicating finite-state machines. J. ACM **30**(2), 323–342 (1983). https://doi.org/10.1145/322374.322380

11. Bravetti, M., Carbone, M., Zavattaro, G.: Undecidability of asynchronous session subtyping. Inf. Comput. **256**, 300–320 (2017). https://doi.org/10.1016/j.ic.2017.07.010

12. Castagna, G., Gesbert, N., Padovani, L.: A theory of contracts for web services. ACM Trans. Program. Langu. Syst. **31**(5), 191–1961 (2009). https://doi.org/10.1145/1538917.1538920

13. Castellani, I., Dezani-Ciancaglini, M., Giannini, P.: Asynchronous sessions with input races. In: Carbone, M., Neykova, R. (eds.) PLACES. EPTCS, vol. 356, pp. 12–23. Open Publishing Association (2022). https://doi.org/10.4204/EPTCS.356.2

14. Courcelle, B.: Fundamental properties of infinite trees. Theoret. Comput. Sci. **25**, 95–169 (1983). https://doi.org/10.1016/0304-3975(83)90059-2

15. Dagnino, F., Giannini, P., Dezani-Ciancaglini, M.: Deconfined global types for asynchronous sessions. CoRR abs/2111.11984 (2021). https://doi.org/10.48550/arXiv.2111.11984

16. Demangeon, R., Honda, K.: Nested protocols in session types. In: Koutny, M., Ulidowski, I. (eds.) CONCUR 2012. LNCS, vol. 7454, pp. 272–286. Springer, Heidelberg (2012). https://doi.org/10.1007/978-3-642-32940-1_20

17. Fournet, C., Hoare, T., Rajamani, S.K., Rehof, J.: Stuck-free conformance. In: Alur, R., Peled, D.A. (eds.) CAV 2004. LNCS, vol. 3114, pp. 242–254. Springer, Heidelberg (2004). https://doi.org/10.1007/978-3-540-27813-9_19

18. Gay, S., Hole, M.: Subtyping for session types in the pi-calculus. Acta Informatica **42**(2/3), 191–225 (2005). https://doi.org/10.1007/s00236-005-0177-z

19. Hennicker, R.: A calculus for open ensembles and their composition. In: Margaria, T., Steffen, B. (eds.) ISoLA 2016. LNCS, vol. 9952, pp. 570–588. Springer, Cham (2016). https://doi.org/10.1007/978-3-319-47166-2_40

20. Honda, K., Yoshida, N., Carbone, M.: Multiparty asynchronous session types. In: Necula, G.C., Wadler, P. (eds.) POPL, pp. 273–284. ACM Press (2008). https://doi.org/10.1145/1328897.1328472

21. Honda, K., Yoshida, N., Carbone, M.: Multiparty asynchronous session types. J. ACM **63**(1), 9:1-9:D67 (2016). https://doi.org/10.1145/2827695
22. Kobayashi, N.: A type system for lock-free processes. Inf. Comput. **177**(2), 122–159 (2002). https://doi.org/10.1006/inco.2002.3171
23. Lange, J., Yoshida, N.: On the undecidability of asynchronous session subtyping. In: Esparza, J., Murawski, A.S. (eds.) FoSSaCS 2017. LNCS, vol. 10203, pp. 441–457. Springer, Heidelberg (2017). https://doi.org/10.1007/978-3-662-54458-7_26
24. Mostrous, D., Yoshida, N., Honda, K.: Global principal typing in partially commutative asynchronous sessions. In: Castagna, G. (ed.) ESOP 2009. LNCS, vol. 5502, pp. 316–332. Springer, Heidelberg (2009). https://doi.org/10.1007/978-3-642-00590-9_23
25. Murgia, M.: A note on compliance relations and fixed points. In: Bartoletti, M., Henrio, L., Mavridou, A., Scalas, A. (eds.) ICE. EPTCS, vol. 304, pp. 38–47. Open Publishing Association (2019). https://doi.org/10.4204/EPTCS.304.3
26. Murgia, M.: A fixed-points based framework for compliance of behavioural contracts. J. Logic Algebraic Methods Program. **120**, 100641 (2021). https://doi.org/10.1016/j.jlamp.2021.100641
27. Padovani, L.: Deadlock and lock freedom in the linear π-calculus. In: Henzinger, T.A., Miller, D. (eds.) CSL-LICS, pp. 72:1–72:10. ACM Press (2014). https://doi.org/10.1007/978-3-662-43376-8_10
28. Pierce, B.C.: Types and Programming Languages. MIT Press, Cambridge (2002)
29. Stolze, C., Miculan, M., Di Gianantonio, P.: Composable partial multiparty session types. In: Salaün, G., Wijs, A. (eds.) FACS 2021. LNCS, vol. 13077, pp. 44–62. Springer, Cham (2021). https://doi.org/10.1007/978-3-030-90636-8_3

Author Index